FOUNDATIONAL FALSEHOODS OF CREATIONISM

Pitchstone Publishing
Durham, North Carolina
www.pitchstonepublishing.com

Printed in the USA

10 9 8 7 6 5 4 3 2

Library of Congress Cataloging-in-Publication Data

Names: Ra, Aron, author.
Title: Foundational falsehoods of creationism / Aron Ra.
Description: Durham, North Carolina : Pitchstone Publishing, 2016. | Includes bibliographical references.
Identifiers: LCCN 2016014000| ISBN 9781634310789 (pbk. : alk. paper) | ISBN 9781634310802 (pdf) | ISBN 9781634310819 (mobi)
Subjects: LCSH: Creationism. | Evolution (Biology)
Classification: LCC BS652 .R24 2016 | DDC 231.7/652—dc23
LC record available at https://lccn.loc.gov/2016014000

You can't prove evolutionism.

Yes, actually I can.

Prove it then!

**I can't prove it right here and now,
but if you'll allow me some time,
I'll create a presentation that will
prove evolution to your satisfaction.**

*You can't prove evolution,
because to do that,
you'd have to disprove God.*

**Whether God exists
or not is irrelevant.**

CONTENTS

PREFACE

My name is not Aaron; it's Aron. It's not pronounced like "errand" but more like "aren't." If you take out the "o," and read it as "Arn," you'll probably say it right; although it's better if you read it like the initials, R. N. It isn't the same as the Hebrew name either. This same name occurs independently in Korean and Indian languages too; mine is Scandinavian. I think it's a convergent derivation of "Arndt" or something like that—maybe Arne. I'm not sure. I have to explain my name most every day, so I may as well start there.

My surname is no terrible secret, but it isn't my preference either. When I signed up for the Usenet group Talk.Origins, I needed a handle and quickly decided on Aron-Ra. Why? I wanted to give a nod to Amen-Ra, a composite of the Egyptian air god and sun god, also known as Amun-Re, whom I see as a template for the god of Western monotheism.

I've always been an irreverent sort of person, with a preference for science over superstition, and ever since I first connected to the Internet, I've spent an unhealthy amount of my time online obsessing over the topic of origins, particularly as relates to the evolution-creationism debate. I worked in a call center back then, one with very low demands that allowed unlimited overtime and did not restrict or interrupt my browsing while taking calls. So I basically just read the Internet all day every day.

The Usenet group Talk.Origins pitted theologians against professional scientists, giving me an opportunity to listen to the best arguments on both sides of the debate. I read the best points of either philosophy. Then I would post my questions and watch them go at it. I compared their answers, and whatever counterarguments followed those. Then I looked up all their references to see who was right about what. This started me on a learning curve well beyond anything I ever did in college. My first attempt at higher education was as a naive and ill-prepared kid. I was a fine arts major, and not a serious student. But in my thirties I studied Internet archives intently, several hours a day, every single day—with increasing interest, and that is no exaggeration. The pace picked up once I was moved to a graveyard shift and hardly got any calls at all. Then I studied with greater concentration. I didn't read physical books, but after a few years of this, I had read countless scientific articles and relevant publications, as well as many of the ancient myths and scriptures of various religions, and many other works and documents I had been advised to read. I had developed such a taste for it that I eventually went back to college as a geoscience major, studying paleontology. Three years into that, I had to quit school again due to a concurrence of life-changing events with my family and career at that time; much the pity, as I would rather spend my life learning than pretending to be learned.

I started this journey as someone who still believed in supernatural things such as souls, psychics and psionics, parapsychology, cryptozoology, extraterrestrial UFOs, and other such nonsense. My position was more or less typical of your average politically apathetic undereducated American who sees this stuff supported on "science" and "history" channels and everywhere else in our mass media. I didn't set out to change my own mind. I wasn't trying to dispel all the pseudoscience I had been led to believe since childhood. I wasn't trying to convince myself of anything either. I just wanted a better understanding of how things really are. I still do.

My country was once admired on the world stage for the very things we ourselves now largely reject. We've fallen victim to anti-intellectualism. A lifetime ago, someone said that when fascism comes to America, it would be wrapped in a flag and carrying a cross. That day has come. Nationalism comes from the same source as other ethnocentric prejudice like racism. We question science, ridicule experts, make fun of genius, and show a complete disdain for tolerance and multiculturalism, as if any of that was "bad." We're incurious and we overtly express hatred of political correctness. We're even opposed to progress! So it's no wonder we're not making any.

The Religious Right is responsible for a lot of that, and creationism is a tool they use to spread their infection. As fortune would have it, when the Texas State Board of Education became world famous for politicizing classrooms, revising and censoring science, sex ed, and social studies—deliberately misleading other people's kids—I found myself in the geographically and intellectually best place to voice my objections and get involved, which I felt compelled to do.

How do I reach people? One of my best friends from high school is now an ordained Southern Baptist minister and the principal of a young-earth fundamentalist Christian school teaching that the Bible is the only source of truth in our world. He recommended that I read *Evolution of a Creationist* by Dr. Jobe Martin, someone touted as a creation scientist. Turns out this "scientist" has a degree in dentistry. His book revealed that he has no idea what evolution even is. I wrote a 19,000-word rebuttal to the second chapter of his book, proving that he got literally every sentence in the chapter wrong. Seriously, every single sentence! The problem is no one would ever read such a lengthy detailed newsgroup post—not even the friend for whom I wrote it. He said I should have written a book instead. So I guess he can take partial credit for this one. I tried educating educators, but that didn't work. I needed to teach people directly. I realized that in order to understand evolution by conventional means, one typically should have been raised in an environment where critical thinking, curiosity, literacy, and scientific methodology are supported and endorsed early in one's life. In addition, one must have an in-depth and very specific education in order to understand the extent of bizarre criticisms against science levied by the laity. Such an education requires a significant commitment of time, effort, and money. But faith-based psycho-babble is shared for free everywhere by people who don't know what they're talking about and don't care to. You don't have to understand it either, as long as you profess your belief.

The only way I could see to counter that was to make a presentation suitable for those with minimal education and a short attention span. In my years on Usenet, I noticed that the same few creationist arguments are constantly repeated, and that these are the basis for all the erroneous arguments of evangelical pseudoscientists. I tried posting brief explanations of this list of errors on Christian discussion forums, but I found the moderators kept deleting them. So I made a series of videos addressing each of these foundational falsehoods of creationism.

At that time, YouTube imposed a maximum limit of eleven minutes on the length of any user-uploaded video. So I had to pack a lot of information into my videos, flashing text, illustrations, and animations in strobe effect as I narrated at a quick pace. No one understood everything that was being presented, but I got the point across. That series was very popular. It changed my life, and I have hundreds of emails in my inbox from others who say that it changed their lives too.

Finally, I met a book publisher at the American Humanist Association's national convention in New Orleans. The publisher wanted me to convert those videos into text. I was hesitant because it's easier to go from a book to a movie than the other way around. The publisher asked me to say everything I said in that series, but to also "flesh out" the arguments and evidence and references that I had to flash so fast before—so that readers could better understand the depth and breadth of this topic as I see it. This book is the fulfillment of that request.

THE 1ST FOUNDATIONAL FALSEHOOD OF CREATIONISM

"Evolution = Atheism"

In 2004, a Gallup poll reported that 35% of Americans believed that evolution was supported by the evidence. Another 35% said evolution was not supported by the evidence, and 25% admitted they didn't know enough to say. What this seems to imply is that only 35% actually knew what evolution is and what the evidence for it is. The 35% who say that evolution is not supported by the evidence should be grouped with the 25% who don't know enough to say (many of whom probably don't even know what evolution is). Thus, it is actually 60% who don't really know enough to say, but less than half of them would admit that.

At that time, Gallup said these figures had not significantly changed in decades. The situation has not improved. In 2012, Gallup reported that 46% of Americans still believe that God created humans in their present form; 32% say that God guided human evolution; and only 15% say that man evolved with no assistance from any deity. Put another way, the division in the American population between those who agree with evolution and those who believe in creationism is still roughly half-and-half.

The U.S. population seems pretty evenly divided over whether the human species is biologically related to other animals or whether we were "specially created" as part of a flurry of miracles. Even our collective politicians—seemingly all of them—are wrapped up in this controversy, yet it's hard to find even one of them who knows what it's about.

In the 2008 presidential race, Senator John McCain was one of very few presidential candidates to express any degree of understanding of what evolution even is, so it was especially ironic that he chose Sarah Palin as a running mate—someone who doesn't even know what science is. Sadly, the next presidential race four years later included:

- Michele Bachmann, who got her law degree from the televangelist Oral Roberts University in Tulsa, Oklahoma;
- Ron Paul, who rejected evolution as being "just" a theory;
- Newt Gingrich, who expressed his fear that his grandchildren would inherit "a secular atheist America dominated by radical Islamists";
- Rick Santorum, whose decade-long crusade to usurp science was actually one of the primary catalysts driving me to activism in defense of secular political values;
- and Rick Perry, my own state's governor, who seemed determined to destroy education altogether at every level, along with the economy, the environment, any remaining socially conscious benefit programs, and anything else he could possibly bring to ruin.

Soon after the first of two failed presidential runs, Perry returned to his office as governor. Texas then revised the state's Republican Party Platform to include the following:

> Knowledge-Based Education—*We oppose the teaching of Higher Order Thinking Skills* (HOTS) (values clarification), *critical thinking skills and similar programs* that are simply a relabeling of Outcome-Based Education (OBE) (mastery learning) which focus on behavior modification and have the purpose of ***challenging the student's fixed beliefs*** and undermining parental authority. [Emphasis mine]

Yes, they really just came right out and said that, and they didn't even know that they should have been embarrassed by it. Apparently, the dominant party in my state believes that knowledge and understanding are bad, that citizens should obediently believe whatever they're told, and that their beliefs should remain "fixed"—that is, rigid, unreasonable, and not to be corrected even when they are known to be wrong. How can there be any justification for that?

Just to be sure we know which beliefs they want to remain "fixed" and unchallenged, the Texas Republican Party Platform also specifically opposed the teaching of evolution.

> Controversial Theories—We support objective teaching and equal treatment of all sides of scientific theories. We believe *theories such as life origins* and environmental change *should be taught as challengeable* scientific theories subject to change as new data is produced. Teachers and students should be able to discuss the strengths and weaknesses of these theories openly and without fear of retribution or discrimination of any kind. [Emphasis mine once again]

Some translation is required here. By "objective" they mean "subjective," and by "equal treatment of all sides" they mean the application of a double-standard in favor of their side only. By "theories of life origins," they mean to include evolution—the study of biodiversity—along with abiogenesis, Big Bang cosmology, plate tectonics, and everything else that isn't permitted or indicated by the Bible. What they call "controversial" is not controversial at all outside of religious dogmatism. So they want to portray scientific theories as if they were unsupported speculation, which they can treat as "challengeable." By this, they mean that mere high school teachers should be allowed to condemn levels of scholarship far beyond their own education rather than teach anything that they themselves refuse to accept or even comprehend.

The seemingly fair-minded reference to "strengths and weaknesses" is actually a subterfuge frequently employed by right-wing politicians to undermine evolution in the classroom. Discussions of the "strengths" of evolution are inhibited, to say the least; they are certainly never promoted by them and are often countered by parental protest. Many teachers avoid mentioning evolution at all for fear of inciting controversy. Even in the rare instances that a teacher understands the subject, and can adequately present it, this language allowing the "strengths and weaknesses" to be discussed permits religiously indoctrinated students to recite pseudoscientific nonsense, disrupting the lesson in an attempt to refute it. If nothing else, it discourages most science-minded teachers from even trying to teach evolution properly, while empowering creationist teachers to cite long lists of alleged weaknesses that are contrived, concocted, and promoted by religious organizations committed to the preservation of biblical literalism. These organizations also invariably endorse some degree of abstinence from (and raise an objection to) the general philosophy and methodology of science.

Teachers and students really should be able to discuss the strengths and weaknesses of all scientific theories. That could uniformly happen if religious fundamentalists didn't get so angrily defensive in any environment where their beliefs may be subjected to equal treatment by knowledgeable persons. However, these discussions would be much more appropriate in a college setting among educated adults rather than in high schools, where no one (including often the teacher) is likely to know what they're really talking about. Most high school teachers are in no position to condemn the work of expert specialists in any field of study, and students should never be encouraged to reject the findings of advanced scholarship—at least until they've been taught what those theories actually are, and what the evidence actually is.

Even if some theories really are controversial, they still cannot be rejected until they are adequately understood. How can you know where the flaws in any theory are if you don't know what a given theory is, or what it actually says? Defenders of the faith fear that children who are taught the science behind evolution will likely accept it, so instead the children are deceived, distracted, misinformed, and misdirected however possible in order to prevent them from understanding undesired concepts.

Why is it that there is such concern in so many schools (kindergarten through twelfth grade) about teaching evolution, yet there is complete consensus among scientists all over America and the rest of the world that evolution is the backbone of modern biology and a demonstrable reality historically as well?

In 1987, *Newsweek* famously reported that "by one count, there are some 700 scientists with respectable academic credentials (out of a total of 480,000 U.S. earth and life scientists) who give credence to creation-science." This was based on a list of signatures of scientists who "doubted Darwinism." That list was provided by the Discovery Institute, an "intelligent design" (ID) think tank dedicated to undermining evolution. Actually, the statement that bore these signatures was this one:

> We are skeptical of claims for the ability of random mutation and natural selection to account for the complexity of life. Careful examination of the evidence for Darwinian theory should be encouraged.

Phrased this way, I might have signed the statement myself even though I have no doubt whatsoever that all life evolved through natural processes without any supernatural influence. However, creationists seem to think that anyone who is skeptical about science has also rejected science to some degree, and must therefore believe in some sort of creation. That list of "scientists" also included medical doctors, mechanical engineers, and many others who aren't scientists at all—but there were some actual "earth and life" scientists in that group. The list was promoted as a "scientific dissent from Darwinism" and was meant to imply that a growing number of scientists were moving away from naturalistic theories, as if evolution were a theory in crisis. This is a lie that has been repeated many times over the last hundred years or so. It is a lie because all other polls consistently show the opposite trend over that same period.

The implication as *Newsweek* saw it is that the total number of collective geologists and biologists of any specialty who still believe in supernatural creationism over evolution via natural selection is only 0.14%, barely more than one-tenth of one percent. So if 99.44% equals "pure" (as it does in the silver trade), then Bill Maher was right when he said that evolution is supported by the *entire* scientific community.

Amusingly, one response to this "scientific dissent from Darwinism" was the National Center for Science Education's "Project Steve." This was a much longer list composed exclusively of actual scientists, all of whom support evolution and all of whom are named Steve or some variant thereof—Stephen, Stefan, Stephanie, Stevie, Estaban, etc. People who share variants of this name are estimated to represent only about 1% of the U.S. population. So the fact that this list has hundreds more names on it than the list of alleged dissenters of Darwinism is another jab at the silliness of creationist statistics.

Here is the statement endorsed by Project Steve:

> Evolution is a vital, well-supported, unifying principle of the biological sciences, and the scientific evidence is overwhelmingly in favor of the idea that all living things share a common ancestry. Although there are

> legitimate debates about the patterns and processes of evolution, there is no serious scientific doubt that evolution occurred or that natural selection is a major mechanism in its occurrence. It is scientifically inappropriate and pedagogically irresponsible for creationist pseudoscience, including but not limited to "intelligent design," to be introduced into the science curricula of our nation's public schools.

This sentiment has been echoed or paraphrased by many professional scientific and academic organizations across the country and around the world. If all those who signed the Discovery Institute's petition had seen this statement instead or been asked to examine both, then the weakness of the alleged dissent from Darwin would be that much more obvious, at least to degreed experts.

What about your average person in the street? Most people really don't understand science—what it is, how it works, what hypotheses and theories are, or even the purpose behind it. Sadly, even those on your school faculty or state board of education often need an education themselves before they can be trusted to govern how or what our kids will be taught.

Of course I'm referring to the infamous right-wing Republican-dominant voting bloc, which made our state's board of education into a source of international embarrassment. Back in 2005, shortly before all of that had blown up to the absurd proportions that it did, I had an opportunity to try reasoning with these people. I had been arguing evolution versus creationism in various web formats for many years by then, and I was convinced that I could change the mind of any reasonable person who still didn't understand the scientific perspective.

Of course, what I usually encountered were people who were determined to be unreasonable, who didn't even care what the truth really is, and who would stoop to any means necessary to preserve their precious beliefs. This requires a more confrontational challenge. Often such people would say that neither evolution nor creationism could be proven and that they both had either the same evidence or no evidence, and so both required an element of faith. To show how wrong that position was, I frequently repeated the following proposal:

> I can prove that biological evolution is the truest, best explanation there is for the origin of our species, and that it is the only explanation of

> biodiversity with either evidentiary support or scientific validity. I can prove this even to your satisfaction over the course of a dozen mutual exchanges. The only trick to that is that you must properly address every point or query, ignoring none. If you repeatedly ignore direct questions, you will default this discussion, and I will be under no obligation to continue.

It wouldn't make much sense to do this in private messages, not if I meant to hold someone accountable to make a point for the common good, so the forum was always a public venue where the content would not likely be altered or deleted. Sometimes I would have moderators in case my contender repeatedly ignored direct questions (as creationists habitually do), then someone else could show an objective consensus as to when the discussion had become pointless to continue.

The trick is to get my opponent to think. Instead of simply telling them anything which they are trained to summarily dismiss without consideration, I would engage their minds with scientific quandaries and insist that they provide their own answers to them. Watching someone squirming to evade that is to watch an exercise in hilarious futility.

I didn't have any defined 12-step program. I never followed an outlined progression. I just wanted there to be an accessible limit of exchanges in which I should be able to present my case well enough even against ingrained misconceptions—enough posts to prove the point beyond reasonable doubt, but few enough that the more intrepid antagonists might take the bait.

Whenever I have seen creationists say that they could validate their position, it often involved substantial investment with high stakes, huge financial commitments, and no time limits on elaborate trials. These overly inconclusive and unfalsifiable challenges were designed to discourage all comers. So I did the very opposite: I tried to make it as simple and straightforward as anything could be with a minimal investment of time, where the onus was entirely on me, and where I could only "win" if the contestant admits that I have made my point. How could anyone confident in their own position possibly be intimidated by that?

Surprisingly, I've gotten very few takers in the decade or so that I've been posting that challenge. Most dropped out early on, while others simply refused to risk it at all. Seems they either are *not* confident in their own position or are not able to question their faith to that degree. One person who initially accepted my challenge deleted the entire exchange from his bulletin board shortly thereafter. In a post to another board that I wasn't expected to find, I saw where he explained to someone else that he had to end the discussion when he realized I was making sense. Another would-be opponent initially feigned objectivity, but soon admitted that he would rather take a bullet in the ear than give up his faith. So much for being objective.

In order to prove any point beyond such unreasonable doubt, you really have to know your own argument. You have to get your facts straight. You have to know how to present them in the right sequence, and get your quarry to acknowledge each point in progression. If you can do that, you can corner them so that they know their position is indefensible. At the same time, they will also know that they can't admit that openly even if it's already clearly obvious to everyone else, especially when their emotional investment in and financial dependence on doctrinal obligations prohibits them from conceding the point. Then when your opponent can neither change the subject nor escape, the result is usually a sort of psychotic meltdown. Rather than admit what should be obvious to all by then, your rival will often collapse into a childish emotional tantrum. This only happens when their belief is based on faith. Someone whose beliefs are instead based on reason is typically able to reconsider their position honestly if given good reason to. Even if it might be disappointing, they're not afraid of truth and wouldn't hide from it.

Only once has anyone actually taken my challenge and seen it all the way through. She was an old-earth creationist posting on Christian Forums under the username ConsideringLily. After our discussion, she accepted evolution, but at great personal cost, seeing as she was subsequently estranged from her husband, friends, and family, and even lost her job. This dearth of Christian love was just part of the impetus that drove her to reject all her remaining religious beliefs. Up to then, she had been a teacher in a fundamentalist Christian school, but within a year she was teaching science in a public school. ConsideringLily was transformed into the atheist activist LilandraX.

She was also the one who brought me the opportunity to address the Texas State Board of Education. She had a side job at that time as a newspaper reporter, which put her in touch with Mark Ramsey, then head of the Greater Houston Creation Association. Ramsey was a mechanical engineer from the oil fields of west Texas. He claimed to be a former evolutionist, although he couldn't even adequately define what evolution is. He also said that "scientific evidence" convinced him that the earth was miraculously created only 6,000 years ago, but he refused to say publicly what that scientific evidence was. Somehow, this evidence was described as personal and confidential, raising a few questions about how scientific it could be. He also fronted a group called Texans for Better Science Education, whose mission, according to its website, was to "teach the controversy"—that is, to teach both the "strengths and weaknesses" of evolution.

Some translation is required here. By "teach the controversy," he means to pretend there is a controversy. By "better" science, he means antiscience, because he is wholly opposed to scientific methodology. For example, his creation association actually proclaims that whenever reality conflicts with our interpretation of the Bible, then reality is wrong. So by "education" he obviously means indoctrination. Despite his promise, he also refuses to acknowledge which "strengths" of evolution even exist.

So let's see his summary of allegedly well-known scientific weaknesses of evolution:

- No explanation for the origin of useful complex information contained in the DNA double helix
- No explanation for the origin of life itself from dead chemicals
- No explanation for the Cambrian explosion of life
- No explanation for the massive, persistent, and systematic gaps in the fossil record that are nearly devoid of transitional forms
- No explanation for the irreducibly complex nature of cells and other biologic systems, to name a few

Apparently, our teachers are supposed to tell students that whenever we haven't figured something out yet, we should stop our research and assume God did it—as if that counts as an explanation. If we don't yet know exactly how the first living cells formed, that somehow negates everything we do know to be true about evolution after that. But worse, creationists want to mislead our kids into thinking that every kind of life appeared all at once, ignoring all the evident stages of progression stretched across time, and all the apparent predecessors found in earlier strata. They want to teach as fact outright falsehoods easily disproved, as well as pseudoscience already publicly exposed in a court of law.

We will expand each of these categories and examine them in more detail later on, but our impressions of them will not change. I learned that Mark Ramsey proudly showed his summary to some members of the State Board of Education. They shared his agenda of undermining science in public school classrooms, and they were already doing noticeable damage. That's when I decided I should speak up and do what I could to help minimize the damage.

To adequately understand evolution, you not only have to understand how to be scientific (which is the real trick for most people), but you also have to know something about cellular biology, genetics, anatomy, and geology (particularly paleontology), as well as environmental systems, tectonics, atomic chemistry, and especially taxonomy—which most people don't know squat about at all. Most people who accept evolution also tend to know a whole lot about cosmology, geography, history, sociology, politics, and, of course, religion. That is because those who respect academic scholarship and the acquisition of knowledge are not satisfied that anyone should simply believe whatever is told to them. We prefer that others should know and care why we think what we do. That's also true in the opposite case; we care to know *why* others think the way they do. That's what "understanding" means.

But to believe in creationism, you don't have to know anything about anything, and it's better if you don't, because creationism relies on ignorance. It is *not* based on honest research! It is a scam, a con job exploiting the common folk and preying on their deepest beliefs and fears. Creationist apologetics depends on misrepresented data and misquoted authorities, out-of-date and out-of-context, and uses distorted definitions, if it uses definitions at all.

There are basically two types of creationists: the first are the professional or political creationists—these are the activists who lead the movement and who will regularly deliberately lie to promote their propaganda. The second type are the innocently deceived followers, commonly known as "sheep." Somehow they don't consider this a derogatory term because they already identify themselves this way! Keep that in mind the next time you think about the children's hymn "I Am Jesus' Little Lamb."

I've known lots of intellectual Christians, but I can't get any of them to actually watch the televangelists because they either already know how phony they are or they don't want to find out. But that only allows this radical fringe to claim support from the masses they also claim to represent. So there's nothing to stop them. Professional creationists are making money hand over fist with faith-healing scams, or by bilking little old ladies out of prayer donations, or by selling books and videos at their circuslike seminars where they have undeserved respect as powerful leaders. All of them feign knowledge they can't really possess and profess absurdities they clearly don't know anything about, and some of them claim degrees they've never actually earned.

For example, a famous evangelist and convicted fraud Kent Hovind famously paid around $100 for a mail-order doctorate in education from a bogus diploma mill. (Patriot Bible College's campus looks like it is no more than a mobile home parked in the desert.) Hovind's illicit credentials were parodied in an episode of *The Simpsons*, in which an expert witness for creationism claimed to be a scientist by virtue of his "PhD in Truthology from Christian Tech."

Were it not for this con, these frauds and charlatans would have to go back to selling used cars, wonder drugs, and multilevel marketing schemes. They will never change their minds no matter what it costs anyone else. So it is obviously the "sheep" whom I'm attempting to reach, so that they might not be sheep anymore and will stop feeding fuel into that manipulative movement. Because it's one thing to believe in something that *might* be true, like some notions of a "higher power" (even though such can neither be substantiated nor tested in any objective way), but it is a whole other matter to willfully deceive others into believing things (like creationism) that are definitely *not* true—especially when we can also prove that those doing this know their assorted arguments are bogus and know they're lying to our children and that they hope to continue doing so under the guise of education.

I thought that the best way to enlighten the creationists on the Texas Board of Education was to have a debate between myself and Mark Ramsey. Dr. Steven Schafersman, head of Texas Citizens for Science, cautioned me against that, saying that any scientist who debates a creationist would give creationism the illusion of credibility and simultaneously diminish the reputation of that scientist. I told Dr. Schafersman that I was an undergrad rather than a professional scientist, so Ramsey would essentially be debating a former tattooist with no reputation to ruin. However, I still made the same mistake that many actual scientists have when debating creationists: I naively believed that once I disproved their allegations, properly address their criticisms, and showed them what the facts really are, they would have a moral obligation not to teach those falsehoods in school. What I had yet to learn was that not everyone cared to know or accept what is true.

Creationism extorts support through peer pressure, prejudice, and paranoid propaganda, and sells itself with short, simplistic slogans, which appeal to those who don't want to think too much or are afraid to question their own beliefs. Worst of all, it actually forbids critical inquiry and promotes anti-intellectualism, and it is based on a series of foundational falsehoods.

First and foremost among these is the idea that accepting evolution requires the rejection of theism, if not all other religious or spiritual beliefs as well. But while every relevant poll over the last century indicates that more than 95% of the global science community accepts evolution over millions of years, polls also show that as many as 33% of these "evolutionist" scientists also hold that the god they were raised to believe in is still somehow involved and helping to guide that process.

I knew better than to have a live debate, because that's the only time when creationists have the advantage; it takes longer to refute a lie than it does to tell one, and they can tell as many lies in one sentence as there are words in that sentence. (I've actually seen that done a few times.) Plus, they will pull out blind assertions without citation. Since there is no way to look these up on the fly, then there is no way to prove how, why, or even if they're wrong on that point. They'll make too many erroneous assertions per minute to even keep track of and they'll declare you the loser if you can't refute every single one of them in the same amount of time. This would still be impossible even if you already knew off the top of your head what was wrong with everything they said; even if you knew in advance everything they were going to say, and came prepared to refute it all, you're simply not allowed enough time to correct every error because there are so many and each requires some degree of re-education.

This creationist debate strategy is known as the Gish Gallop, named for Duane Gish of the Institute for Creation Research, who exemplified this style. That's why debates of this type should be in writing; any claim made in a blog-type forum can be examined by both sides before either one posts a reply. Bullshit won't fly in such a format, so it is not possible that any young-earth creationist could beat a competent advocate of science. People say I'm arrogant for saying that, but it's consistently true.

The tricky part of a written debate is getting the intended audience to actually read the posts and comment on them to show that they understood what they read. This is compounded by the fact that creationists will *not* be held accountable, so to counter both of these problems, I suggested that we have a half-dozen moderators, three of them creationists and three of them evolutionary scientists. I also insisted that all of the moderators be Christian.

For decades, those behind the creationism movement have tried very hard to portray the illusion that one cannot accept evolution and still believe in God. They know better, but they still want you to believe that evolution is atheist—that the only two choices are evolution without God or God creating without evolution. That's been their central claim since the creationism movement began. But this supposed controversy never was about whether there is a god. Most people believe there is a god, and they believe he is in control of all the seemingly random events of our lives. This is also true of most of the people in the general population who accept evolution. Most of them believe in God as well, and they believe that God is in control of evolution, and that evolution—like every other system in nature—is part of God's design.

Of the hundreds of different and often violently conflicting denominations of Christianity, the largest of them by far is Catholicism and the oldest is Orthodoxy. Both of these have stated support of evolution and denounced creationism. Pope Benedict XVI once described evolution as "an enriching reality," and he described creationist contests against it as "absurd." Two of the popes before him advised Christians around the world to consider evolution to be "more than a hypothesis" and not to fear acceptance of it as being any challenge to their faith in Christ. Following Benedict, Pope Francis also said that God created beings that evolve, and that God had allowed humans to develop according to internal laws.

So when a Wisconsin school board passed antievolution policies in 2004, hundreds of reasonable Christians got together on the Clergy Letter Project. (Different letters were composed for Christian clergy, Jewish rabbis, Buddhist sangha, and Unitarian Universalist ministers.) It was their version of Project Steve, except that instead of a poll of scientists endorsing evolution, it was a poll of religious believers also endorsing evolution. It read:

> We the undersigned, Christian clergy from many different traditions, believe that the timeless truths of the Bible and the discoveries of modern science may comfortably coexist. We believe that the theory of evolution is a foundational scientific truth, one that has stood up to rigorous scrutiny and upon which much of human knowledge and achievement rests. To reject this truth or to treat it as "one theory among others" is to deliberately embrace scientific ignorance and transmit such ignorance to our children. We believe that among God's good gifts are human minds capable of critical thought and that the failure to fully employ this gift is a rejection of the will of our Creator. To argue that God's loving plan of salvation for humanity precludes the full employment of the God-given faculty of reason is to attempt to limit God, an act of hubris. We urge school board members to preserve the integrity of the science curriculum by affirming the teaching of the theory of evolution as a core component of human knowledge. We ask that science remain science and that religion remain religion, two very different, but complementary, forms of truth.

As of October 25, 2015, the Clergy Letter Project had over 13,000 signatures. That's an order of magnitude more than Project Steve, which itself was hundreds more than the Discovery Institute's list of "dissenters from Darwinism."

The early pioneers of evolutionary science were all initially Christian (including Darwin), and many leading proponents of modern evolutionary science are still Christian today. For example, microbiologist Ken Miller describes himself as a very traditional Catholic; he was also an expert witness who testified against intelligent design creationism in *Kitzmiller v. Dover.* Another outspoken proponent of evolution is Robert T. Bakker. He was not only one of the leading and most recognizable paleontologists in the world, but he also happens to be a Bible-believing Pentecostal preacher—though he interprets Genesis differently than literalists would. In his book *Bones, Bibles and Creation*, he says that to treat the Bible as though it were common history is to degrade its eternal meaning. One of the earliest geneticists, Theodosius Dobzhansky, was an Orthodox Christian who many times professed his belief that life was created by God. But he famously said that "nothing in biology made sense except in light of evolution." All these men agree that even if there really is a god, and even if that god is the Christian god, and even if that god created the universe and everything in it (which they all believe), evolution would still be at least mostly true and creationism would still be completely wrong.

So on the science side for our debate, I selected geneticist Jill Buettner MSc as my first moderator. She was my own biology teacher, and she had been cited in the journal *Nature* for her contribution to the human genome project. I also knew she was active in her church, smoothing the division between fact and faith. I chose her based on comments she had made to me relating to religion: "I am a Christian and I *can* accept that Noah's Ark was a folk tale told by mouth until it was written down around Moses' time—it is not a firsthand account! Only literal Bible readers get bogged down trying to prove that the Creation story, Adam and Eve, and Noah's Ark are absolute fact."

I also chose her because of comments she made to me relating to her field of scientific specialty:

> The evidence of taxonomic relationships is overwhelming when you look at the comparisons between the genomic (DNA) sequences of both closely related and even distantly related species. The DNA of yeast and humans share over 30% homology with regard to gene sequences. Comparison of the human and mouse genome shows that only 1% of the genes in either genome fails to have an orthologue with the other genome. Comparison of nongene sequences, on the other hand, shows a huge amount of divergence. This type of homology can be explained only from descent from a common ancestor. The probability of these things being a coincidence, which I guess would be the argument of creationism and intelligent design, is statistically so small as to be negligible.

My second moderator was geologist Glenn Morton PhD, a former young-earth creationist and former contributor to *Creation Research Society Quarterly*. Nowadays he is a theistic evolutionist, something he describes as a necessary adaptation for a Christian working in petroleum exploration. He wrote an intriguing account of his departure from the young-earth creationist perspective and how his former associates reacted to that. He also created a meme with an article explaining his own confirmation bias, which he called "Morton's Demon."

My last science-minded moderator was the famed paleontologist Dr. Robert T. Bakker, a long-haired, bearded, grizzled old prospector with two Ivy League doctorates. He was famous as a science adviser for the original *Jurassic Park* movie. Actor Thomas Duffy's character in *The Lost World* was based on him. But Bakker also lived a relatively secret life as a Pentecostal preacher. He was very familiar with the radical right-wingers on the Texas Board of Education, and he referred to them as "bigots."

With these people on the panel, I figured that much of the contentions in defense of religion should be negated since all these evolutionary scientists were Christian. Plus, any absurdity posed on genetics, geology, or the fossil record would immediately be shot down with expertise, so that the pseudoscientists would know what they're up against. Sadly, I did not yet know what *I* was up against.

Of all the developed nations throughout "Christendom," only the United States has a significant number of creationists, and they're the minority even here! Every other predominantly Christian country tends to regard creationism as an incredulous (if not insane) radical fringe movement that is an almost exclusively American phenomenon. It is not taken seriously anywhere else in the West. Poll after poll continue to reveal that around the world most "evolutionists" are Christian, and most Christians are evolutionists. So evolution is not synonymous with atheism, and creationism isn't synonymous with Christianity either. Most creationists aren't even Christian! There are millions more Muslim and Hindu creationists than Christian ones.

No matter which religion they claim, creationists can be collectively defined as the fraction of religious believers who reject science—not just the conclusions of science but its methods as well, and I mean *all* of them, from uniformitarianism and methodological naturalism to the peer review process and the requirement that all positive claims be based on testable evidence. These people rely instead on blind faith in the assumed authority of their favored fables. In all cases, creationism is an obstinate and dogmatic superstitious belief, which holds that members of most seemingly related taxonomic groups did not evolve naturally but were "created" magically—that plants and animals were literally poofed out of nothing fully formed in their current state, unrelated to anything else despite all indications to the contrary.

The first two moderators on the antiscience side of our debate were my intended audience. Terri Leo was Lilandra's primary contact on the Texas

State Board of Education and my primary target. Leo was so shamelessly opposed to any progressive position that her critics referred to her as "the Sultana of the Texas Taliban." She once stopped the state board's dues payments to the National Association of State Boards of Education because the national association was opposed to the harassment of homosexuals. So here is a woman who apparently condones harassment on the basis of sexual orientation—a hate crime.

On his science blog, Pharyngula, Prof. P. Z. Myers said of her:

> She's a real piece of work. She's been working like a maniac to gut textbooks; she's even tried to get publishers to add little 'facts' like "Opinions vary on why homosexuals, lesbians and bisexuals as a group are more prone to self-destructive behaviors like depression, illegal drug use, and suicide." She's a perfect example of anti-science, anti-intellectual, intolerant bigotry, and yet there she is on the state board of education. . . . Leo would be a hilarious clown if she weren't so dangerous.

I invited Leo only because she said that science classes shouldn't present theories as fact. Refusing to accept what "theory" really means, she also denied the theory of gravity, saying that it is a law instead. She should have been taught better than this in middle school. She also said that no transitional species have ever been found. All these comments are easy to refute, so it should be no problem showing where she was mistaken so that she would admit her error and change her position accordingly. Right?

The next moderator for supernaturalism was a mutual choice, again from the State Board of Education. Mark Ramsey wanted Don McLeroy to moderate opposite my scientists. At that time, I thought McLeroy was easily the single least intelligent member of the whole board. In a public statement, he said that "most of the books we are considering adopting claim that Nothing made a spider out of a rock." McLeroy's knowledge of science is so out-of-date he professed to believe that our solar system is the center of the universe, and even thinks that this Copernican view is still accepted by scientific consensus today. Apparently McLeroy's textbooks were printed in the 1500s. It shouldn't be surprising that our highly questionable governor later promoted McLeroy to chairman of the State Board of Education—makes perfect sense.

In a failed attempt to ease my concerns over that, a few emails came to me explaining how some other members of the board were even less competent than McLeroy. For example, Vice Chairman David Bradley famously said, "This critical thinking stuff is gobbledygook. Students need to be able to jump to their own conclusions."

Yes, he thought that jumping to conclusions is a good thing and that rational thinking is bad. Bradley's biases also prompted him to reject an algebra textbook because he didn't like the illustrations and references to women's suffrage, biology, and the Vietnam War. It seems that if it ain't sexist, nationalist, and creationist, he doesn't wanna know about it. He also rejected an advanced placement environmental science textbook because it promoted environmentalism, which he and other board members considered to be "anti-American and anti-Christian." That's right. Environmental conservation is anti-Christian. As Ann Coulter wrote in a piece titled "Oil Good; Dems Bad," "The ethic of conservation is the explicit abnegation of man's dominion over the Earth. The lower species are here for our use. God said so: Go forth, be fruitful, multiply, and rape the planet—it's yours. That's our job: drilling, mining and stripping. . . . Sweaters are the anti-Biblical view. Big gas-guzzling cars with phones and CD players and wet bars—that's the Biblical view."

Cynthia Dunbar, another board member, said that Barack Obama was a terrorist who would declare martial law within the first six months of his administration, and she made no apology when he didn't. She also wrote a book wherein she refers to public education as a "tool of perversion" that is unconstitutional and tyrannical, given that secular education does not endorse her extreme religious or political views.

Bradley and Dunbar are both college dropouts who homeschool their kids—no doubt to shelter them from "perversions" like sex education, advancement of civil rights, progressive equality, and all forms of modern understanding. Yet none of that seems to be enough to disqualify their participation on the Texas State Board of (alleged) Perversion.

Gail Lowe, who normally chaired board meetings, said she would never support the adoption of an environmental science book that attributes global warming to human activity. She believed that all such statements followed a political agenda rather than "solid objective science." The irony is that hers was the political agenda; her promise to preserve and promote the values of Christian conservatives required her to deny all the solid objective science that kept contesting her conclusions. For instance, she said that "the

National Academy of Sciences has still stated [evolution] is not a fact, and we don't believe evolution ought to be taught as a fact."

Of course, the actual position of the National Academy of Sciences is that evolution is both a fact and a theory. Gail refused to correct her statement because she refused to accept that evolution is at all factual.

Ken Mercer was another board member who criticized evolution and promoted intelligent design. He was also opposed to sex education, and he advocated discrimination against gays. There were many other creationists on the board too. In fact, most of the members were, and they all voted as a bloc right along party lines as if they were strategically positioned to do so—because they were.

It turns that many of these creationists were given substantial campaign donations all from the same source: James Leininger, a former physician and successful businessman with more than two hundred million dollars. He was known as the Daddy Warbucks of the Republican Religious Right in Texas. This sinister saboteur of science and sanity (also known as "God's paymaster") donated tens of thousands of dollars to elect Rick Perry as governor, and he donated a couple hundred thousand dollars just to seat right-wing Republicans on the State Board of Education. That includes $35,000 to Ken Mercer and $20,000 to Don McLeroy. Compare those to equivalent amounts donated to the governor's campaign, and you can see where Leininger's interests lie. Better-suited and more appropriate candidates not backed by such powerful special interests would have to fund their own campaigns, and would be hopelessly outgunned—especially when you remember that serving on the SBOE is an unpaid voluntary position! However, McLeroy described that position thusly: "Sometimes it boggles my mind the kind of power we have."

This is no paranoid conspiracy theory; this is a matter of verifiable public record. It was also a successful strategy as indicated in the following email sent to me from one of the board members on the left of the domineering majority:

> Unfortunately, the State Board of Education is controlled by the religious right. They will vote against the body of research that supports the teaching of evolution in our science books. The only ones that will vote with the Science professionals will be Mary Helen Berlanga, Bob Craig, Mavis Knight, Rene Nunez, Pat Hardy, and maybe Lawrence Allen. All the

> others will vote for the inclusion of "intelligent design" and "creationism" in our science books. You can bring a million experts in the field of science and research scientists will tell them otherwise, but, they have the votes and they will do to the Science TEKS [Texas' Essential Knowledge and Skills] what they did to the English Language Arts and Reading TEKS: vote against the body of research supporting the professionals in that field.
>
> Until we can replace David Bradley out of Beaumont, and he does have an opponent in November and replace Cargill (The Woodlands) who has a write-in candidate running against her, there is no hope of doing what is right for public education.

Clearly it would be impossible to reason with any of these purposefully appointed paid volunteers whose insidious mission was no secret; it was an all-out crusade. In a televised interview on ABC News, Don McLeroy actually admitted (by acknowledging in the affirmative) that his intention as an elected official was to impose his own religious and political views onto other people's children.

None of the board were as notorious as McLeroy. When he announced his plan to teach his imagined weaknesses of evolution to high school students, a coalition of more than 800 Texas scientists endorsed the following response on the website TexasScientists.org:

> The fact that biological populations evolve is not in question. Evolution is an easily observable phenomenon and has been documented beyond any reasonable doubt. The "theory" part of evolutionary theory concerns the experiments, observations, and models that explain how populations evolve. At this level of introductory instruction, it is ludicrous to think about teaching what some people disingenuously call "weaknesses." We teach what is known and has been supported by a huge body of scientific research.

To this McLeroy casually replied, "Evolution is vital to understanding a lot of the sociological sciences, not the biological sciences." As if a dentist knows more about this than all the experts in the field. Remember he's the one who famously said, "I don't agree with these experts; somebody's gotta stand up to experts."

The only thing McLeroy ever said that I agree with was, "It is wrong to teach opinions as fact." However, that is exactly what he does. He is a fourth-grade Sunday school teacher who definitely does teach his own personal, unsubstantiated, usually refuted, and often-bigoted religious opinions to other people's children. McLeroy considers his opinions to be factual, and he considers actual fact to be a matter of opinion. A man without humility or shame at his own hubris, he explained on his website his intent to oppose not only evolution, but also an even bigger target.

> In the words of Phillip Johnson it is "metaphysical naturalism" or "materialism" or just plain old "naturalism"; it is the idea that nature is all there is. Modern science today is totally based on naturalism. In all of intelligent design's arguments against both Darwinian evolution and the chemical origin of life, it is their naturalistic base that is the ultimate target. The important aspect of Darwinian evolution is its naturalistic claim that all life is a result of purposeless, unintelligent, material causes.

McLeroy clarified his stance in an opinion piece to the *Austin American-Statesman*:

> If science is limited to only natural explanations but some natural phenomena are actually the result of supernatural causes, then science would never be able to discover that truth—not a very good position for science. Defining science to allow for this possibility is just common sense. Science must limit itself to testable explanations, not natural explanations. Then the supernaturalist will be just as free as the naturalist to make testable explanations of natural phenomena. The view with the best explanation of the empirical evidence should prevail.

In a Sunday school lesson at Grace Bible Church, he encouraged fourth-grade students to

> *Keep chipping away at the objective empirical evidence*, and keep pointing out their deductive reasoning depends on the premise "Nature is all there is." Remind them that they may be wrong. [Emphasis mine]

McLeroy obviously does not understand that the word "miracle" is defined as an event or phenomenon that cannot be accounted for by science because it defies physical laws. If it defies the laws of physics, then it is physically impossible by definition. Thus science generally does not consider the impossible to be possible unless otherwise indicated, which in this case it is not. In fact, if you look up the words "miracle" and "magic" and compare them in a handful of different dictionaries, you'll see that they share much the same definition; they are the same thing.

Miracle:
1. Literally, a wonder or wonderful thing; but appropriately,
2. In theology, an event or effect contrary to the established constitution and course of things, or a deviation from the known laws of nature; a supernatural event. Miracles can be wrought only by almighty power.

Magic:
1. The art or science of putting into action the power of spirits; or the science of producing wonderful effects by the aid of superhuman beings, or of departed spirits; sorcery; enchantment. [This science or art is now discarded.]

—*Webster's 1828 Dictionary*

Miracle:
An event that appears inexplicable by the laws of nature and so is held to be supernatural in origin or an act of God.

Magic:
a. The art that purports to control or forecast natural events, effects, or forces by invoking the supernatural. The practice of using charms, spells, or rituals to attempt to produce supernatural effects or control events in nature.
b. The charms, spells, and rituals so used.

—*TheFreeDictionary.com*

Miracle:
an effect or extraordinary event in the physical world that surpasses all known human and natural powers and is ascribed to a supernatural cause.

Magic:
the art of producing a desired effect or result through the use of

incantation or various other techniques that presumably assures human control of supernatural agencies or the forces of nature.

—*Dictionary.com*

The only difference between miracles and magic is who does it. A boat may be considered a ship if it's big enough. When a rich man is neurotic, we call him eccentric. When a V.I.P. is murdered, it's an assassination. When a god performs magic, he's working miracles.

Again, if you look at a consensus of sources, citing only those definitions in the relevant context, then magical miracles or miraculous magic is defined the same; both are an evocation of supernatural forces—be it by charms, spells, incantations, or prayers, all of which are found in the Bible—to effect control of nature in ways that are inexplicable by science because they defy the laws of physics. It would also be fair to say that anything that the laws of physics cannot account for or permit is physically impossible. Therefore miracles are impossible by definition.

I'm sure McLeroy doesn't realize that either. He thinks it is possible to have a supernatural explanation that is testable, but the definition of that word holds that the supernatural cannot be measured or tested. Only natural explanations can be testable, verifiable, or potentially falsifiable. Neither does McLeroy understand what the word "empiricle" [sic] means. But he should at least be able to grasp that "chipping away" at empirical evidence cannot be the best explanation of that evidence. He is right only in that the view with the best explanation of the empirical evidence should prevail, and indeed it has. Creationism lost that contest back in the nineteenth century and has been losing continuously ever since.

Moving from bad to worse, my opponent in the proposed debate chose, as his third moderator, Mark Cadwallader of Creation Moments online ministries. Cadwallader claimed to be a "government scientist" because he has a master's degree in chemistry and once testified before the House of Representatives about hazardous materials. However, he somehow displayed even less competence than your average layperson on every topic we brought up, and he complained that I used too many big words. His funniest moment had to be when he said, "We are not eukaryotes because we were never eukaryotes, get it?"

That's when I had to explain to Mr. Government Scientist that a eukaryote was any organism with nucleic cells—that is, cells with a nucleus. Then he argued that we weren't organisms either, because he thought the word "organism" applied only to single-celled microbes. We had to argue about that too, and he refused to ever accept when he was wrong about anything, no matter how academic the point.

Lilandra, who had put this debate together, had only recently abandoned her belief in special creation. I think she was still evaluating those long-held beliefs, and may have used this venue to compare how each side presents its case. Unfortunately for her failing faith, my opponent Mark Ramsey was being extremely difficult about the whole thing even before it started. The most amusing episode occurred when I posted a list of definitions for many of the technical terms that invariably come up in all of these types of discussions. Ramsey didn't object to the definitions I presented, but he wouldn't agree to them either. Instead, he said that we should leave all these terms undefined so that our readers could decide what *they* think these words should mean. He actually said that! The only reason I can imagine for him to take a position like that is that he intends to move the goalposts by misusing each of these words, trusting that his audience of undereducated laypeople wouldn't understand scientific jargon any better than he does. Otherwise he would lose before we begin, because I would force him to adhere to actual definitions rather than the straw men that all his arguments depend on.

Cadwallader waited until the debate began to argue over the definition of evolution. He said that textbooks defined evolution as "the godless origin of life from nonlife," or something very similar to that. I found a half-dozen textbooks, most of them current in Texas schools at that time; all of them defined evolution as "descent with inherent genetic modification," or "variation of allele frequencies in reproductive populations over time," or some paraphrased but essentially similar version of that. Cadwallader refused to accept the consensus definition from any number of sources, regardless of authority. He insisted that most textbooks typically pushed his definition even though he couldn't point to a single one that ever did. He also objected that genetics and evolution were even related topics. For him, evolution equals irreligious materialistic atheism, and he would not concede where he was wrong on any (much less every) point.

I opened the debate with a complete dismantling of every last one of Ramsey's alleged weaknesses of evolution point by point. From my perspective, I won the debate in the first post, because there was no way he could defend anything that he still wanted to teach. So his first post largely ignored everything I had already said. He said at one point that there were very few transitional species that he was aware of, but that he still wanted to teach that there were none at all because he said it was important for students to *believe* there were none. He admitted that he knew better, but he intended to lie to students anyway. I have seen many creationists imply that whether we believe something matters more than whether it is true, and Ramsey showed that he has no problem misrepresenting data to deliberately deceive impressionable children rather than educating them properly or honestly.

I thought that all of the moderators would have been outraged by that, but Cadwallader was such a raving loon that they weren't even paying attention. He immediately and repeatedly attacked each of the science-minded moderators for not being "true" Christians, and he accused them of being in league with the devil in some secret Satanic atheist conspiracy against God. I'm not kidding; he really did accuse them of being Satanic Christians and Christian atheists and atheist devil-worshippers. The first time I remember ever seeing the term "bat-shit crazy" was a comment from one of the discussion boards describing Cadwallader's behavior in that debate.

Once Cadwallader started his raving, my geneticist Jill Buettner simply stopped participating. All the idiotic comments my opponents made about genetics and mutations went unchecked, but I can't fault her for walking out when pious monkeys start throwing feces. I just wish she had said something about the panel's inappropriate conduct.

I was initially delighted about having Bob Bakker involved, but he didn't seem to read any of the posts in that debate either. All the nonsense Ramsey claimed about the fossil record was posted without moderation. Bakker wrote quite a lot in that forum, but he hardly addressed the debate at all. Instead he used my venue to get on his soapbox and rant about his disdain for atheists. He did this knowing that I was an atheist, and knowing that he was there at my invitation because of my respect for the man. We had a very encouraging conversation on the phone when I first invited him, so I certainly didn't see this coming.

The most disappointing participant of all, however, was Glenn Morton. His hatred of atheists seemed to dwarf that of everyone else on the panel. Suddenly he had nothing but vitriol for Lilandra and me. I suspect that he was even sending personal messages to Ramsey trying to help him behind the scenes. The defense of accurate science education didn't even matter to him anymore, and he wasn't at all ashamed about that. He said that he just didn't want an atheist to win a debate against a Christian—even if the Christian is wrong and lying, and a dangerous threat to the education of millions.

Years later, Glenn Morton again demonstrated that religious biases matter more to him than either honesty or academic accuracy. Late in 2012, he deleted all his files from his own website simply so that atheists wouldn't make use of the cache of facts he had previously posted. Morton explained his reasoning in a disturbing diatribe wherein he revealed the depths of his right-wing prejudice, especially his hatred of atheists. Worst of all, he considers it acceptable to teach children alleged facts that we can all prove—and that *he* can easily prove—to be certainly wrong. He says religion has a right to be wrong, and that means it's okay to lie to children under the guise of teaching them. He says creationism is factually wrong and he considers it a detriment to Christianity as a whole, but he still prefers its teaching to a science education based only on actual facts because he somehow considers such an education to be tantamount to a totalitarian forced conformity. This was a man once respected by theists and rationalists alike. It is hard to see how anyone could respect him anymore.

At the close of our debate, it seemed that Terri Leo had read only one comment from my final post and clearly did not understand it. Nor could she have, since I was never allowed to explain it. All parties walked away from that debate prematurely in the third round with both sides claiming victory. Don McLeroy may not have known the debate was even going on until after it was over. So the two people I was trying to reach both remained wholly oblivious to my efforts. The debate was a complete failure on all fronts.

Lilandra and I had intended this to be a well-documented confrontation of legitimate science versus pseudoscience, fact versus fraud, but it degraded immediately into a flame war of pure prejudice from dogmatic believers, all allied against me alone. What's more, Lilandra paid to have the debate hosted on a site that simply disappeared a year or so later, so that the only record of this encounter are the comments about it that can still be found on places like ChristianForums.com. In my foolish innocence, I really thought that men of science who cared about academics would—as a matter of honor—be able to put aside what I consider to be their relatively trivial religious beliefs for the sake of honest factual academic accuracy. I was sadly mistaken.

Creationists may side either with Western Abrahamic religions, in which there are conflicting versions of the same tales, or with one of many Eastern religions where the sacred stories of creation are much older, completely different from their Abrahamic counterparts, and dedicated to other gods and pantheons. But in every case, the proposed "creator" is supernatural, meaning that it is not a part of perceptible reality. It is therefore undetectable by any testable means and can only be assumed to exist for subjective emotional reasons, or as a result of cultural indoctrination rather than because of any measurable evidence or logical rationale. In other words, there's no way to say if it's really there. Worst of all, there's also no way to distinguish anyone's gods or ghosts from the imaginary beings some primitive folks just made up either. This doesn't mean no god exists, but it *does* mean that science can't say anything about them, because even if gods are real, they still don't appear to be and apparently don't want to since all the holy books demand they be believed on faith alone. As there is nothing anyone can verify and thus actually know to be correct about gods, then science is unable to make any comment about them at all, because science can only ever investigate things with demonstrable evidence that can be tested or measured. The supernatural by definition *cannot* be tested or measured, indicated, or falsified. In all instances, the supernatural is indistinguishable from the imaginary.

From the creationist's perspective, the method or mechanism of creation which these mystical beings use is nothing more than a golem spell, where according to ancient Semitic folklore, clay statues are magically animated with an enchantment. Or it's an incantation in which complex modern plants and animals are "spoken" into being. That's right, magic words which cause fully developed adult animals to be conjured out of thin air. Or a god simply wishes them to exist, so they do.

That's it! There really is nothing more to it than that: just pure freakin' magic by definition. Remember that the next time you hear anything from a creation "scientist."

So for those who believe in God, the question really is how God created—whether he did so through one of many inextricably integrated natural systems he seemingly designed, or whether he simply blinked, wiggled his nose, wished upon a star, and said "*avra kehdabra.*"

THE 2ND FOUNDATIONAL FALSEHOOD OF CREATIONISM

Sacred Scriptures Are the "Word of God"

When believers argue over any of the many things that contradict their religion, they often challenge us to decide, whom are we going to believe? The alleged "word of God"? Or that of men? As if human inquiry had no chance against the authority they imagine their doctrine to be. But when they say "men," they're talking about science, and when they refer to the "word of God," they're talking about myths written *about* God—by men.

If there really is an intelligent and purposeful creator, then it would have to be he (assuming it is not a goddess, or a god of no gender) who constructed the fossil record revealing evolutionary history, who conceived the genetic patterns which also trace that same course, and who must have added all the other lines of evidence that point to the evolutionary conclusion exclusively and in brilliant detail. Why else would all these things exist? It's as if he were trying to tell us something! Men couldn't create any of those things. But men can tell stories, whoppers in fact. And it was men who wrote all the scriptures pretending to speak for God.

Every one of the world's supposedly holy doctrines of any religion describe themselves as being written by men, not gods. Sure, the men were reportedly "moved by" or inspired by their favorite gods, or perhaps taking dictation from angels, but they were written by lowly, imaginative yet imperfect mortals nonetheless—not by angels, and certainly not by gods.

It is easy for most Christians to charge that the Book of Mormon was written by a man, and that his work didn't really have the supernatural editorial he claimed. They might say the same about the Bahá'í Kitab-i-Aqdas, and the Qur'an as well, but they don't examine the works attributed to the apostles with the same consideration. The Jewish perspective is that Christians are contorting the Tanach, the Jewish Bible. Muslims say the whole of the Bible, both the originally Jewish foundation and the Christian revisions, were corrupted by human editors, but they say the Qur'an remains perfectly flawless—unless of course it's a translated version not in the original Arabic. New interpretations of scripture are often promoted as "getting back to the original message," while the orthodox of course declare those same works to be heresies. Consequently, all these conflicting divisions of the Abrahamic god have been at war with each other continuously, each since their inception. This is not a compelling testament to any omniscient infallible all-father simultaneously associated with infinite truth, love, and peace.

If there really was one true god, it should be a singular composite of every religion's gods, an uber-galactic super genius, and the ultimate entity of the entire cosmos—neither Krishna nor Christian, but both and beyond either one. If a being of that magnitude ever wrote a book, then there would only be one such document—one book of God. It would be dominant everywhere in the world with no predecessors or parallels or alternatives in any language, because mere human authors couldn't possibly compete with it. And you wouldn't need faith to believe it, because it would be consistent with all evidence and demonstrably true, revealing profound morality and wisdom far beyond contemporary human capacity. It would invariably inspire a unity of common belief for every reader. If God wrote it, we could expect no less.

But what we see instead is the very opposite of all that. Instead of only one religion leading to one ultimate truth, we have many different religions, some with no common origin, all constantly sharding into ever more deeply divided denominations seeking conflicting "truths," and each somehow claiming divine guidance despite their ongoing divergence in every direction.

The Jewish Torah, the Christian gospels, the Qur'an of Islam, the Kitab-i-Aqdas of Bahá'u'lláh, the Hindu Vedas, the Avestas of Zarathustra, the Adi Granth of the Sikhs, the Mahabarata's Bhagavad Gita, the Book of Mormon, and the Urantia Book are all declared by some of their devotees to be the "absolute truth" and the "revealed word" of the "one true god," and believers of each say the others are deceived. The only logical probability is that they are all deceived, at least to some degree.

None of these have any particular advantage over the others. None of them have any evidentiary support, and none of them are historically verifiable. All of them require faith and apologetics as well, because they also contain inconsistencies, absurdities, and primitive notions once held true but which have since been disproved. So they can claim no evidence of divine wisdom. Many of them promote heinous atrocities in place of morality, and some claim to be validated by prophecies now fulfilled, where each may also contain prophecies that failed to come to pass as predicted.

For example, Ezekiel prophesied that Nebuchadnezzar II, the king of Babylon, would conquer the city of Tyre and destroy it utterly, breaking down all the walls, streets, and towers, and killing or driving away everyone therein. Then God was supposed to step in and cause the city to sink beneath the deep and be lost forever. It would never be found again, because it would be unrecognizable—just an uninhabitable bare rock in the midst of the waters and no more than a place for fishermen to spread their nets.

Biblical literalists are forbidden to admit that the Bible could ever be wrong about anything, which is why so many apologists are employed full time making up excuses trying to conceal the many errors of scripture. Being desperate to defend this prophecy, some have argued that it may have taken centuries for this to happen, and that it took "many nations" to do it, but that this prediction was eventually fulfilled to the letter. Or some will say that it was sort of fulfilled depending on how you look at it.

I maintain that to believe this prophecy actually came true, one must employ both colorful rationalizations and strategic omissions. For sheer amusement, I will illustrate these by citing the apologists themselves. While I discuss the conditions and the situation, I will also compare the explanations given by two sets of defenders of biblical prophecy. According to Padfield.com, which represents the Church of Christ,

> After the destruction of Jerusalem and the carrying away of her king Zedekiah into captivity, "Nebuchadnezzar took all Palestine and Syria and the cities on the seacoast, including Tyre, which fell after a siege of 13 years (573 B.C.)" (E. A. Wallis Budge, *Babylonian Life And History*, p. 50). The

> inhabitants of Tyre fled to a rocky island half a mile offshore. The walls on the landward side of the island were 150 feet high. "The channel between Tyre and the mainland was over twenty feet deep, and frequently lashed by violent south-west winds. Their fortifications, they believed, would resist the strongest battering-ram yet devised. The city-walls stood sheer above the sea: how could any army without ships scale them? Shore based artillery was useless at such a range." (Peter Green, *Alexander of Macedon 356–323 B.C.*, p. 248).

So the Church of Christ says that Nebuchadnezzar captured Tyre as predicted, but that the Tyrians fled to "a rocky island." That sounds like a successful campaign until you realize that the island is more than just rocks. It has a sheer impenetrable wall fifty yards high around a city, and the name of that city is Tyre. Their own citation describes "the channel between Tyre and the mainland," meaning of course that Nebuchadnezzar never took the city at all. Tyre was an island city visible from an associated settlement on the mainland. Here is how Ezekiel 26 distinguishes the two establishments in the New American Standard Bible:

> (3) therefore thus says the Lord GOD, 'Behold, I am against you, O Tyre, and I will bring up many nations against you, as the sea brings up its waves. (4) 'They will destroy the walls of Tyre and break down her towers; and I will scrape her debris from her and make her a bare rock. (5) 'She will be a place for the spreading of nets in the midst of the sea, for I have spoken,' declares the Lord GOD, 'and she will become spoil for the nations. (6) 'Also her daughters who are on the mainland will be slain by the sword, and they will know that I am the LORD.'

The New Revised Standard Bible refers to Tyre's "daughters on the mainland" as "daughter-towns in the country." In either case, we can see that the target is the island city, not the "daughter-towns." The mainland settlement was supposed to be taken in addition to the primary city on the island, the one with the sheer wall, which Nebuchadnezzar never breached. He did slay the daughters on the mainland, but that's all he ever managed to do. The first failure of this prophecy is that all the rest of that stuff that the Bible says Nebuchadnezzar himself was supposed to do was actually done by Alexander the Great a couple centuries later. According to BibleArchaeology.org, which is maintained by Associates for Biblical Research,

> Nebuchadnezzar, king of Babylon, besieged Tyre for 13 years (585–572 BC), but the precise historical facts of its outcome are still unclear. He evidently did not conquer the city, but it may have surrendered conditionally to him. Both Jeremiah (27:3–11) and Ezekiel (26:7–14) spoke of this event. Apparently both Tyre and Sidon surrendered to Nebuchadnezzar, based on a fragmentary Babylonian administrative document which mentions the kings of Tyre and Sidon as receiving rations from the royal Babylonian household.

So the Associates for Biblical Research concede that Nebuchadnezzar failed where the Church of Christ say he succeeded, and the only way they can justify that admission is if they interpret negotiated treaties to count as some form of conquest. They also have no reason to assume that Tyre made any such treaties at that time, and Ezekiel 29:18 states very clearly that no such treaty was ever made. As for the fragmentary Babylonian document, archaeologist James B. Pritchard's classic anthology *The Ancient Near East in Pictures Relating to the Old Testament* is cited, but neither of the biblical citations given by the Associates for Biblical Research support their conclusion at all. Jeremiah 27:3–11 says that God had given all the land to Nebuchadnezzar, and that he would drive the Tyrians out so that they would perish, which is essentially the same prophecy made in Ezekiel 26:

> (7) For thus says the Lord GOD, "Behold, I will bring upon Tyre from the north Nebuchadnezzar king of Babylon, king of kings, with horses, chariots, cavalry and a great army. (8) He will slay your daughters on the mainland with the sword; and he will make siege walls against you, cast up a ramp against you and raise up a large shield against you. (9) The blow of his battering rams he will direct against your walls, and with his axes he will break down your towers. (10) Because of the multitude of his horses, the dust raised by them will cover you; your walls will shake at the noise of cavalry and wagons and chariots when he enters your gates as men enter a city that is breached. (11) With the hoofs of his horses he will trample all your streets. He will slay your people with the sword; and your strong pillars will come down to the ground. (12) Also they will make a spoil of your riches and a prey of your merchandise, break down your walls and destroy your pleasant houses, and throw your stones and your timbers and your debris into the water. (13) So I will silence the sound of your songs, and the sound of your harps will be heard no more. (14) I will make you a

> bare rock; you will be a place for the spreading of nets. You will be built no more, for I the LORD have spoken," declares the Lord GOD.

It is difficult to envision such petty threats as these coming from the creator of a trillion galactic clusters. Notice there is no hint of any treaty here, and there is no implication of any conditional surrender either. Nor could such have even been possible in this passage. Verse 7 and the first half of verse 8 are all that came true as predicted. The second half of the eighth verse through the twelfth verse could be said of Alexander the Great, but not Nebuchadnezzar, so the prophecy fails. The argument of some apologists is that the sentences beginning with "he" refer to the king of Babylon, but that the sentences beginning with "they" refer to a series of later invasions by Macedonians and even Muslims as far into the future as the crusades. But of course this prophecy worded as it is doesn't allow for Greece or anyone else to step in centuries after the failure of Babylon. Verses 13 and 14 never came true at all, nor did any of the subsequent verses of this same prophecy:

> (15) Thus says the Lord GOD to Tyre, "Shall not the coastlands shake at the sound of your fall when the wounded groan, when the slaughter occurs in your midst? (16) Then all the princes of the sea will go down from their thrones, remove their robes and strip off their embroidered garments. They will clothe themselves with trembling; they will sit on the ground, tremble every moment and be appalled at you. (17) They will take up a lamentation over you and say to you, 'How you have perished, O inhabited one, From the seas, O renowned city, Which was mighty on the sea, She and her inhabitants, Who imposed her terror On all her inhabitants! (18) Now the coastlands will tremble on the day of your fall; Yes, the coastlands which are by the sea will be terrified at your passing.'" (19) For thus says the Lord GOD, "When I make you a desolate city, like the cities which are not inhabited, when I bring up the deep over you and the great waters cover you[.]"

Actually, the reverse happened here. Alexander the Great built a huge causeway joining the island to a peninsula with the mainland. None of that nonsense with "princes of the sea" ever happened. The coastal kingdoms continued to deal with the Tyrians pretty much as before, and for more than a thousand years after this. The city was never rendered desolate, or uninhabited, and the waters never covered it over either.

> The modern city of Tyre is of modest size and is near the ancient site, though not identical to it. Archaeological photographs of the ancient site show ruins from ancient Tyre scattered over many acres of land. No city has been rebuilt over these ruins, however, in fulfillment of this prophecy.

So say Lane T. Dennis and Wayne Grudem in *The ESV Study Bible*, as quoted by the Church of Christ. What do the Associates for Biblical Research say?

> Tyre also recovered from Alexander's devastation. In 126 BC, now a peninsula extending into the Mediterranean, Tyre became a Roman province and later the capital of Rome's Syria-Phoenician province.

There seems to be some disagreement here. Did the city recover or not? According to Christian apologists George and Ray Konig in their book *100 Prophecies*,

> In Ezekiel 26:14, the prophet says the Phoenician city of Tyre would be destroyed and never be rebuilt. This was fulfilled when Alexander the Great conquered Tyre in 332 BC. His conquest brought an end to the Phoenician Empire. The empire never recovered from the attack. And so, it could never rebuild Tyre. Other nations and empires have built and rebuilt cities on or near the original Phoenician site.

So saith the novices, but what about the scholars? According to Edward S. Ellis and Charles F. Horne, PhD, in an excerpt from *The Story of the Greatest Nations and the World's Famous Events, Vol. 1* (quoted on PublicBookShelf.com),

> Tyre was still their [Phoenician] leader; for she rose phoenix-like from the destruction caused by Alexander. Only eighteen years after his successful assault we read of the stubborn Tyrians enduring another siege from one of the Greek generals who succeeded to his empire. This time Tyre made profitable terms of peace after holding back the besiegers for fourteen months.

Here we finally hear of the treaty that the Associates for Biblical Research might have been talking about, but this led to a profitable peace, not a conditional surrender. This practice continued as Tyre persisted throughout the stories of the New Testament, according to Acts 12:20. So the city was never taken by Nebuchadnezzar, and did recover after Alexander. It may not have recovered completely, but it recovered immediately. It retained and regained some of its power, was never made desolate, and it never sank beneath the waves either, right? The Church of Christ quotes Gleason L. Archer from *Encyclopedia of Bible Difficulties*,

> In point of fact, the mainland city of Tyre later was rebuilt and assumed some of its former importance during the Hellenistic period. But as for the island city, it apparently sank below the surface of the Mediterranean.... All that remains of it is a series of black reefs offshore from Tyre, which surely could not have been there in the first and second millennia BC, since they pose such a threat to navigation. The promontory that now juts out from the coastline probably was washed up along the barrier of Alexander's causeway, but the island itself broke off and sank away when the subsidence took place; and we have no evidence at all that it ever was built up again after Alexander's terrible act of vengeance. In the light of these data, then, the predictions of chapter 26, improbable though they must have seemed in Ezekiel's time, were duly fulfilled to the letter—first by Nebuchadnezzar in the sixth century, and then by Alexander in the fourth.

Wait, the island actually did sink beneath the waves? Nothing left but an uninhabitable bare rock in the midst of the deep? Are you quite sure about that? Let's ask the Associates for Biblical Research,

> Today Tyre is a depressed city that suffered greatly during Lebanon's civil war and Israel's subsequent occupation of southern Lebanon. The modern isthmus that joins the island to the mainland holds streets of houses and shops. There is a picturesque fishing harbor on the north side of the isthmus, adjoining a lively souq. The administrative center for a number of nearby villages and towns, Tyre has a number of unplanned squatter settlements. As important as any industry to modern Tyre are the Greek and Roman archaeological remains which cover the ancient mainland city of Palaetyrus, the accumulated isthmus and the island city.

That's what I thought; Tyre was rebuilt many times, but it sustained no damage from Nebuchadnezzar, because that prophecy failed multiple ways. The island remains above water, the ruins of the old city have been found again, and the city was always inhabited and still is. It was never just a "bare rock" jutting out of the sea, nor is it an unrecognizable place that is only good for fishermen to spread their nets. How many different ways can this prophecy fail?

The probability of any possibility increases as more time is allowed for it. It's a safe bet that whenever any government shows its arrogance, bragging that it is the greatest military power on earth, someone is going to take it down eventually, even if it doesn't happen for hundreds of years. You don't need a prophet to see that coming, and you don't need a miracle to bring it about either. Neither would any omnipotent being need any succession of human armies to do his bidding. Nor would anything truly divine allow itself to be seen as endorsing any of this petty madness. No god worthy of worship would be associated with such character flaws as vanity, jealousy, vengeance, or wrath. God should be above such deadly sins and would not be encumbered by them. That's why even if some or all of the Bible's prophecies had come true, it still wouldn't indicate the existence of any actual god—and certainly no deity deserving of devotion. At best, you would have some frail fallible faker using obvious clues to make a lucky guess. At worst, you'd have some propagandist trying to extort fear to exert submission to control. From Ezekiel 26,

> (19) For thus says the Lord GOD, "When I make you a desolate city, like the cities which are not inhabited, when I bring up the deep over you and the great waters cover you, (20) then I will bring you down with those who go down to the pit, to the people of old, and I will make you dwell in the lower parts of the earth, like the ancient waste places, with those who go down to the pit, so that you will not be inhabited; but I will set glory in the land of the living. (21) I will bring terrors on you and you will be no more; though you will be sought, you will never be found again," declares the Lord GOD.

If only Ezekiel knew about Google Maps, right? Then he would be able to see the city that *was* rebuilt where people *still* live on an island that *never* sank, and he would know he is a *fraud*.

That's the problem with everything claiming to be God's word: it always turns out to be the word of some delusional reactionary zealot speaking without any real authority about things he doesn't actually know anything about. The only reason a prophet really needs to claim a god is so he can dwell in a state of make-believe and assume clout that he doesn't deserve. It's rather like when a harmless bull snake is cornered and tries to mimic a rattler; that's all a prophet is. If you're gonna call it a prophecy, give us the who, where, when, and how, and don't make excuses when most of that prediction turns out to be wrong.

Ezekiel's 29th entry is an unambiguous error without defense from apologetics. This is partly because it includes an admission that he also botched his previous prediction.

> (8) Therefore thus says the Lord GOD, "Behold, I will bring upon you a sword and I will cut off from you man and beast. (9) The land of Egypt will become a desolation and waste. Then they will know that I am the LORD. Because you said, 'The Nile is mine, and I have made it,' (10) therefore, behold, I am against you and against your rivers, and I will make the land of Egypt an utter waste and desolation, from Migdol to Syene and even to the border of Ethiopia. (11) A man's foot will not pass through it, and the foot of a beast will not pass through it, and it will not be inhabited for forty years. (12) So I will make the land of Egypt a desolation in the midst of desolated lands. And her cities, in the midst of cities that are laid waste, will be desolate forty years; and I will scatter the Egyptians among the nations and disperse them among the lands." (13) For thus says the Lord GOD, "At the end of forty years I will gather the Egyptians from the peoples among whom they were scattered. (14) I will turn the fortunes of Egypt and make them return to the land of Pathros, to the land of their origin, and there they will be a lowly kingdom. (15) It will be the lowest of the kingdoms, and it will never again lift itself up above the nations. And I will make them so small that they will not rule over the nations. (16) And it will never again be the confidence of the house of Israel, bringing to mind the iniquity of their having turned to Egypt. Then they will know that I am the Lord GOD." (17) Now in the twenty-seventh year, in the first month, on the first of the month, the word of the LORD came to me saying, (18) "Son of man, Nebuchadnezzar king of Babylon made his army labor hard against Tyre; every head was made bald and every shoulder was rubbed bare. But he and his army had no wages from Tyre for the labor that he had performed

> against it." (19) Therefore thus says the Lord GOD, "Behold, I will give the land of Egypt to Nebuchadnezzar king of Babylon. And he will carry off her wealth and capture her spoil and seize her plunder; and it will be wages for his army. (20) "I have given him the land of Egypt for his labor which he performed, because they acted for Me," declares the Lord GOD.

Never happened. God couldn't give Egypt over to Nebuchadnezzar just like he couldn't give Tyre to him. Remember that Judges 1:19 shows how we're talking about a provincial god only whose military capabilities can be bested even by the simple machines of the Iron Age, so the demands made in Ezekiel 29 are clearly either beyond his god's ability or his god simply didn't want to do as he was foretold. The important thing here is that there was never any point in biblical or postbiblical history when neither man nor beast dwelt in Egypt, or when Egypt was ever completely barren for any duration at all. This prophecy fails completely, and it was very specific about when that was to happen and who was involved. Once again, Nebuchadnezzar was the instigator, and this prophecy should have come to fulfillment during the reign of Pharaoh. It never happened to any degree whatsoever. This prophecy is an absolute failure built on a precedent of earlier errors already admitted.

These people don't learn from their mistakes, and they don't take responsibility for them either. Nor will they apologize when they're wrong, which is often. When I was still in high school, I remember Pat Robertson (former presidential candidate and leader of the Christian Coalition) declaring that a planetary alignment would pit the gravity of the sun on one side of the earth with the combined gravities of all four Jovian planets on the other. This he said would rip the earth in half. There was no apology when the planets aligned on March 10, 1982, and both hemispheres held unflinchingly together.

For another modern example, in 1997 all 150 members of God's Salvation Church sold everything they owned back home in Taiwan, and all of them bought neighboring houses less than two miles from my own home in Garland, Texas. Why? Because their leader, Hon-Ming Chen, known to members as Teacher Chen, promised that God himself would personally land a flying saucer at the site of the church (which was a converted suburban home) in order to save that church from imminent nuclear war. God's spaceship was due to land on March 31 at 10:00 A.M.

Central Time. God would take human form—conveniently that of Chen himself. Why not? Chen had already declared that his two children were the reincarnations of Jesus and Buddha. When asked what they would do should God fail to appear, Chen said that his dozens of followers would all be free to move back to Taiwan, as if the move to Texas hadn't already cost them everything.

When Harold Camping prophesied the rapture would occur on May 21, 2011, there were many who sold everything they had and gave it all away, leaving themselves penniless and destitute in their advancing years. Some committed suicide. One woman tried to murder all three of her own daughters in order to save them from Satan during the impending rapture. When the world miraculously kept existing beyond the expiration date, Camping said that he didn't understand it. But days earlier, his roster was leaked to the press showing that he was still scheduling appointments and programming for days after the time when he said the world would be destroyed. So he already knew business would go on as usual. Camping made millions on that scam in which many of his followers lost everything. Alleged prophets simply have no accountability.

In addition to the many other failed prophecies in the Bible, there are also examples of prophecies in other religions that claim to have been written beforehand, the same as biblical prophecies were said to be, but were just as likely to have been written later and inserted into new editions of the old texts. Here is one such example from the Bhagavada Purana (Srimad Bhagavatam):

> Then in the beginning of the Kali yuga, the supreme lord Krishna will appear as Buddha, the son of Anjana, in the province of Gaya, for the purpose of deluding the atheists who are envious of the faithful theists. (Canto 1, chapter 3, verse 24)

We also have newer scriptures claiming fulfillment of prophecies from elder scriptures, yet those earlier prophecies are now missing from the complete and inerrant word of God, if they were ever there in the first place. Jesus does this a couple times himself. In Luke 24:46 and again in John 7:38, he speaks of "what was written" and is now fulfilled. Presumably he should be referencing the Tanach or Old Testament, but he doesn't say *where* it is written, and no one has ever been able to find it. If Jesus is God, as the

Christians contend, and God wrote the book, and if Jesus said that this or that prophecy should be there, then why aren't they?

This isn't the only time he got something wrong, either. For example, Jesus' second coming was supposed to be back when some of his apostles were still alive, meaning he's about 1,950 years overdue. Of course, the wanna-believers must then deny that the Bible really says that:

In Mark 9:1, Matthew 16:28, and Luke 9:26–27, Jesus says, "there are some of those who are standing here who will not taste death until [they see Jesus returning in his kingdom]." This has been interpreted symbolically, to imply that a Christian disciple would only die in the flesh but not in the spirit. There are several problems with that. First, since the soul is alleged to be immortal, that would give no time limit at all and would render the preface of Jesus' prophecy meaningless. Second, Jesus said they would not even "taste" death, which more than accounts for the death of the flesh. Regardless of the posthumous promises of an eternal afterlife, Christians, pagans, and atheists alike may find themselves writhing on the floor clutching their chests, trying to take that last breath. That's what death tastes like. No religion saves us from that, and that's the only part of death worth being afraid of. As Mark Twain wrote in his autobiography, "Annihilation has no terrors for me, because I have already tried it before I was born—a hundred million years—and I have suffered more in an hour, in this life, than I remember to have suffered in the whole hundred million years put together."

Finally, Jesus also specified that only *some* of his listening audience would still be alive by then, and that negates all the other symbolic interpretations I have ever heard. Because if only some of the apostles but not all of them would be dead, then we must be talking about natural death "in the flesh," and that means we're talking about events due to occur within a few decades of that time at most. In Mark 13:30 and Matthew 24:34, Jesus specifies that he is talking about "this generation," referring specifically to those individuals standing with him at that time. He says they "will not pass away" until a great tragedy occurs in which people flee to their rooftops or the hills, when the sun and moon go dark and the stars fall from the sky (as if that were even possible). They will then see a more powerful Jesus returning in clouds. So it is clear that Jesus meant to imply that he would return before all his original disciples died. That means the second coming of Christ should have been in the first century CE. After waiting for another twenty centuries since then, it's safe to say we've been stood up; he ain't coming.

Some have argued that maybe Jesus was referring to the transfiguration that immediately follows. But that wasn't even a week later, and that neither fits the descriptions of "coming in the clouds at the right hand of power" nor the prophesied apocalypse either. As Ken Harding writes in BibleTrash.com,

> This apologist basically says: "It can't mean those living at the time of Jesus, because he would not have said that." They say that "this generation" means the generation that's alive during the tribulations. Let us take a good look at this 'explanation.' First, the claim that 'generation' could mean race, family, or the nation or tribe of Israel.
>
> What are the Greek words for *Nation, Tribe, Family* and *Generation*? Generation is "genea," the root of genealogy. "Family" is "patria." "Tribe" is "phule." "Nation" is "ethnos," as in ethnic. Next, we need to look up these words as they appear in the New Testament, and cross-reference the Greek words with the English words. I have done this. **Every single** occurrence for *Nation* that I looked up gave the word "ethnos." **Every single** occurrence for *Generation* that I looked up gave the word "genea." When the writers meant *nation*, they wrote *ethnos*. When they meant *generation*, they wrote *genea*. They were apparently very clear in this. They never used "patria" or "phule" in any of these instances. [Emphasis in original]

So this is one of several prophecies of the messiah that never came true, yet should have been fulfilled a long time ago if it was ever to come true at all. There are plenty of other examples of that too.

In order to understand these, it is important to remember that the Old Testament is Jewish, not Christian. So when Jewish authors spoke of "The Lord," they weren't talking about Jesus. The god of Abraham is known to the Jews as Abba (the father), Yahweh (YHWH), or Yohvah (Jehova). This is the same deity whom Islam refers to as Allah (the god). When Muslims talk about God, they too are talking about the god of Abraham, YHWH the father, not Jesus the self-proclaimed demi-human son. Since Jesus himself was supposedly born and raised in Jewish culture, he naturally referred to his god as his "father." The Christian tradition that arose based on the character of Jesus is a very different religion from Judaism (from which it emerged), and rabbinical scholars have very different interpretations of the prophesied messiah. According to JewFAQ.org's article "Mashiach: The Messiah,"

> The mashiach will be a great political leader descended from King David (Jeremiah 23:5). The mashiach is often referred to as "mashiach ben David" (mashiach, son of David). He will be well-versed in Jewish law, and observant of its commandments. (Isaiah 11:2-5) He will be a charismatic leader, inspiring others to follow his example. He will be a great military leader, who will win battles for Israel. He will be a great judge, who makes righteous decisions (Jeremiah 33:15). But above all, he will be a human being, not a god, demi-god or other supernatural being....
>
> The mashiach will bring about the political and spiritual redemption of the Jewish people by bringing us back to Israel and restoring Jerusalem (Isaiah 11:11-12; Jeremiah 23:8; 30:3; Hosea 3:4-5). He will establish a government in Israel that will be the center of all world government, both for Jews and gentiles (Isaiah 2:2-4; 11:10; 42:1). He will rebuild the Temple and re-establish its worship (Jeremiah 33:18). He will restore the religious court system of Israel and establish Jewish law as the law of the land (Jeremiah 33:15).

So from the mainstream Jewish perspective, Jesus doesn't qualify as the mashiach because he didn't do any of the things the mashiach was supposed to do. The organization Jews for Judaism has published a handbook titled *The Jewish Response to Missionaries*, which explains why the Jesus as messiah story didn't fit the needed criteria and is not biblically supported from the perspective of the original tradition:

> In an accurate translation of the Jewish Scriptures, the word "Moshiach" is never translated as "Messiah," but as "anointed." Nevertheless, Judaism has always maintained a fundamental belief in a Messianic figure. Since the concept of a Messiah is one that was given by G-d to the Jews, Jewish tradition is best qualified to describe and recognize the expected Messiah. This tradition has its foundation in numerous biblical references, many of which are cited below. Judaism understands the Messiah to be a human being (with no connotation of deity or divinity) who will bring about certain changes in the world and who must fulfill certain specific criteria before being acknowledged as the Messiah.

These specific criteria are as follows:

1. He must be Jewish. (Deuteronomy 17:15, Numbers 24:17)
2. He must be a member of the tribe of Judah (Genesis 49:10) and a direct male descendent of both King David (I Chronicles 17:11, Psalm 89:29–38, Jeremiah 33:17, II Samuel 7:12-16) and King Solomon. (I Chronicles 22:10, II Chronicles 7:18)
3. He must gather the Jewish people from exile and return them to Israel. (Isaiah 27:12-13, Isaiah 11:12)
4. He must rebuild the Jewish Temple in Jerusalem. (Micah 4:1)
5. He must bring world peace. (Isaiah 2:4, Isaiah 11:6, Micah 4:3)
6. He must influence the entire world to acknowledge and serve one G-d. (Isaiah 11:9, Isaiah 40:5, Zephaniah 3:9)

All of these criteria for the Messiah are best stated in the book of Ezekiel chapter 37:24–28:

> And My servant David will be a king over them, and they will all have one shepherd, and they will walk in My ordinances, and keep My statutes, and observe them, and they shall live on the land that I gave to Jacob My servant . . . and I will make a covenant of peace with them; it will be an everlasting covenant and I will set my sanctuary in their midst forever and My dwelling place shall be with them, and I will be their G-d and they will be My people. And the nations will know that I am the Lord who sanctifies Israel, when My sanctuary is in their midst forever.

As Jews for Judaism states, "If an individual fails to fulfill even one of these conditions, he cannot be the Messiah."

Note that these are not things the messiah is supposed to do only after he comes back from the dead. According to the Jewish Virtual Library, "One talmudic source does apparently attribute immortality to Messiah (Suk. 52a), and the Midrash (mostly later) singles him out among the immortals of Paradise." So the Jewish messiah only gets one life, because he is immortal—he *can't* die. Yet he is supposed to accomplish all these qualifications *before* he can be called "messiah." Bearing this in mind, let's look at Isaiah 7:14, one of the Christian's favorite prophesies so often quoted as being fulfilled by Christ: "Behold, a virgin shall conceive, and bear a son, and shall call his name Immanuel."

This single comment taken out of context is often presented as though Isaiah foretold the coming of Jesus the Christ. Jesus was never referred to as Immanuel, except by a few who lived after him—and who were only doing so to force fulfillment of this erroneous prophecy. But that's not the reason the prophecy fails; there is a whole lot more wrong with it than just that, such that it cannot possibly be a reference to Jesus. Anyone who says it is must not have continued reading beyond that point.

Viewed in context, we have the king of Judah in around 730 BCE, fearing that he may eventually be conquered by combined Syrian and Israeli forces. And we have YHWH commanding his prophet Isaiah to deliver this message to the worried king: "Take heed, and be quiet; fear not, neither be fainthearted, for. . . . It shall not stand, neither shall it come to pass." (Isaiah 7:4–7)

In order to ease the king's worried mind, the prophet offers a sign, a prophecy that the king can see fulfilled in his own time, to watch unfold as foretold: "The maiden is with child and will soon give birth to a son. Curds and honey he shall eat, that he may know to refuse the evil and choose the good. For before the Child shall know to refuse the evil and choose the good, the land that you dread will be forsaken by both her kings." (Isaiah 7:15)

The mother-to-be is described as *almah* (maiden), not *bethulah* (virgin). Christian apologists argue that their translation was not in error, but despite the objections, there is no indication in the context that there should be anything unusual or miraculous about the birth of the child in question, nor of the child himself. Further, how could he be Jesus, son of YHWH, God himself in the flesh, and yet there is some period wherein this child can't yet distinguish right from wrong? So there are two more reasons why this prophecy can't possibly refer to Jesus.

This prophecy also offers a time limit, even if it is somewhat vague. In Hebrew, the verse reads in the present tense "is with child," not "will conceive and bear a child." So whoever it is, she is already pregnant, and the prophecy would be fulfilled once her baby is old enough to make moral choices. By that time, the king of Judah would see that the kings of Syria and Israel weren't a threat to him anymore. That is the crux of the prophecy, that the kingdoms of Syria and Samaria would fall before this fetus could reach the age of reason. It could not have been meant to occur many centuries too late to ease the king's looming concerns for his own life. No, this could only have been a sign meant for fruition in his own time, because it was to show the king that the professed prophet of "God's word" could be trusted.

Amusingly there is still a good deal more wrong with this prophecy. To ensure his own success, Isaiah impregnated a maiden himself, and he brought witnesses. Kinky. She bore him a son, which could have helped him along—except that Isaiah decided not to name the child Immanuel. Curiously, though, he repeated the same prophecy again upon the birth of the boy with the wrong name. But it gets even worse than that! As Farrell Till writes in *Prophecies: Imaginary and Unfulfilled*, Isaiah made the prophecy to assure King Ahaz that the Syrian-Israelite alliance would not prevail against him, yet the Bible record shows that the alliance not only succeeded but did so overwhelmingly. Second Chronicles 28 reports that Ahaz's idolatrous practices caused "Yahweh his God" to deliver him "into the hand of the king of Syria." (v:5) (This king was the Rezin of Isaiah 7:1.) The Syrians "carried away of his a great multitude of captives" and took them to Damascus. (v:5) Simultaneously, the Israelites attacked Judah under the leadership of Pekah (the same Pekah of Isaiah 7:1), and in one day 120,000 "valiant men" in Judah were killed and 200,000 "women, sons, and daughters" were "carried away captive." (vv:6–8) The battle casualties included Maaseiah, Ahaz's son; Azrikam, the governor of the house; and Elkanah, who was "next to the king." (v:7) If these results were Isaiah's idea of Syrian and Samarian failure, one wonders what kind of drubbing the alliance would have inflicted had Isaiah prophesied its success.

(Note: Former Church of Christ minister and missionary turned atheist Farrell Till was a theologian who specialized in biblical errancy. He did an extensive refutation of not only this but also many other biblical prophecies, showing a network of interconnected failures on the part of several prophets. As a sad coincidence, Farrell Till died in the midst of my writing this very chapter. Please review his life's work at infidels.org.)

So, there is no possible link to Matthew 1:23, meaning that passage is wrong too. What Isaiah said would come to pass didn't, and what God himself said would *not* come to pass did. How much more could any prophecy possibly fail? Yet this is the example Christians most often bring up whenever they want to show me a prophecy that they think should convince me that it actually came true! Really? If this is the best you've got, why keep trying?

However, that's still not the best *I've* got. There is another example that is even better than this, where YHWH laughably fails to fulfill his own prophecy. Exodus 17:14 reads, "Then the LORD said to Moses, 'Write this in a book as a memorial and recite it to Joshua, that I will utterly blot out the memory of Amalek from under heaven.'"

Yes, record it, recite it, announce it, so that all will memorize the name that I want them to forget. This passage is a biblical example of the Streisand Effect, an Internet phenomenon where an attempt to suppress information actually publicizes and perpetuates it. One would think that God would be just a bit smarter than that, but of course the Bible doesn't just have fortune-tellers who are no better than a Magic 8 Ball. It also has God saying a whole lot of really crazy things that no superior being would ever say.

Many religious believers seek to justify their faith on the fulfillment of prophecy, but that is actually the least of the criteria they should be looking for. None of the world's "holy" books contains anything indicating divine guidance, nor can their influence be said to be omniscient, inerrant, omnipotent, and just all at the same time. Devotees of the Tanach, the Bible, the Qur'an, and the Vedas all claim that their tomes contain knowledge that would be unobtainable without divine revelation, but none of the specific citations bear that out.

The Bible has no moral guide acceptable as such according to human laws or social ethics. The practices endorsed in many of the sacred scriptures are almost entirely criminal by modern human standards. There is no coherent or useful revelation of the hereafter either; the descriptions of that are just as vague, internally inconsistent, and mutually contradictory as we would expect if they were written by different people who were just making it up as they went along with no apparent comprehension of anything.

One thing that is obviously lacking from most if not all of the scriptures is any hint of higher understanding or wisdom. It's not just that they tell us to do wrong by everyone and call it "righteous," which these books do; it is also that they tell us "truths" that we know to be wrong on many levels. Remember when Jesus said the stars would fall from the sky? He apparently did not know what a star is; he never recognized that the sun is a star, and he certainly didn't understand the size of a star or its distance in relation to the earth, because it's just not possible for any star to "fall from the sky"—yet he said they would.

In later chapters I will review several of the absurdities of the Bible as they relate to science. Otherwise for further clarification on these points, I will turn to Donald Morgan, a former born-again evangelical Christian who was at one time a member of the Board of Elders of a well-respected Bible church, and for a time, chairman of its Christian Education Committee. He attended many Bible studies, Bible seminars, Sunday school sessions, etc., and studied the Bible itself to the point that he began to see its many problems, as well as the many problems with Christian theology. He now identifies as a nontheist who nevertheless finds the Bible a fascinating book. Drawing on his extensive familiarity with scripture, this "atheist theologian" compiled a list of several hundred absurdities, atrocities, and contradictions in the Bible, as well as examples of highly questionable judgment on the part of an allegedly infallible creator. These compilations and a lot more useful information relevant to an in-depth study of this topic are available online at infidels.org. Similar lists have been compiled by other authors listing ridiculous inconsistencies in the Qur'an, as well as in Vedic scriptures.

I focus on the Bible only because I was raised immersed in predominantly Christian society, and Christian dogma is what I know best. Whenever I question the historical accuracy or inerrancy of the Bible, those who revere it turn not to the points being addressed, but to the interpretations of them by biblical scholars. Religionists tend to defer to authority, as long as the authority agrees with them. That's why when Penn & Teller explored the historicity of the Bible on their show *Bullshit!*, they interviewed Dr. Paul Maier, professor of ancient history at Western Michigan University. As a documentary author, seminar lecturer, and professional theologian defending the historical accuracy of the Bible, Dr. Maier was as much an authority on scripture as any Christian could ask for. That's why the following comment from him is such an important reference:

> I'm not saying that now I have proven the Bible is authoritative, is accurate historically. Of course not. You *still* have to have faith.

So there you have it. This is the first of several citations this book will show proving that faith—in the opinion of leading theologians—is a belief that is held in the absence of evidence or proof. Here also we see that whatever else biblical literalists want to claim, scholars of religious history admit that all of the scriptures of any religion were written by human hands and were thus subject to the interpretations, impressions, and perspectives of their primitive and often prejudiced and politically motivated authors, and they cite this as the explanation behind many of the contradictions in those books, especially those in the Bible.

In Richard Dawkins' video documentary *The Root of All Evil*, pastor Ted Haggard told him, "The evidence I can present is we've got a book written over 1,500 years by forty different authors on one subject, and it doesn't contradict itself." This is one of many things Haggard said that is laughably and demonstrably wrong on every single point. Haggard was himself a living contradiction, preaching against homosexuality and drug use until he was exposed as a gay meth addict. Regarding the Bible, Haggard should have said that no one really knows how many authors there were nor *who* they were, since most of them were anonymous, and that works attributed to illiterate Romans were actually written in Greek decades too late to count as eyewitnesses.

There are other examples where scripture was written in a language and by a culture with which the original "author" could not have been familiar. Many of the works attributed to Moses were actually compiled from earlier works from different regions and religions. It's not one book either, but dozens, and should actually include many more than are currently there. More importantly though, these books talk about very different subjects from different perspectives, and very different beliefs. And yes, they *do* contradict each other, and not just on theology. They also contradict science and even the interpretations of modern right-wing religious politics.

For example, American Christian conservatives have taken a very radical "pro-life" stance. That's already a bit ironic for biblical literalists, considering that the Bible keeps pushing the idea that killing everyone for practically any reason is such a great method of judgment and punishment. Some Christian activists will even kill the doctors who perform abortions. The contradiction is that the Bible actually condones abortion, and this is with God's direct endorsement and involvement.

Remember, the Bible is Jewish, at least the Old Testament is. But the Tanach is not the only ancient book of Jewish laws. Jewish belief holds that the basic laws of the written Torah were given by God to Moses together with the oral Torah, a tradition eventually recorded in the Mishnah, the Talmud, and the Midrash. If this is so, then these can be used to enhance our understanding of the Tanach from the Jewish perspective.

Now while today's right-wing Christian fundamentalists hold that life begins at conception, and therefore an abortion done at any stage for any reason at all should be considered murder, the Babylonian Talmud Yevamot 69b states that "the embryo is considered to be mere water until the fortieth day." For a while after that, it still isn't considered a fully living being. Rashi, the great twelfth-century commentator on the Bible and Talmud, states clearly of the fetus in Mishnah Oholot 7:6: "*lav nefesh hu*"—it is not a person, not until it is born.

> If a woman was in hard travail [such that her life is in danger], the child must be cut up while it is in the womb and brought out member by member, since the life of the mother has priority over the life of the child; but if the greater part of it was already born, it may not be touched, since the claim of one life cannot override the claim of another life.

Note that the above passage describes a partial-birth abortion, and explains how to perform one in accordance with Hebrew religious law. Jewish tradition holds that the fetus is not considered a separate person until the head or most of the body has passed through the birth canal. This is concordant with Mosaic law because they're both Jewish traditions believed to be of the same source. The book of Exodus (21:22–23) illustrates how the Bible does not consider a fetus to have value equal to that of a human life.

> If men struggle with each other and strike a woman with child so that she gives birth prematurely, yet there is no injury, he shall surely be fined as the woman's husband may demand of him, and he shall pay as the judges decide. But if there is any further injury, then you shall appoint as a penalty life for life.

Dr. Richard Carrier offered me some further clarification in correspondence. "The word rendered 'injury' here actually more vaguely means 'evil.' So the passage says in Hebrew 'if her child goes out without evil, then' a fine will be levied, but 'if with evil, then you shall give a life for a life.' So what does 'evil' mean? We have to look at what the Jews themselves understood it to mean—after all, it is their language. The Greek Septuagint translation, produced over a century before the time of Jesus by a committee of rabbinical experts, makes the issue very clear. It translates this passage as 'if the woman's child goes out and is not fully formed,' then a fine is levied, but 'if the child is fully formed,' then the man who struck the blow pays with his life. Therefore, the ancient pre-Christian Jews understood their own text as referring only to killing the baby in both cases—but only killing a 'fully formed' baby warrants death, whereas any other pre-formed baby can be killed and it is a mere property crime."

There are other references to the "breath of life," where the movement of air is akin to spiritual essence. The traditional belief is that when newborns take their first breath, they are infused with the spirit and become a living being. According to Rav Moshe Feinstein, considered a supreme authority in Orthodox Jewry,

> Once the head appears, however, and the child is able to breathe independently, he is treated as an entity separate from the mother. He is now independent of the mother's circulatory and respiratory systems. We grant him the full rights and privileges of an adult. The most important of these privileges is the right to life.

From the fortieth day of conception until birth, the fetus is considered to be part of its mother—not a separate entity, but equal to one of her limbs. Sanhedrin 80b of the Talmud describes the fetus as "*ubar yerech imo*"—the fetus is as the thigh of its mother.

Now bear all of this in mind as you read the fifth chapter of the book of Numbers. When a man suspected a woman of infidelity, he was to take her to the tabernacle, the tent where God was supposed to live. Therein the priest would take "dust" from the floor and mix it with water in a bowl. (Remember that this was also where all the animals were brought to be sacrificed, which might have impacted the quality of the "dust" used.) He would also write a curse on a scroll and wash the ink off into the bowl too. Then they would force the woman to drink the cursed potion of bitterness. If she was faithful and falsely accused, then nothing would happen—but if she was guilty of infidelity and *had* cheated on her husband, then the curse would cause her belly to swell and her "thigh" to either "rot" or "fall away," depending on which translation you read.

Obviously they're not talking about one of her legs. We already know that an unborn fetus was considered to be one of the mother's thighs, and that it would be referenced that way. That's why instead of saying "thy belly will swell and thy thigh will fall away," the New Revised Standard Version of the Bible says that the curse will cause the uterus to drop and the womb to discharge. It actually says that God himself will personally cause this to happen.

> Let the priest make the woman take the oath of the curse and say to the woman, "The LORD make you an execration and an oath among your people, when the LORD makes your uterus drop, your womb discharge." (Numbers 5:21)

Now there is no clear indication that the accused woman is necessarily pregnant, nor is there anything to imply that she is not pregnant. All we have is that the woman is suspected of having cheated with another man. Each of the leading Bible versions hold that she definitely had sex with a man, but that she wasn't caught in the act and there were no witnesses against her. So how could we be so certain that she had sex with a man, and what reason would her husband have for believing she had cheated on him? The most obvious answer to both questions would be if she becomes pregnant when she shouldn't be.

This is a time before birth control, when men might be away for more than a month at a time. This was also a culture wherein a man might have many wives and concubines, and he might not "favor" all of them. This culture also made a big deal about knowing when a woman was on her menstrual cycle, to the point that there were prohibitions against even approaching women during those periods. So if an estranged woman skips one or more months with no bloody rags to show, then she would be suspect.

This test is no more sensible nor reliable nor realistic than any medieval test to confirm whether someone was a witch. The point is that the only way to fail this test was for the uterus to drop, for the womb to discharge, and the fetus to rot and fall away. In other words, the only way to fail this test was for the curse to cause a miscarriage. Whenever someone deliberately causes a miscarriage, we call that an abortion.

I have seen steadfast objection to this interpretation from people insisting that the woman in question could not have been pregnant at this time because they can't admit that the Bible says what it does or that it means what it says. So they argue that it says an innocent woman would still be able to conceive children. At best, that *might* mean that the innocent woman didn't happen to be pregnant and literally had nothing to lose. It could also mean that her fetus survived if there was one. But there is nothing in this chapter to imply that the guilty woman is not pregnant, nor that she could not be pregnant when this test is applied. Maybe not every woman accused of adultery was, but at least some of them surely would be, given the nature of the charge against them, yet I get no answers when I pose the question of what would happen if a guilty woman did happen to be pregnant when this test says it would have left her "barren." The only defense I have yet heard is that a pregnant woman wouldn't be allowed to take this test, but no such prohibition exists. Nor would it be possible back then to know if she were pregnant if it were in the first forty-five days or so. That also ignores the only way the Bible says anyone could fail this test: if she cheated on her husband, the fetus would be aborted. That's what it says.

Remember this the next time some right-wing extremist says that abortions should be treated as murder even for underage rape victims, or in cases of extreme medical necessity. Remember that God says it's okay to terminate a pregnancy just because an insecure or underperforming husband gets jealous.

Notice also that the last line of this chapter says that the man will be guiltless but that the woman shall bear her iniquity. There is no provision wherein the woman can pass the test and be owed any compensation or apology. If she does pass, she simply won't be punished any more than she already has been. There is no punishment against a husband for falsely accusing her, and no recourse for her if the roles are reversed. She is his property, little more than livestock, and there is no equality for her.

> Then shall the man be guiltless from iniquity, and this woman shall bear her iniquity. (Numbers 5:31)

Now we know what happened in biblical times when a man suspected his wife of cheating. If she appeared to be pregnant and he didn't think he impregnated her, that meant trouble for her. But what if a wife suspected her husband of cheating? What would be her first clue? Well, there would likely be a naked girl in her husband's bed, crying on bloody sheets. What could the woman do in that case? Nothing. The new girl would be yet another wife, perhaps her husband's new favorite. Otherwise, the girl might just be a new toy, something they used to refer to as a "concubine." Either way, it didn't matter. The guy could do whatever he wanted, and his wives, girlfriends, slaves, and prisoners of war all had to submit to him.

The new girl might not even have been a teenager yet. In Hebrew tradition, men are legally separated from boys at thirteen years of age plus one day. Similarly, women are distinguished from little girls at twelve years plus one day. The rules laid out in the Talmud are supposed to be the recently written records of what had reportedly been an oral tradition since the time of the Pentateuch (the five books of the Bible attributed to Moses). If this is so, and it offers any indication of Hebrew practice in biblical times, then that also offers another reason not to consider the Bible as any sort of moral guide.

Looking again to the book of Numbers, Moses (acting on God's authority) orders his people to pillage a village, take all their booty, and burn the rest to the ground. His men had been told to murder everybody, but they bring back women and children along with the rest of the livestock. God's own appointed prophet then becomes enraged that his men were merciful. So he orders them to hack all the little boys to death and put all their mothers to the sword in what would have been a scene of senseless horror—but if that wasn't evil enough, Moses had a different plan for the little girls.

> Now therefore kill every male among the little ones, and kill every woman that hath known man by lying with him. But all the women children that have not known a man by lying with him, keep alive for yourselves. (Numbers 31:17–18)

Presumably these "women-children" would all have been preteen girls, according to the tradition of the Talmud, being less than twelve years old. Some of them would be murdered too, if they were not virgins—and let's pause for a moment to consider how Moses' men were supposed to determine that. What defensible implication could there possibly be in keeping only little girls alive if there is also a stipulation that they must be virgins?

Every defense I have ever heard for this passage attempts to paint the females "among the little ones" as though they were acceptably mature by modern standards, even though the New American Standard Version refers to the daughters as "girls." The New Revised Standard Version calls them "young girls," while the King James Version calls them "women children" and Young's Literal Translation calls them "infants." In any case, it's clear we're talking about kids kept alive only because of the purity of their sex.

Defenders of the faith also reject that there was any impure intent in keeping only virgin girls alive "for yourselves." Any sexual innuendo is dismissed as immoral, because God couldn't be immoral, right? Worst of all? The fact that each of these girls were so deeply and horrifically traumatized by being invaded, raided, captured, and enslaved, and by having their privates inspected by the very savages who were still butchering everyone they love all around them—that was *not* acknowledged as immoral! Everyone who attempts to defend the divinity of these scriptures must try to justify all this terror. They make excuses that these victims somehow deserved this, as if that were even possible. How immoral is *that*? Remember too that Mosaic law also holds that if a betrothed woman is raped in the city, she must be murdered if no one heard her scream (Deuteronomy 22:23–29). And if a maiden is raped, then she might be forced to marry her rapist and is forbidden to divorce him. How could anyone believe that these are the mandates of a superior being?

The Bible is clearly not the word of God, and mainstream Christians know it already. How do we know that they know? First off, most of the people who consider themselves Christian haven't even read the Bible, which you'd think they would have if they really thought it was written by God. Secondly, even those who have read the Bible still put up Christmas trees, even though they are expressly forbidden by God himself speaking through the prophet Jeremiah:

> Thus saith the LORD, Learn not the way of the heathen, and be not dismayed at the signs of heaven; for the heathen are dismayed at them. For the customs of the people are vain: for one cutteth a tree out of the forest, the work of the hands of the workman, with the axe. They deck it with silver and with gold; they fasten it with nails and with hammers, that it move not. They are upright as the palm tree, but speak not: they must needs be borne, because they cannot go. Be not afraid of them; for they cannot do evil, neither also is it in them to do good. (Jeremiah 10:2–5)

Christians also often get religious tattoos, even though Leviticus 19:28 expressly forbids that too. This is why nonbelievers criticize Christians for "cherry-picking" passages from the Bible that they claim to live by while ignoring other passages that they say don't count or don't matter somehow. For example, Christians won't hesitate to cite the book of Leviticus, where it says that homosexuality is an abomination (18:22), and that homosexuals should be put to death (20:13). But if it is mentioned that eating shellfish is also an abomination (11:9–12), as is cross-breeding cattle, mixing crops, or wearing wool at the same time as cotton (19:19), then the excuse is that Jesus has applied special rules so that most of the Old Testament commandments no longer apply—even when they're taken from other chapters of the same book as the ones that *do* apply!

It doesn't matter to them that Leviticus 3:17 says it will be a "perpetual" rule that no one be permitted to eat either blood or fat—ever. That means it's not temporary; it's gonna stay the same forever, and it ain't ever gonna change. That's why Jesus said in the New Testament that Christians are to obey every "jot and tittle" of all 613 of those Old Testament commandments (Matthew 5:17–19). So if you like your steak juicy and rare, get over it. You can't have lobster or bacon, either. You're supposed to keep a bunch of holidays no one even knows how to celebrate anymore, according to the second edition of the "Ten" Commandments written by God himself in Exodus 34. You're also obliged to kill anyone you see working on the Sabbath. Just wait till after the Sabbath to do it, lest you do work on the Sabbath yourself.

Oh, and remember that female slaves are half-price according to Leviticus 27. But if you can't afford one of your own and you rape another man's slave, she'll be the one who gets punished for it, because women get punished for everything. You won't be, but you will have to give him your best sheep. Either that, or you can pay the slavemaster for his damaged goods. Remember now, God said these rules are perpetual, and he sees the future and everything, so the shekel-to-livestock exchange rate isn't going to change. Now how many slaves do you need?

If these really were the laws of God, then they would apply equally to all cultures at all times, and no god worth worshipping would ever have posted any of these rules. But the Christian Apologetics & Research Ministry explains that these despicable mandates only applied to one culture, the Jews, God's chosen people—and then, only at one point in history. So God's word is neither timeless nor universal; instead, it is a collection of temporary transient provincial indefensible absurdities that are not only misogynist, but racist as well.

The Bible doesn't just contrast with all modern sensibilities; it has many internal inconsistencies too, including quite a lot of clear contradictions—hundreds of them, in fact. Project-Reason.org compiled a list of some 439 contradictions of various types just in the context of the work itself. Here are just a few choice examples:

We know that Judas Iscariot died in the potter's field. But who owned the field, and how did he die? Judas sold Jesus out to the chief priests either before Passover (Matthew 26:14–25, Mark 14:10–11, Luke 22:3–23) or after Passover (John 13:21–30). The priests paid him thirty pieces of silver, which Judas used to buy the potter's field (Acts 1:18). But he also gave the same money back to the priests, and they used the same money to buy the same field themselves (Matthew 27:3–7). They didn't buy it from him either, because they bought it after he died, in the version where he never bought anything. In the version where Judas threw the money back at the priests, he then went to the field that they eventually bought and hanged himself. In the alternate reality where Judas owned the field, he went there and fell "headlong," busting his guts open. The implication is that he dove off a cliff or something similar. But however you do it, it's really hard to fall "headlong" when your head is tied to a tree.

Likewise, Genesis 1 has trees, birds, and other assorted animals being created before humans, and then multiple men and women were created together at the same time. But Genesis 2 has only one man alone being created before trees, birds, or other animals, and then one single woman being created after all of that.

It's not just that the New Testament contradicts the Old Testament (which it does), and it's not just that different gospel writers contradict each other (which they do), and it's not just that different chapters of the same book contradict each other too; it's also that any single alleged author also contradicts his own writing, as if every book was written by more than one person and later contributors didn't pay enough attention to the whole. There is just no other way to explain how there are so many contradictions. Referring to Genesis, Dr. Maier says, "Well I wouldn't call them contradictions as much as commentaries, the one on the other. Again, let's point out, we probably do have two different authors here, whose work was blended together then, in an editorial revisioning, somewhat."

Dr. Maier was right about there being more than one author for Genesis, and it was definitely not an eyewitness account!

The scholarly consensus is that the Bible was written by many human authors from different generations, different nations, and even different religions, some of them pagan. These include the Sumerian King List, the creation myth of Enûma Eliš, the Epic of Atrahasis, the Epic of Gilgamesh, the Persian Avestas of Zoroaster, the Egyptian Book of the Dead, the religion of Amen-Ra, the astrology of Helel bin Shahar, and the collective legends of many different neighboring religions just within the Iraqi floodplain alone. Some of these count among the oldest documents ever recovered, and date to as far back as 2200 BCE, with oral traditions going back much further. Gilgamesh, for example, lived around 2700 BCE, and unlike Moses, he's widely accepted to be a historical figure. Archaeologists even announced in 2003 they may have discovered his tomb.

A word on the evolution of Genesis: The Akkadian cuneiform tablets of Enûma Eliš are the oldest of all creation myths, dating as far back as the early Kassite era in the eighteenth century BCE. That tale is told in seven tablets, accounting for seven generations of gods. On the sixth tablet, we read that the sixth divine generation created man to complete creation so that the seventh generation could rest. Sound familiar?

These first people were created out of the blood and bones of one of the gods. In the Babylonian Epic of Atrahasis from at least the seventeenth century BCE, the goddess, Mami (also known as Nintu and Ninhursag), tells the other gods that they should create man, again to complete creation so that the gods could rest. It does not actually say, "let us make man in our own image," but that seems to be implied. According to this fable, the first people were created out of clay figurines. Seven male and seven female they made them. Another important parallel is that in this version, they also used the blood of one of the gods mixed into the clay. Except that this time, they sacrificed that god and the clay figurines were washed in his blood. Some themes that you thought were original to Jesus are more than a millennium older than that.

In a Sumerian legend, the god Enki trespassed on the sacred garden and ate of its forbidden fruit. The goddess Ninhursag cursed him for it and he fell. Then she forgave the fallen immortal and bore seven daughters to cure his wounds. One of them, Ninti, was "lady of the rib," for she was born to close the wound to his side. Does that sound familiar? These are the ancient original versions of the fables now found in the much more recent compilation of Genesis.

What may have happened (and this is my own postulation) is that the shifting kingdoms of Sumer, Babylon, Akkad, and Chaldea once boasted and shared the earliest syllabic text ever invented. They lived in sophisticated city-states and sent their children to formal schools where they were taught how to read and write in cuneiform. This was the use of a special tool for pressing letters of a sort into clay tablets, which were then baked into permanent documents. These people were fully literate until the fall of the Mesopotamian empire, after which the schools were closed, education was abandoned, and the ability to read cuneiform was lost. This would have been roughly concordant with Hammurabi's initial construction of the Marduk ziggurat, now more popularly known as the Tower of Babel. The base of the tower is still there, but the loss of literacy may be the only element of truth that fable has. For a millennium afterward, all the old stories were kept alive by memorization and rehearsal.

No longer "set in stone," these originally unrelated collective legends were subject to embellishment and cultural integration. Very often even the names of the characters were changed. For example, the character now known as Noah was originally known as Ziusudra in Sumer, Atrahasis in Akkad, and Utnapishtim in Babylon. Each of these early accounts vary from each other, and from the Bible as well. Yet so many verbatim passages are shared by most that they are all clearly talking about the same event that the Bible does: a local flood centered on the city of Shuruppak at the end of the Jemdat-Nasr period around 2900 BCE. So ancient Mesopotamian mythology evolved over many centuries of occasional enhancement such that the old legends were adapted for and integrated with the emerging culture of Judaism. By the time the Phoenicians wrote the newer versions of some of these stories down again some three thousand years ago, they were "blended together then, in an editorial revisioning."

Some experts now recognize four different sources just for the Pentateuch, the five books of "Moses." According to the Wellhausen documentary hypothesis, these forgotten contributors are now referred to as the Yahwists (J), from circa 950 BCE in the southern Kingdom of Judah, the Elohists (E) c. 850 BCE in the northern Kingdom of Israel, the Deuteronomists (D) c. 600 BCE in Jerusalem during a period of religious reform, and the Priestly writers (P) c. 500 BCE by Kohanim (Jewish priests) in exile in Babylon.

Modern scholarship doesn't credit Moses as the author of anything. One reason is that Moses couldn't possibly have posthumously written about his own death and remembrance as detailed in Deuteronomy. Worse, Moses evidently never even existed as described. As senior rabbi David Wolpe said in a 2001 Passover sermon at Sinai Temple in Los Angeles,

> The truth is that virtually every modern archaeologist who has investigated the story of the Exodus, with very few exceptions, agrees that the way the Bible describes the Exodus is not the way it happened, if it happened at all.

It seems that the character now known as Moses is a compilation of a few different predecessors from elder mythos. The most obvious of these is Hammurabi, the Babylonian lawgiver. He received the Stele Law Code from the sun-god Shamash some five hundred years before Moses was supposed to have lived, which is generally estimated to be around 1250 BCE. Otherwise the principle difference between these two figures is that the Stele Law Code of Hammurabi is an actual artifact on display at the Louvre, while the Ark of the Covenant is one of those things that only seems to exist in movies.

Egypt provided another precursor in the form of Djadjamankh, chief ritual priest of the Pharaoh Snefru, from the 4th Dynasty of the Old Kingdom in the twenty-fifth century BCE. One of the five tales included on the Westcar Papyrus details a voyage wherein Snefru took a score of buxom beautiful girls, stripped them all naked, and enjoyed the view of his crew as they each took an oar and rowed his long boat across a lake. One of these girls accidentally dropped a turquoise bauble over the side and got so upset about losing it that the Pharaoh called on his high priest for help. Djadjamankh then chanted an incantation spell that folded half of the water in the lake onto the other half, as though he were looking under a blanket. Thus the Pharaoh could appear a hero to a topless rowing maiden.

Okay, so it's not quite Cecil B. DeMille, but it does cause one to wonder how the story of Moses parting the Red Sea could be considered original in the same land that had already dreamt this satirical farce up well more than a thousand years earlier.

In their book *Deceptions and Myths of the Bible*, Lloyd and Elizabeth Graham describe parallels between Moses and the much more serious story of Sargon.

> This myth woven about the legendary Sargon I, 2750 B.C., strikingly resembles the early history of Moses, that is, his infancy. This part is given

> only by the Elohist, "... when she, Moses' mother, could not longer hide him, she took for him an ark of bullrushes and daubed it with slime and with pitch, and put the child therein, and she laid it in the flags by the river's brink" (Exodus 2:3). And on the tablets of Kouyunjik, Sargon tells his story.
>
> 4. My mother, the princess, conceived me; in difficulty she brought me forth.
> 5. She placed me in an ark of rushes with bitumen my exit she sealed up.
> 6. She launched me in the river which did not drown me.
> 7. The river carried me to Akki, the water-carrier, it brought me.
> 8. Akki, the water-carrier, in tenderness of bowels, lifted me ...
>
> In appreciation, Sargon named his capital Agadi, called the Semites Akkad, and Akkad was near the city of Sippara. Now note that Moses' wife was "Zipporah."

At this time, it is important to note that while there definitely are many valid claims of strong parallels between different religious characters, not all such claims have been adequately researched or verified before publication, and many can't be sourced. Some are based on earlier speculations that in turn are based on documents or cultural artifacts that have either been lost or misinterpreted—or that never existed to begin with. For example, the Grahams list a series of parallels between Moses and Mises, another character alleged to have multiple origins. However, the citations the Grahams provide for these parallels do not include the passages or references they described. Consequently, I and others have been unable to verify any of the alleged parallels with regard to Mises. Two glaring examples of this sort of misinformation are Kersey Graves' book *The World's Sixteen Crucified Saviors* (1875) and Peter Joseph's compilation of conspiracies in the 2007 movie *Zeitgeist*. In both instances, I was able to verify very few of the many claims made about the assemblage of myths that culminated in the character of Jesus the Christ. I was unable to find any information at all for several of Graves' list of crucified saviors, and what I did find included allegations that I could show were definitely not true. Some of my associates have independently arrived at the same conclusion.

Dr. Richard Carrier (a notable scholar in this area) wrote the following review of *The World's Sixteen Crucified Saviors*:

> Although I have not exhaustively investigated this matter, I have confirmed only two real "resurrected" deities with some uncanny similarity to Jesus which are actually reported before Christian times, Zalmoxis and Inanna, neither of which is mentioned by Graves or John G. Jackson (another Gravesian author—though both mention Tammuz, for whom Inanna was mistaken in their day). This is apart from the obvious pre-Christian myths of Demeter, Dionysos, Persephone, Castor and Pollux, Isis and Osiris, and Cybele and Attis, which do indeed carry a theme of metaphorical resurrection, usually in the terms of a return or escape from the Underworld, explaining the shifting seasons. But these myths are not quite the same thing as a pre-Christian passion story. It only goes to show the pervasiveness in antiquity of an agricultural resurrection theme, and the Jesus story has more to it than that, although the cultural influence can certainly be acknowledged. . . .
>
> The only case, that I know, of a pre-Christian god actually being crucified and then resurrected is Inanna (also known as Ishtar), a Sumerian goddess whose crucifixion, resurrection and escape from the underworld is told in cuneiform tablets inscribed c. 1500 B.C.E., attesting to a very old tradition. The best account and translation of the text is to be found in Samuel Kramer's *History Begins at Sumer*, pp. 154ff., but be sure to use the third revised edition (1981), since the text was significantly revised after new discoveries were made. For instance, the tablet was once believed to describe the resurrection of Inanna's lover, Tammuz (also known as Dumuzi). Graves thus mistakenly lists Tammuz as one of his "Sixteen Crucified Saviors." Of course, Graves cannot be discredited for this particular error, since in his day scholars still thought the tablet referred to that god (Kramer explains how this mistake happened).

The oldest book of the Bible that is not obviously adapted from previous polytheism is Job, and it supposedly hails from as early as 1500 BCE, though other scholars say it came much later. There is a scholarly consensus that Genesis was compiled (probably by Ezra) from several unrelated oral traditions less than 2,500 years ago. Other documents filtered in at about the same time, all of which attributed to authors, mostly between the sixth and second centuries BCE. So the Bible is nowhere near as old as believers say it is! The Dead Sea Scrolls are the oldest archaeological texts known in all of Western monotheism, being radiocarbon dated to between 335 BCE and 122 BCE. Yet they're each centuries younger than Zoroastrianism, Buddhism, Hinduism, and various Hellenist, Druidic, Chinese, and Egyptian religions. Curiously, nearly all of these major religions either began around 600 BCE or endured major revisions at around this time. This flowering of philosophies also included atheism and the emergence of scientific naturalism among Ionian Greeks.

Christianity began with the Gnostic faith, and then the Docetics and Ebionites, all in the first few centuries of the Christian formation. In summary, Docetics believed that Jesus was entirely a god, an immortal spirit whose physical form was an illusion. He could never have actually died, so he couldn't have been resurrected. The Ebionites were renunciate Jews who believed that Jesus was a fully human messiah and prophet, but of natural origin with no supernatural aspects. Gnosticism is actually a pre-Christian concept of "knowledge through faith," which essentially means convincing yourself to be absolutely certain of things that are really just idle speculation, or pretending to know what you don't really know. A window into this type of deranged philosophy was given by seventh-century Greek theologian St. Maximus the Confessor: "That mind is perfect which, through true faith, in supreme ignorance supremely knows the supremely Unknowable."

This particular brand of antilogic represents the greatest flaw in creationism and, as I will explain, is simultaneously its only strength.

The completely different perspectives of Jesus held by the Gnostics, Ebionites, and Docetics were eventually combined into a kind of compromise called Orthodoxy, which taught that Jesus was both human and divine, a physical manifestation and an avatar of God. Other Christian subsets around at that time (such as the Luciferians) were overrun and discredited by further biblical revisions.

One of these revisions relates to the king of Ugarit, who supposedly lived some 3,500 years ago. As his followers were the principle competition with the emerging religion of Moses, scribes working on the New Testament chose to demonize *Ba'al ZeBul*, the "Lord on High," by distorting his name to *Beelzebub*, the "Lord of the Flies." So the Bible has been deliberately and deceptively altered for both religious and political reasons.

The rest of what became the New Testament was finally canonized by the fourth century after a series of committee decisions at a convention in Nicea. According to both Christian (Dr. John Ankerberg) and atheist (Prof. Bart Ehrman) scholars of biblical origins, four gospels were accepted and sixteen more were rejected, as if the facts of the matter could be evaluated or dismissed by popular vote. As Dr. Maier notes,

> The way the canon developed was by what was being read on Sunday in the centers of Christianity. What do you read on the second Sunday after Easter in the church in Jerusalem? What's the church of Rome reading at this time? And they found that again and again, they were zeroing in on the same stories in the gospels. And so the core of the canon kind of developed from the usage of the early church.

So the council simply accepted as gospel whatever few relevant stories the uneducated masses happened to like at that time. But they still opted to remove more than a dozen books from the Bible even though they were still referenced by other books they chose to include. Among the rejected items were the writings of both apostles and prophets.

Removed from the Old Testament:

1. The book of the Covenant (Exodus 24:7)
2. The Wars of the Lord (Numbers 21:14)
3. The book of Jasher (Joshua 10:13:2 and Samuel 1:18)
4. The book of Statutes (1 Samuel 10:25)
5. The book of the Acts of Solomon (1 Kings 11:41)
6. The book of Nathan (1 Chronicles 29:29 and 2 Chronicles 9:29)
7. The book of Gad (1 Chronicles 29:29)
8. The Prophecy of Ahijah and Visions of Iddo (2 Chronicles 9:29, 12:15, and 13:22)
9. The book of Shemaiah (2 Chronicles 12:15)
10. The book of Jehu (2 Chronicles 20:34)
11. The Acts of Uzziah written by Isaiah (2 Chronicles 26:22)

12. The Sayings of the Seers (2 Chronicles 33:19)
13. The Prophecies of Enoch (Jude 14)

Removed from the New Testament:

1. An Epistle of Paul to the Corinthians (1 Corinthians 5:9)
2. An Epistle of Paul to the Ephesians (Ephesians 3:3)
3. An Epistle of Paul to the Laodiceans (Colossians 4:16)
4. The Prophecy that Christ should be a Nazarene (Matthew 2:23)
5. The Predictions known to the scribes in Jesus' day that Elias must restore all things before the coming of Christ (Matthew 17:10)

Why would God's word refer us to other books that were some human's word or that are no longer available? Who were the real editors here? The Bible often names human authors, but how could it make such an admission if God were the real author and editor of this haphazard jumble of fables, parables, and psalm lyrics?

The Bible was very definitely written by men, and not superior men either—far from it! This is why so much of it can be shown to be historically and scientifically dead wrong about damned-near everything from back to front. We're talking about people who believe in talking snakes and donkeys (Genesis 3 and Numbers 22:1–35), in incantations (Genesis 1:3–29 and Ezekiel 37:1–10), in blood sacrifice (Genesis 4:4 and 31:54; Lev 1:9 and 9:18; 2 Kings 16:15; Ezekiel 39:17) and human sacrifice (Leviticus 27:28–29; Judges 11:29–39; Numbers 3:11–13 and 31:31–40; Exodus 22:29 and 34:19; Ezekiel 20:26), in ritual spells (Numbers 5 and Leviticus 14), in enchanted artifacts (1 Samuel 5:6–9; Exodus 7:8–12; 1 Samuel 5:69 and 6:19), in pyrotechnic potions (Numbers 5:20–26), in astrology, (Genesis 1:14–15; Job 38:32; Isa 14:12–14; Luke 21:25, Matt 12:32 and 28:20), and in the five elements of witchcraft (each employed in a single spell in Leviticus 14).

They thought that if you use a magic wand to sprinkle blood all over someone, it will cure them of leprosy. We're talking about people who think that rabbits chew cud (Lev 11:6), that bats are birds (Deuteronomy 14:11–18 and Lev 11:13–19), that whales are fish (Jonah 1:17 compared to Matthew 12:40), and that pi is a round number (1 Kings 7:23 and Chronicles 4:2).

These folks believed that if you display striped patterns to a pregnant cow, it would bear striped calves (Genesis 30:37–43). How could anyone say that who knows anything about genetics? Obviously the authors of this book didn't. If the Bible had been written by a supreme being, then it wouldn't contain the mistakes that it does. If it was written by a truly superior being, and meant to be read as a literal history, then the Bible wouldn't contain *anything* that it does.

As a moral guide, it utterly fails, because much of the original Hebrew scriptures were written by ignorant and bigoted savages who condoned and promoted:

- Animal cruelty (Joshua 11:6 and 2 Samuel 8:4)
- Incest (Genesis 4:1–17, 9:1, 19:30–38, 20:11–12; 2 Samuel 13:1–15)
- Slavery (Numbers 31:31–35; Leviticus 25:44–45; Exodus 21:27; Ephesians 6:5; 1 Timothy 6:1–2)
- Abuse of slaves (Exodus 21:7, 20–21; Luke 12:47–48)
- Spousal abuse (Numbers 5:5:31; Deuteronomy 22:13–21, 28–29)
- Child abuse (Genesis 22; Deuteronomy 21:18–21, 23:2, 20:30, 22:15, 23:13–14; Psalm 137:9)
- Child molestation (Numbers 31:17–18) (Also condoned in the Talmud: Sanhedrin 55b, 69a–69b, Yebamoth 57b, 60b, 12b, Niddah 44b)
- Abortion (Amos 1:13; 2 Kings 15:16; Hosea 9:11–16, 13:16; Numbers 5:5–31)
- Pillage (Genesis 34:13–29; Deuteronomy 20:13–14; Numbers 31:7–12)
- Murder (Exodus 2:12, 21:15, 22:17; Leviticus 20:9–10, 13, 27; Deuteronomy 13:13–19, 17:12; Proverbs 20:20; Judges 9:5, 11:29–39, 14:19; 2 Samuel 18:15; 1 Kings 2:24–25, 29–34, 46, 9:27, 10:7; 2 Chronicles 21:4; Ezekiel 20:26)
- Cannibalism (Leviticus 26:29; Deuteronomy 28:53; Isa 49:26; Jeremiah 19:9; Ezekiel 5:8–10; 2 Kings 6:29)
- Genocide (Genesis 6:11–17, 7:11–24; Exodus 17:13, 32:27; Numbers 21:3, 21:35; Deuteronomy 2:33–34, 3:6, 7:2, 20:16; Joshua 8:22–25, 10:27–40, 11:8–23; 1 Samuel 15:3, 7–8)

- Prejudice against:
    - Race (Exodus 23:23; Ezra 9:11–14, 28; Numbers 21:35; Deuteronomy 3:6, 7:1; Matthew 15:22–28)
    - Nationality (Leviticus 25:46; Joshua 6:21–27; Matthew 11:21–24)
    - Religion (throughout the Bible, but especially in 2 Kings 10:19–27)
    - Sex (Genesis 38:16–24; Judges 9:53–54, 19:22–29, 21:10–12; Deuteronomy 21:10–14, 22:23–24, 28–29, 25:11–12; Leviticus 12:1–8, 14, 15:19–30, 18:19, 19:20, 21:29, 27:3–7; Numbers 1:2, 20:13–15, 30:3–16, 31:14–18; Zechariah 14:1–2)
    - Sexual orientation (Deuteronomy 22:5; Leviticus 18:22–23, 20:13)

Why? To justify their own inhumanity by claiming to do the will of God. But creationists still continue to ignore all of that. Some of their sites even admit that wherever reality conflicts with the Bible, then reality must be ignored! And why is that? Because if creationists didn't have their beloved books, they wouldn't have a god either. One *is* the other in their world.

Ironically, the faithful reject the "works" of God as worshipping creation over the creator. But then they prop up the words of men before God, *as* God, and even insist that disproving their supposedly "holy" books would somehow disprove God too—and not just their version of God, but everyone else's version of God as well. Creationist Christians think that if the Bible is wrong, then God lied. They cannot accept that God could exist if the Bible is wrong because they can't distinguish doctrine from deity. So it is a form of idolatry wherein the believers worship man-made compilations as though those books were God himself—because they think it is *his* word. But God never wrote nor dictated any of the scriptures of any religion. Everything men chose to reject from or include in the supposedly inalterable word of whatever god was conceived, composed, compiled, translated, interpreted, edited, and often deliberately altered and enhanced by mere fallible men.

THE 3RD FOUNDATIONAL FALSEHOOD OF CREATIONISM

Human Interpretation Is "Absolute Truth"

"What Krishna has said 5,000 years ago in the Bhagavad Gita has stood the test of time. You can read it today and it is still perfectly valid. Your scientific theories will come and go but the absolute truth will not change."

—Madhudvisa dasa Swami, July 14, 1995

The third foundational falsehood of creationism is the assertion that any human's understanding of their various internally conflicting and intercontradictory faith-based beliefs should or even could be considered infallible, inerrant, or completely accurate.

I hadn't yet turned thirteen when a girl asked me to walk her home after school. She was charming enough and she seemed to have a romantic interest in me, so I took the bait. As we were walking and talking, she asked me what religion I was. At that time, I knew almost nothing about religion. I knew that different beliefs were dominant in other parts of the world, and that the mythos of other cultures were based on legendary accounts of heroes and deities that had nothing whatsoever to do with the stories in the Bible or the god of Abraham. I might have been somewhat henotheistic at the time because I thought that all these prophets of foreign dogma were glimpsing portions of the supernatural realm, each through their own peephole, and they were trying to properly interpret their mystic visions just like the rest of us. Everyone says their religion is the one true faith, but none of them can actually prove it, so we don't really know. It didn't make any sense to believe that any one religion was completely true and that the others were totally false. Instead, I thought it far more likely that all religions might be partially true, but that none of them would be absolutely true.

This was also why I never declared myself to be Mormon like my mother and grandmother had. I couldn't do that because I didn't know the scope of what Mormons believed. I hadn't studied it, so how could I say whether I believed everything they do? More importantly, I hadn't studied any other religions either, so how could I know whether one of them might make more sense to me? But one thing I had figured out by then was that when someone asks what my religion is, it is a prelude to an uncomfortable confrontation.

We moved a lot when I was a kid, sometimes several times every year. By that point in my life, I had lived in eight different states. Mormon families were often the dominant demographic when we were in small towns in Arizona, but in the metropolis surrounding Los Angeles (where the Mormons didn't own and control everything), I noticed quite a different reaction to that label. So I tried to answer her in the most reasonable way I possibly could. "*Well, my family is Mormon . . .*" The sentence obviously didn't end there, but I still didn't make it to the "but" before she was already pleading with the sky: "Oh God, why was I given a Mormon?"

In retrospect, I think it's interesting what I didn't notice then. People commonly say things like "Oh God, why is this happening?" but when they do that, they usually know that they're only talking to themselves. She didn't. She actually thought she was conversing with some invisible entity listening in. I missed that, having immediately taken offense. I wanted to say, "You weren't *given* a Mormon. I'm *not* a Mormon, and I wasn't *'given'* to you either," but before I could say any of that, she turned to me and asked, "Do you know what Mormons believe?"

This was a conversation I'd had a few times before. When they think that I'm Mormon, they would then ask whether I knew that Mormons actually believe [insert absurdity here]. They were essentially asking whether I knew what I believe, as if I wouldn't already know that better than they would or could. But it gets worse, because then they would tell me that Mormons believe all kinds of crazy things that no Mormon I know has ever mentioned.

Now I'm fully aware that Mormons really do believe some truly outrageous nonsense—every religion does—but it wasn't the same kind of crazy that their critics charged. Once or twice I tried to invite these people into my house to tell my mother that she believes in all this stuff (that I'm sure she'd never even heard of either), but that invitation had always been refused. I wonder why.

If I hadn't been raised by a Mormon living in a non-Mormon culture, I might have missed out on an important lesson regarding religious prejudice. What I learned is that if you want to know what a Mormon believes, ask a Mormon; don't ask a Baptist, because you won't get a straight answer that way. Of course the same applies to any other sociopolitical or religious perspective too. You can ask a Baptist what Baptists believe. But if you ask another Baptist the same thing, you might get a different answer, one that won't match.

Continuing with the story, we finally get to her place, and her whole family is home. She knew they would be, but I hadn't expected that. Turns out she didn't want to be alone with me. She had been previously instructed to lure me (or anyone really) into the house and turn me over to them. Her parents and siblings were very interested in meeting me and talking with me, and they seemed nice enough for about a minute or so. But then they surrounded me and suddenly started yelling into my face that they were all going to vanish into thin air "any minute now" (this was in 1974), being called up to heaven, but that I would find myself embroiled in a worldwide lake of fire—and that the only way I could save myself from eternal torment was to break down in shame, repent the sins of Mormonism, and confess my belief in their horrible raving insanity instead.

One of them was writhing on the couch. I think she was speaking in tongues. The others seemed to be undulating like natives dancing around a totem. But despite all their convulsive evangelizing, I stood there frozen in silent alarm, tears streaming down my face. I didn't know how else to react, since I was only a child and this was the first time I had ever been in a madhouse.

I went home as soon as I could get out the door. I shared that harrowing experience with my mother, and was surprised that she was not surprised. She was quite confident in the superiority of her faith. It's all those other religions who act like that, because (of course) they don't know "the truth" like she does. Isn't it odd that adherents of every other belief system also say the same thing about her belief, with equal confidence and the same justification? Through faith, the same method of study will lead like-minded initiates to an ever-widening variety of completely different conclusions—impossibly, all of these are asserted as "the truth."

My mother never indoctrinated me or took me to church. She didn't think it was necessary. I guess she always assumed that I just believed whatever she told me, even though I questioned all of it. So she unwittingly allowed me to grow up as a freethinker. But now she challenged me to investigate her religion. She insisted that I read her "Pearl of Great Price," the Book of Mormon.

The Book of Mormon is similar to the Qur'an and also the Kitab-i-Aqdas of the Bahá'í Faith. All three of these books are based to some degree on the stories in the Bible, both Old and New Testaments. They are each presented as an addendum to the Bible, the final word from the last great prophet—whether that prophet happens to be Muhammad or Joseph Smith, who reportedly took dictation from angels, or Bahá'u'lláh, who reportedly acted as God's messenger directly. But how could all these prophets of the same god disagree with each other on so many interpretations at once?

It wouldn't be a fair evaluation to read the Book of Mormon without reading these others too, so that I could have a proper comparison. But what if the Bible they're all based on turns out to be wrong? If the foundation fails, wouldn't that cause all the houses on it to fall? If the Bible is correct, would it matter what any of the subsequent books had to say if they differ from that?

Every new sect is a heresy of whatever precedent denomination it is based on or related to. I have often heard representatives of offshoot denominations making the same claim of unique authority as several other relatively new denominations have. In short, the older, more established orders have "corrupted" God's "truth" to such a degree that some new belief system had to be created to get back to the "true" way—the one that God had originally intended but that is not recorded anywhere and likely requires some very strained and particular interpretation. Mormons, Muslims, and Jehovah's Witnesses all make this claim.

Alternately, some sects might also profess to be older than the dominant denominations, but say that their newly emerging cult was somehow the keepers of the secret truth that has been kept hidden for all these centuries. Of course none of this could really be the case if any of these doctrines were really established by an omnipotent, omniscient, and infallible immortal.

If people have a tendency to corrupt an original truth such that it generates all these other variants, then it would not logically make sense to examine each of these end nodes hoping to find the whole and uncorrupted "absolute" truth therein. The completely correct truth shouldn't be some unique belief held only by one particular cult, especially not a new one. Rather, since each religion derives the basis of their beliefs from some other religion, then logically whatever truth there is to any of them should be found in the elements they all hold in common. I knew better than to compare each of the leaves of this tree. I had to go to the root, so I refused to read the Book of Mormon since I hadn't read the Bible first.

I expected everything in the Bible to be "coded" such that primitives would accept and repeat it, but that only the enlightened would truly understand it, so I read it the way I think most people do, imagining what deeper meanings each verse might have apart from whatever they actually said. This was the 1970s. People did that with music lyrics too and found hidden messages all the time.

Did you know that Paul McCartney died in a car crash in 1966, and that his death was covered up, and that John Lennon had quietly replaced him with an exact double? Someone who not only looked like Paul but also sang with the same voice? That's why McCartney is shown crossing Abbey Road barefoot in a suit; he was the corpse. Lennon (in white and looking like Jesus) represented the divine, whom everyone else follows. Ringo was dressed as a minister, and George Harrison was in denim because he was the gravedigger. That's why Paul was referred to as "the walrus," which was either a Scandinavian symbol of the dead or a Greek word meaning "corpse." Not really, but this is what many fans understood from their interpretation of a complex network of obscure unrelated and enigmatic lyrics in the Beatles' later albums.

Charles Manson had a very different interpretation, especially with reference to the Beatles' unlabeled record popularly known as the *White Album*. One track was a montage of overlapping sounds entitled "Revolution #9." Manson took that to be a reference to Revelation 9, and he thought that the Beatles were the "locusts" with "the hair of women" that are mentioned in that biblical chapter. Manson's murderous cult was founded on his impression that he was the Son of Man, that the Beatles were angels of the apocalypse, and that they were sending him secret messages encoded in their music—messages that could only be understood by those who were "spiritually enlightened."

Like music lyrics, this is apparently how the Bible is supposed to be read too—it could potentially be made to say anything and to endorse either side of almost any argument, regardless of whatever message was originally intended. When you get into the various dreamlike visions and prophecies in Revelation, they do read a lot like the meaningless nonsense lyrics written by John Lennon on acid.

The first insurmountable problem I had with the Bible was that every part of it was so weirdly illogical and grotesquely immoral that it was obviously not inspired by any superior being. The Bible is written remarkably badly, especially for something that was so revered and intently studied by the very people who compiled and composed it. I was told that the Bible was the "word of God." That meant that if I read it, I would find a level of wisdom, morality, and understanding far beyond the scope of the most brilliant or compassionate man. That is what I fully expected when I cracked open that book. Of course what I found instead were only the insane ravings of superstitious primitives apparently trying to justify their own inhuman atrocities by pretending to speak for their god, thereby claiming ultimate authority. The Bible is a deeply repugnant tome that celebrates evil, promotes ignorance, and punishes wisdom as an abominable sin. No deity worthy of worship would want to be associated with that despicable compilation of the worst that men can be. The deeper I read, the worse it got. On that first attempt, I don't think I had gotten out of Genesis before I reached my limit of tolerance. I eventually threw the book across the room in disgust.

The majority of Americans who were once Christian were reportedly polled as saying that reading the Bible was the #1 catalyst that turned them atheist, or that it was most often at least the first step down that path. Many atheists (myself included) actually encourage seemingly reasonable Christians to read the Bible for that reason. Penn Jillette famously said, "We need more atheists, and nothin' will get you there faster than reading the damn Bible."

When I first tried to read the Bible at the age of twelve, I still believed in God. At that time, I didn't know that it was even possible not to believe in God. I knew that atheists existed, but I didn't know what an atheist really was. I was told that atheists secretly believed in God, but that they hated him; either that, or they believed in nothing, as if all atheists were nihilists denying their own existence. I didn't yet know what faith really was either, nor did I need it. Everyone told me the existence of God was a "conclusively proven scientific fact," another lie commonly repeated by defenders of the sacred truth. Questioning the existence of God seemed to me at that time to be no less crazy than questioning whether the United States really had a president, as if it were all an elaborate hoax. You had to be some kind of weird conspiracy nut to think that God wasn't real despite the billions of people who believe in him. Such was my mind-set as a boy.

So I believed in God, but I still had to reject the Bible at least on principle, if not on fact as well, because my presupposition that God is real left only two options: either God is good, and the Bible is completely wrong about him, or if the Bible is true, then God is neither holy, nor just, nor wise, nor loving, nor forgiving, nor anything else that could be considered "good." He was a monstrous amoral dictator, base, capricious, unworthy of respect much less admiration, and thus impossible to worship. Anything that is actually deserving of worship wouldn't want to be worshipped anyway.

So I said a prayer. I've only done that a handful of times in my life, but this was important. I said it alone in the privacy of my room, just like the Bible says in Matthew 6:5–6. I looked at the place where the wall meets the ceiling, and I assumed that's where God would be watching me. I spoke to that place out loud, as if addressing someone there who would actually hear me. In my prayer, I challenged God, demanding that he either provide some explanation to justify all the heinous atrocities and horrible judgments attributed to him in the Bible, or he should distance himself from all of that by showing me that the Bible does not speak for him. I even gave an ultimatum—that if he did not answer my prayer, then he could not have my soul. I knew exactly what that meant when I said it, but I was being sincere. I would sooner damn myself to Hell than be forced to serve the insufferable despot that the Bible portrayed God to be.

In some sense, I can honestly say that prayer was answered. Over the course of my life, it seems that practically every subject of study eventually implies that—even if there is a supernatural aspect to the universe, and even if there really is a god—the Bible is wrong, and it's wrong about everything. The Bible isn't just wrong culturally, politically, scientifically, logically, historically; it is wrong factually, philosophically, ethically, and morally.

So the course of my religious studies soon took me beyond the Bible, beyond all the common concepts of collective Christians, and quickly away from every religion based on the Abrahamic god of Western monotheism. I looked into Eastern religions—Sikhism, Zoroastrianism, Hinduism. I actually enjoyed reading *The Science of Self-Realization* by His Divine Grace, A. C. Bhaktivedanta Swami Prabhupada—until it got into all the mythology. I was interested in transcendental meditation, reincarnation, and all of that, but all the stories of Krishna and Vishnu sounded no different to me than tales of Zeus and Hera. So I skipped over polytheism altogether, and eventually rejected theism in any form. For a while, I considered Druidic, Taoist, and Pantheist spirituality. But once I understood what faith was, I was compelled to discard every belief that was dependant on faith simply because faith was required.

One of the things the faithful do is declare their Truth with a capital T. Yet the beliefs they promote are not supported by evidence and thus cannot be verified. Obviously it cannot be "truth" if it is not true. Nor could we honestly call it truth if we cannot show that it is true. If we're in a courtroom with an eyewitness sworn to tell the truth, we still would not automatically accept that it is the truth until or unless the evidence bears that out. There must be some means of confirmation.

Truth is defined as "that which is concordant with reality." That means reality itself is not truth. But statements about reality can be—if they're true. So we can't call it "truth" unless we can show that it is concordant with reality. With that sort of situation in mind, if truth is defined as, or determined by, whatever we can show to be true (as I think it should be), then there is no truth in any religion. The greatest weakness of every religion is that all of their claims fall into only one of two categories: faith-based beliefs that are either not evidently true or evidently not true. This means there is no reason to believe a particular religion's claims and there are good reasons not to believe them. Despite all their apparent veracity, no religion can honestly show that their beliefs are any more accurate than those of every other religion. How do they respond to this? By asserting their convictions all the more vehemently, with each one claiming theirs is the "absolute truth." But what they now assert with a facade of undeniable certainty is still highly questionable, because it's really only blind speculation. Without supportive evidence or means of confirmation, it can't be more than that.

If you gather representatives of a dozen different mutually exclusive religious beliefs, where no two of them can be completely correct at the same time yet all of them say theirs is the absolute truth, then logically only one of them at most can be telling the truth at all, and the greater probability is that every one of them is wrong to at least some degree. Yet this is the situation with many of the world's religions. Even if they're all partially or mostly accurate, the assertion of absolute truth is still rendered a false statement, because it is an absolute. It therefore requires only a single exception to prove it wrong, and there are many such exceptions.

Every religion claims to be exceptional too, but virtually every trait that any of them hold as unique is also being promoted by another religion in exactly the same way, and for the same reason. Case in point, Christians sometimes assert a greater conviction by saying that theirs is not merely a religious belief, but also a "personal relationship," such as would not be possible unless Jesus were the one true god. However, George Harrison, the guitarist for the Beatles, was a Bhakti Hindu. He believed in a personal god, and he said that if one chants the mantras with devotion, Lord Krishna would visibly appear and speak to him in an audible voice.

> If there's a God, I want to see Him. It's pointless to believe in something without proof, and Krishna Consciousness and meditation are methods where you can actually obtain GOD perception. You can actually see God, and Hear Him, play with Him. It might sound crazy, but He is actually there, actually with you.

Many pagans are similarly convinced of having met their deities too. For example, a cat fancier in Texas whom I have known for decades insists that he began worshipping Bast only after the Egyptian cat-headed goddess dramatically appeared as a physical manifestation so that he could see her, feel her embrace and hear her voice. He said she had personally chosen him, and bade him to become her disciple. So of course, he did.

I have never met any Christian whose "personal relationship" with Jesus compares to either of these examples. The Abrahamic god never seems as personal, tangible, nor apparently real as it often seems with the spiritual beings and deities of other faiths. Every religion offers its own proofs, and they're often of the same type. They may be described as life-changing revelations of secret truths or the conversion experience awakening a new awareness, or they may purport to evoke powers and abilities beyond those of mortal men.

For example, the traditional Chinese religion is a mixture of Confucianism, Taoism, Buddhism, polytheism, and ancestor worship. Devotees of this blend of traditions claim direct communication with their gods and spirits, and they credit these entities for making them capable of remarkable feats of faith. Most notable among these are the extreme body and face piercings paraded through the annual vegetarian festival in Phuket, Thailand, in which devotees pierce all manner of bizarre objects through their cheeks, tongues, and skin. These objects can be anything from axes, swords, and guns to wrenches, lampposts, and even bicycles. They impale themselves this way and make a public spectacle of it in an attempt to shift evil spirits from others unto themselves. This annual tradition began as a response to a deadly epidemic that swept through the tin-mining community in the 1800s. The people decided they could stop the epidemic by showing homage to the nine emperor gods, Kiu Ong Iah. The epidemic ended around the same time the practice began, and believers now say these gods protect them from harm in the piercing practice. Remarkably, participants report feeling no pain and experience little or no blood loss, regardless how horrific their spectacular impalements may appear.

Near the end of the fifteenth century in the Punjab region of South Asia, Guru Nanak had a divine revelation that neither of the two great religions was the right path. He was referring to Hinduism and Islam. Christianity went unmentioned because it had no presence in this region, despite its dominance in Europe during the Dark Ages. Instead, Nanak was compelled to follow what he called "God's path." Thus, he became the first master of Sikh Dharma, starting a religion that numbers over 23 million devotees today. Guru Nanak was credited with many miraculous feats; in one instance, he is said to have squeezed blood and milk out of two different pieces of bread, as a demonstration of the inequity of the caste system. When he died, his followers laid flowers on him and covered his body with a shroud. When they pulled back the shroud the following morning, only the flowers remained, for his body had vanished.

Today this region has a lot of religious diversity. If someone reports having a vision of spiritual revelation or a near-death experience, the particular details will typically be consistent with Sikh, Hindu, Muslim, or Jainist belief, whatever the subject is already familiar with and accepts to some degree. In the Orient, Bodhisattvas or traditional Chinese spirits are involved, or there may be a "remembrance" of past lives regardless whether any of this was ever real or not. Neopagans and such will likely recall an "out-of-body" experience. Christian visions only occur among those who were raised in or taught the Christian religion. There is never a revelation of spirituality that the subject did not already know about, which indicates that mystic visions and near-death experiences are likely illusory and only accentuate preconceived notions.

Significant stress to the brain—whether from drugs or loss of oxygen or whatever—has often produced religious experiences, and not always in subjects who were previously religious, but such stress-induced experiences almost always produce effects that match the religious bias of the subject's culture. Those who study many religions equally won't likely "experience" any one of them specifically, and those with no specific religion's mythology will likely not have any religious experience whatsoever.

A great experiment that illustrated this fact involved the "God Helmet." Significant stress on the brain can be safely duplicated without being "near-death," such as being flung in a centrifuge for astronaut training. The so-called God Helmet is a device that electrically stimulates the brain in subtle ways. The most common reaction people have is the idea that another unspecified but intelligent presence is in the room with them. Some thought it was something like a ghost; some thought it was a god. At least one person thought that the other presence was himself having an out-of-body experience. A parapsychologist, convinced that supernatural entities were a matter of psionic fact, described the test as one of the most extraordinary experiences she had ever had. However, famous atheist Richard Dawkins described only feeling slightly dizzy and "quite strange," but otherwise pleasantly relaxed.

This is consistently what happens as anecdotal religious claims are put to the test. They are found to be subjective, implausible, and not likely genuine experiences of anything "supernatural." Yet adherents of all these religions (and some Buddhists as well) talk about their spiritual rebirth whenever they accept whatever version of whichever deity into their lives. Varying forms of spiritualism that are otherwise different all demonstrate or claim to practice transcendental meditation and astral projection, psychic viewing, levitation, aura reading, spell casting, divination, and various forms of spiritual healing—all of which leave the initiate believing these experiences are real. If the ritual is right and the believer believes "hard enough," he can even be convinced of almost anything, even lycanthropy!

Every religion boasts their own miracles, artifacts, and prophecies proving theirs is the truest faith. So it's no surprise that Christians say the same things about their versions of God too. No religion is significantly different from any other in this respect. But whatever else may be going on, when men claim revelation from God, it usually means they've decided to promote their own biased and baseless opinions as if they were divinely inspired. So it's not like any true god, ghost, or spirit is really guiding all these people the way they all insist that he, she, they, or it is.

If any god exists, and it happens that there's only one of them, then surely every spiritually enlightened and visionary holy man from any nation or tribe should be able to sense it, if men can sense such things at all. And their scribes would write the scrolls seeking to make sense of it, however feeble an attempt that may be. Perhaps that's why there are so many different religions—because no man can know the true state of God and there's no way to correct errors of perception to determine which concept is the most accurate. So there's no way to know what is really true. Following a belief based on faith is a matter of the blind leading the blind.

Whenever one walks away from any faith to pursue evidence instead, it doesn't matter which faith they came from; they will be guided toward the same one-and-only position that is held by the worldwide scientific community. That is the truth of nature and the nature of truth: there can be only one truth, and only one version of it. But rather than coming together (as everyone's search for the one truth should), religions continuously shard further and further apart into more divided factions with mutually exclusive beliefs, and there are as many wrong interpretations as there are people claiming theirs as the "absolute truth."

In reality, there is no such thing as absolute truth. Logically, there can't be; one reason being that "truth" requires validation while "absolute" denotes an unrestricted unconditional ultimate totality, independent of relation and transcending the limits of experience or observation. No mere fallible human can honestly claim knowledge of absolute truth because everything within the capacity of human understanding contains a degree of error, and everything men know to be true is only true to a degree.

Everyone is inevitably wrong about something somewhere. We don't know everything about everything. We don't know everything about anything! And what we *do* know, we don't know accurately on all points nor completely in every detail. Honest people admit this. Anyone claiming to know the absolute truth is not being honest, especially not when they claim to know anything about that which can only be believed on faith.

Even if humans were given genuine revelations by truly omniscient beings, that must still be filtered and interpreted by weaker minds influenced by our limitations, biases, and misimpressions, as well as linguistic and cultural barriers.

Of course that depends on how one defines "absolute truth." The first time I published commentary on this topic, I interpreted that phrase to mean essentially the same thing as a perfect knowledge. That is the context that is generally used when theists plead for the authority of their various scriptures. I guess I was trying to be philosophical, looking for the deepest possible meaning of "truth." What I meant was that the limits of human perception and understanding are such that our knowledge will never be absolutely complete nor flawless enough to be called "absolute."

Many religious types have since criticized me for this, thinking that I had contradicted myself whenever I expressed certainty in other things that were known to science. In November 2011, I was in a debate with Pastor Bob Enyart on his weekly radio talk show, and he pressed me on this same point. It quickly became obvious in the course of our discussion that he wasn't using so lofty a definition as I was, so I asked him to clarify. Turns out, all he thought (and apparently all *any* apologist thought) was that "absolute truth" only meant whether we could know anything for certain. That's it!

It was so simple! Here I was playing way beyond their game and having to repeatedly explain the same quoted paragraph for every newb believer who saw my video explanation of the 3rd Foundational Falsehood of Creationism. When I found out how utterly basic the theists' definition was during that debate, I remember saying, "I wish I knew that before!"

Later on, I went to the Reason Rally in Washington, DC, and met Eric Hovind of Creation Science Ministries. He immediately asked me whether we could know anything for certain, and I answered in the affirmative. (Later he tried to say that I actually said no, but he kept the video available online.) I tried to explain that even if reality itself wasn't real, even if we only exist within a dream of Brahma or if all we perceive is a computer-enhanced hallucination like the Matrix, it wouldn't change the fact that the rules of that reality would still be real to *us*. We're forced to assume that our perception is at least partially reliable, that we really do exist in a real world that is independent of us, and that other minds exist outside our own. Because it would be impossible to function otherwise. Consequently, such ponderings as solipsism are in my opinion meaningless philosophical nonsense.

The point is that if I had met him back when I posted my video on the 3rd Foundational Falsehood of Creationism and he had asked me then whether I believed that humans could claim knowledge of absolute truth, I would have thought that he was asking a different and deeper question, and I would have given him the opposite answer.

Just to clarify—yes, there *are* some things we can and do know for certain. The testable and verifiable fact that the Bible is wrong about Noah's flood, about the Tower of Babel, the firmament, the Garden of Eden, and a host of other things is a matter of scientific certainty. Through an independent international consensus analysis of overwhelming evidence by expert specialists in every relevant field regardless of religion, we can know that certain elements of the Bible's fables have been conclusively disproved beyond redemption. Not even the existence of God could resurrect the Bible anymore. Even though we don't know absolutely everything, and we don't know anything in completely flawless detail, we do still know at least this much with absolute certainty: the Bible, and all of the subsequent books and religions that are dependent on that mythology, are wrong. Sorry.

The Mormons and Muslims both have their lists of issues, and both deserve a firm rebuke on a great many specifics. They're both tripped up by various biblical fables that they may be compelled to defend, and they both also make very similar defenses of alleged artifacts, mathematical conundrums, claims of coincidence, and pleas of supposedly prophetic passages. In each case, these either don't mean what they want them to, or they have been disproved beyond defense. None of their sacred tomes are infallible documents, because they were all written by men who were themselves limited by the times and cultures they came from.

In the history of history itself, no account human journalists have ever given has been absolutely complete, inerrant, and perfectly accurate—especially when there is a desperate emotional bias, such as there is at the source of each of the world's religious books. All of them were written decades or centuries after the alleged events they claim to have witnessed, and they speak of many scenes that no one could have witnessed at all. Take Jesus' prayer at Gethsemane, for example, or Cain's posthomicidal conversation with YHWH. Who was the stenographer in either case? Similarly, much of the Torah and the Bhagavad Gita were originally written as poetry, so the conversations can't have been verbatim unless all the characters really spoke in rhyme.

These tales also include impossible absurdities that aren't even corroborated by any other internal contributors, much less by external records. For example, we know from genetic tests of ancient human remains and multiple dating methods of associated artifacts that the ancestors of today's Native Americans were already in the western continents many millennia before any date ever given for "the Flood." Yet they're still there because the flood never drowned them. Archaeologists and anthropologists have confirmed that the same is true for virtually every other archaic culture around the world. Each of the ancient civilizations were already established and speaking many different continuously evolving languages a long time before the Tower of Babel (around 1750 BCE).

The same goes for the New Testament. According to the gospels, when Jesus died on the cross, there was supposed to be hours of darkness during the day—yet no literate person anywhere in the whole world noticed when the sun shorted out for a while. There was no eclipse at that time, nor any kind of volcanic disturbance either. At the same time, according to the Gospel of Matthew, several long-decomposed saints reportedly crawled out of their graves to wander the streets, yet no one noticed that either. Not even the other gospels mention that! We have documents recording all manner of relatively trivial events and financial exchanges of that period, but not one other person anywhere noticed when a troupe of the undead danced the Thriller in downtown Judea? This is the sort of thing that is noteworthy! The occupying Romans were very competent historians, yet no matter how many witnesses there supposedly were or how many others should have written about it, the only source for any of the fables in the Bible is the Bible itself.

Archaeology certainly doesn't support any of these stories. Instead, we have earlier versions of many of them coming from the myriad myths of polytheism, some of which written by the very ancestors of the biblical authors. As explained in the previous chapter, they apparently conceived all the original but as-yet unassociated elements that were eventually blended together into the fables we now know as Genesis.

These stories can be interpreted wildly differently by anyone who reads them. Some argue that the Bible doesn't really say some of the things we can prove that it does say, while others are convinced that it clearly says things that it doesn't really even hint at anywhere. For example, nowhere does the Bible say that the serpent in the garden was supposed to be anything but a snake. The Book of Revelations describes Satan as "that old serpent," but it doesn't mean *that* old serpent, not the one from the garden. The last book of the Bible is not commenting on the first book from a different author and religion. Some early interpretations cast the serpent as an incarnation of the dark maid Lilith. That's why so many classic renderings of the temptation of Eve show the serpent as a female. She is even depicted that way on the ceiling of the Sistine Chapel and the Cathedral of Notre Dame and by many of the great artists of the Renaissance. Of course the original version of this story said "the serpent who could not be tamed" was a companion of Lilith. We know that the serpent in the Garden can't be secretly Satan, because the first time Satan is mentioned by name is in a story chronologically after this one, and in that, he is said to be walking. Remember that the serpent was cursed to crawl on his belly all the rest of his days, but in the book of Job, Satan is still walking around and chatting with God as if they'd never had a falling out. This is when God had to ask Satan where he's been, because his infallible omniscience obviously didn't know. So if he had to ask that of Satan, who is later described as the Lord of Lies, then why would God believe him? Or didn't God know any better by then?

So there is no literary link ever implied between Satan and serpents, other than the common insult of calling him a snake. Jesus referred to the Pharisees as snakes too, and he said they were descended from Satan, but that doesn't mean they're descended from snakes—even though John the Baptist said they were. Nor does it mean that any of them were in the sacred garden at the alleged time of Adam and Eve. All of this is interpretation that is assumed on tradition but not at all supported in the text.

Another example is the idea that there was no death before "the fall." The Bible doesn't say that. In fact, it says there was death before the fall, because Adam and Eve had to ingest and digest living cells in order to survive—the very definition of what it means to be an animal. The only way around that was to eat of the fruit of the tree of eternal life, which directly contradicts the creationists' interpretation, because it wouldn't need to be there if they already had eternal life. That's why God kicked them out of the garden before they could eat of that tree too.

What if Adam never ate from either tree? Or from any tree? Would he eventually starve to death? In every other biblical book besides Genesis, "the fruit of" whatever serves as a metaphor of the consequences of our choices. In the Garden of Eden it was a choice between the fruit of the tree of eternal life and the fruit of the tree of knowledge of good and evil. This fable was adapted from the legend of Adamah, also known as Adapa or Adamu, from Ashurbanipal's library. In this story, the seed of man was brought before the gods and warned not to eat or drink anything offered to him because the gods might poison him. But instead of giving him the food of death, they offered the food of eternal life, and he refused.

Perhaps the version of the fable in Genesis 3 was originally meant to represent a choice between innocence and responsibility—that once you grow up, you'll never be the same and can never return to childlike innocence. That too is an interpretation. But in any case, the tree of eternal life and the tree of knowledge of good and evil were obviously metaphorical; you couldn't cut either one down and build a boat out of it, because they were not actual deciduous plants!

Everything in the Bible is a matter of interpretation rather than "absolute truth." Of course one interpretation is that without that particular fable about Adam and Eve, there is no original sin, and thus no fall of man, and consequently no need of salvation; at that point, creationist Christianity comes completely apart. Christians who interpret it that way often say that either the scriptures are mere fables and Jesus was a lunatic or a liar, or that the scriptures are absolutely true (despite their many contradictions) and that "Jesus is who he said he was." But there is also some argument over how Jesus supposedly identified himself. That too is a matter of interpretation.

The gospels attributed to Matthew, Mark, and Luke each sought to portray Jesus either as the new Moses, as the new King David, or as the long-awaited messiah. Then a whole lifetime after any date given for the crucifixion, we have the gospel attributed to John. By then, the political climate turned against the Gnostics and Ebionites, and now ran in favor of the emerging Orthodoxy. The next generation of Christians posthumously deified their Christ and confused his identity with that of the father-god of Judaism (and eventually Islam). Thus the Gospel of John broke from the trend of the previous gospels by attempting to portray Jesus as a god. But even in John, Jesus still never claims to be either the creator of the world, nor the same deity that the Hebrews had worshipped all along.

In the fourth century, the Council of Nicea gathered theologians from all over the Roman Empire, trying to interpret what their scriptures meant rather than what they said. One of the central disputes was the identity and divinity of Jesus, a point that had been hotly contested since Christianity began. The majority position, defended by Saint Athanasius, was that although Jesus had always distinguished himself from God, and never implied otherwise in any of the gospels, if one remembers what is said in 1 Corinthians 1:24, and forgets what was just said in Proverbs 9:1, they may interpret the next use of the word "wisdom" in Proverbs 8:12–22 to mean "Jesus." One could then use that to reject what Jesus himself said about not being the same person as God.

The minority position proposed by Saint Arius of Alexandria was that Jesus was a prophet, the most important human, and a sort of demigod—but still a created being, and not the same as Abba, the "father" god who created him. As 1 Timothy 2:5 says, "For there is one God and one Mediator between God and men, the man, Christ Jesus."

For a time, both sides labeled the others heretics. However, the heresy of St. Arius was so ill-received that it reportedly earned him a punch in the face from none other than Saint Nicholas. Yes, *that* Saint Nicholas, legend has it. Merry Christmas—have a knuckle sandwich.

Those who said Jesus was a prophet of God but not of the same essence as God lost the vote, and were banished in order to prevent their ideas from influencing the Christian formation. Amusingly, some of the banished found themselves in the city of Medina. Because of this placement at that time, some scholars of religious history (including my professor, Dr. Mark Hanshaw) suspect their teaching may well have eventually influenced the emerging religion of Islam.

Once men had decided which fables should be canonized, Bishop Athanasius condemned the use of noncanonical tomes in his Festal letter of 367 CE. Thus, other books would be treated as damnable heresies that must be silenced and put out of sight. Consider the writing of Justin Martyr, a second-century apologist:

> Analogies To The History of Christ. And when we say also that the Word, who is the first-born of God, was produced without sexual union, and that He, Jesus Chris, our Teacher, was crucified and died, and rose again, and descended into heaven, we propound nothing different from what you believe regarding those whom you esteem sons of Jupiter.
>
> Be well assured . . . that I am established in the knowledge of and

> faith in the Scriptures by those counterfeits which he who is called the devil is said to have performed among the Greeks; just as some were wrought by the Magi in Egypt, and others by the false prophets in Elijah's days. For when they tell that Bacchus, son of Jupiter, was begotten by [Jupiter's] intercourse with Semele, and that he was the discoverer of the vine; and when they relate, that being torn in pieces, and having died, he rose again, and ascended to heaven; and when they introduce wine into his mysteries, do I not perceive that [the devil] has imitated the prophecy announced by the patriarch Jacob, and recorded by Moses? And when they tell that Hercules was strong, and travelled over all the world, and was begotten by Jove of Alcmene, and ascended to heaven when he died, do I not perceive that the Scripture which speaks of Christ, "strong as a giant to run his race," has been in like manner imitated? And when he [the devil] brings forward Aesculapius as the raiser of the dead and healer of all diseases, may I not say that in this matter likewise he has imitated the prophecies about Christ? . . . And when I hear . . . that Perseus was begotten of a virgin, I understand that the deceiving serpent counterfeited also this.

Where is the story of Jesus being as strong as a giant to run his course all over the world? This is obviously one of the many accounts of Jesus that apparently played into the origin and development of Christianity, but which was subsequently censored and eliminated. So we really have no idea how many borrowed legends Christianity was really based on. But all the stories we still have were apparently adapted from tales originally told about someone else.

I used to think that Jesus was a real person who lived in Judea somewhere between 4 BCE and 40 CE. I know a lot of Christians who say that Jesus "divided time" by being born in the year zero, but there was no year zero. The gospels erroneously attributed to Matthew and Luke say Jesus was born during the reign of Herod the Great, who died in 4 BCE. However, Luke also says he was born shortly after the Census of Quirinius, which was conducted a decade later in 6 CE. But neither account allows that he could have been born between those dates. Pope Gregory VIII produced his own calendar in 1582, and wanted to back-date it to coincide with Jesus' birth. Since Jesus was supposed to have been born both before the year 41 on the Julian calendar and simultaneously after 50 on that same calendar, and there was no way to reconcile that contradiction, it seems that the Pope just chose a date in the middle and called it zero. That was the forty-fifth year of the Julian calendar, and the Gregorian calendar arbitrarily assigned that to be Jesus' birth year. That's why modern Christians think that Jesus "split time" between the creation of the world (which was *not* 2,000 years BCE) and our current days, which are *not* the "end times," regardless of what Dominionists say.

Of course Jesus is a Jewish character, and according to the Jewish calendar, he should have been born around their year, 3758. As I write this, the year is now 5776.

So we have a child born in Nazareth, which was not occupied until the next century. So he couldn't really have been born there. So other than his two different birthdays in different centuries, we also read that he was raised in two different countries. How can we even discern which stories apply to any actual Jesus and which ones were inserted for some other purpose?

By the time of the Nicene Convention, there were many different sects of Christians with very different interpretations that had to be eliminated in favor of the emerging Orthodoxy. Among these different interpretations were the Coptic Gnostics of Egypt. They responded to Athanasius' decree by hiding their sacred texts some 1,600 years ago. They were accidentally rediscovered near the end of World War II, buried in a jar. The 1,200-page compilation was found near the Egyptian city of Nag Hammadi, and is now known as the Nag Hammadi Library. In that collection, there is a markedly different account of Genesis, which I will briefly paraphrase in summary:

After all manner of immortal deities were formed out of infinity, the goddess of wisdom (who is also described as such in the Book of Proverbs) became the primordial light. Chaos was created afterward, as an abyss, a dark and lifeless abortion flowing out of the goddess of shadow, who was jealous of the light. Through Faith, Wisdom conjured a Lord out of this lifeless abortion of darkness and chaos. This Lord was ignorant of his own origin. He had never seen the face of Faith, for she withdrew into the light. So he was left alone in the darkness of the abyss, and thus assumed that he was the only thing in existence. Isn't that great? God was created in ignorance of his own creation, unaware of his creator. So God was an atheist.

We get several paragraphs into this story before we get to the parallel of Genesis 1:1, where the Lord awakens to find himself a lone spirit over the face of the waters. He then divided the water from the dry land, created Heaven as his abode, and made the earth as his footstool.

Immediately after that, the Lord used an incantation to conjure three sons, Yao, Eloai, and Astaphaios. He then created a half dozen more heavens for his subsequent creations and he filled them with armies of gods and archangels.

The heavens and earth were destroyed by the troublemaker from below, but the goddess of Wisdom bound the unnamed antagonist to the same abysmal prison wherein the Hellenists had previously trapped the Titans. After this, the Lord and primordial "father" was honored by all his creations. Thus he boasted, "I am God, and no other exists besides me" (as if quoting Isaiah 45).

This the elder gods took to be boastful impiety. The goddess of Faith said to the Father Lord, "You are mistaken, blind god; there is an immortal man of light who has been in existence before you, and who will appear among your modeled forms; he will trample you to scorn, just as potter's clay is pounded. And you will descend to your mother, the abyss, along with those that belong to you."

She's not talking about Jesus. Because the father Lord then created Israel (the man who sees God) as his "first-born" son. He already had a daughter, Zoe (Eve of Life), who was created to be his teacher. Jesus Christ, who sits in the throne to his father's right hand, was created after that, as the younger brother of Israel and Eve.

The immortal man made of light was introduced to the father Lord as Adam, the "luminous man of blood." After that, we mix in some Greek gods too, and the mythos soon comes to a parallel of Genesis 2. The principle difference here is that the tree of eternal life grows next to the tree of knowledge, in a garden which is located outside the orbit of the moon. Adam magically conjures a half dozen eternities here and there, but is eventually overcome by darkness and chaos. Then the father lord and a few other gods join together to recreate him, this time in their image. The job is a failure and the new Adam is left lifeless, so Eve comes along and breathes into him the breath of life.

There are actually three successive Adams in this story: a spiritual Adam, a soulful Adam, and an earthly Adam, who becomes the father of Men, but not all men. Another group of lords come upon Eve and take out their vengeance by gang-raping her. This was intended to soil her so that she could only bare mortal children. Her firstborn son Abel is described as a rape baby.

The wisest of all creatures was named "Beast," and he instructed the traumatized Adam and Eve to eat of the fruit of the tree of knowledge. When the seven lords came into the garden, they found that their orders had been ignored; because of that, they were powerless against Adam. They cursed him and tested him, but he always persevered.

The Coptic creation myth On the Origin of the World prophesies the eventual death of everything, including the gods themselves, so these Gnostics had a very different interpretation than other early Christians like the Orthodox, Catholics, and Luciferians. (Yes, the Luciferians were a Christian denomination once upon a time, and Jesus the Christ was once known as "Lucifer the light-bringer.")

Bear in mind that the Nag Hammadi Library has been dated to at least 200 CE. That would make it shortly after or possibly contemporary with the Gospel of John. This collection also has its own noncanonical gospel of Thomas, which consists of a single chapter that reads like snippets taken from other gospels. My favorite bit is from the 31st verse: "No prophet is accepted in his own village; no physician heals those who know him." Jesus appears to be playing the part of a first-century faith healer, and seems no more genuine than the charlatans we see on evangelical TV ministries today.

Scriptural expert Bart Ehrman notes that "there is no doubt but that in the Gospel of John, the 4th gospel, Jesus understands himself to be God, and explicitly calls himself divine." Ehrman also points out that Jesus never calls himself this in any gospel except John, but that "Jesus is portrayed as a divine being, a God-man, in all the Gospels." With all due respect to Professor Ehrman's scholarship and to the fact that his is the dominant interpretation among dedicated theologians, I still have to disagree; there has always been doubt about Jesus' divinity, and there still is. Even modern Christians have very different interpretations of Jesus' divinity. Consider the beliefs of the following nontrinitarian Christians:

1. *Ebionites*: Jesus was a purely human prophet of God.
2. *Monarchists, Arians, and Unitarians**: Jesus may not have been technically divine, even if he was both a perfect human and a spiritually immortal being. But either way, he does not share identity with his father, who is identified as YHWH, the true god of the Jews and creator of the world, whom Jesus himself worshipped. Jesus was God's son, and

his messenger, but not his avatar. Being "begotten" by God also means they cannot be coeternal.

3. *Mormons*: Jesus is truly divine because the son of a god would be a god also. But Jesus is not the same entity as his father. The "Holy Ghost" is a third distinct character who hasn't yet lived on earth, and consequently does not have a body yet. Jehova, his begotten son Jesus, and the amorphous "Holy Ghost" comprise a "godhead" operating as a single monotheistic collective similar to the Hindu Trimurti.
4. *Sabbatarians (Binatarians)*: Some denominations are similar to trinitarian Christians except that they recognize only YHWH and Jesus as two aspects of a single being, but do not consider the "Holy Ghost" to be an entity of its own.
5. *The Restored Church of God*: This is another binatarian belief that the godhead temporarily consists of Jesus the Messiah, as the creator and spokesman, and God the Father. Their God Family doctrine states that every human who ever lived may be spiritually born and enter into the godhead.

*I should note that the reference to "Unitarians" above describes a prominent Christian perspective among some of the founding fathers of the United States. Those Unitarian Christians of the eighteenth century should not to be confused with the Unitarian Universalists of today. Those are two very different notions, almost entirely unrelated.

Trinitarian Christians have the impression that Jesus had at some point declared himself to be an incarnation of God rather than the son (or prophet) emanating from God, as if God and Jesus were the same person in two forms. (Actually three forms since trinitarians also include the Holy Ghost.) So trinitarians can say that God the son is Jesus, but they can't say that God the Father is anyone else but Jesus. To them, YHWH and Jehova are just other names for Jesus. Thus Jesus is his own father, because he is the only identifiable personality of the Christian godhead. Consequently, Michelangelo may be the only Christian artist to depict Jesus and Jehova as two different characters.

What trinitarians (and binatarians) describe is the concept of an avatar. Imagine that you're playing a video game: you as your physical self cannot exist within the confines of that game because you exist beyond that reality and outside their scope of time, so you create a character to represent yourself within that game. While that character cannot be all that you are, it is still "you" in the sense that it is a manifestation of your own true self extended into the game. Thus in the trinitarian concept, Jesus the man is a player character for Jesus the god, who is the actual player and who exists beyond the confines the game (which in this case is the material world).

Of course that idea falls apart once you read about Jesus' prayer at Gethsemane: "My Father, if it is possible, let this cup pass from Me; yet not as I will, but as You will" (Matthew 26:39 and Mark 14:36). Why would Jesus need to pray to someone else? And who is Jesus pleading with? Whose will is to be done here? Because it clearly is not the will of Jesus himself. The same goes for when Jesus said, "Father, forgive them" and "into your hands I commend my spirit," and "why have you forsaken me?" He's obviously not talking to himself! Jesus can't believe that he is the same essence or personality as God, even if he did believe himself to be the son of God and thus a god of sorts himself.

I remember having a very long conversation in the office of a Southern Baptist minister, who was once a close friend of mine in high school. I listed several pre-Christian god-men from neighboring religions, all with suspicious similarities to Jesus. The minister dismissed them all, saying that Jesus was the only one among them who ever said he was God. Immediately I pointed out that at least a few of these other characters had loudly and proudly declared themselves to be God in the flesh, but that Jesus never described himself that way, and he was one of the few who did not do so. In fact, the Bible never even mentions the trinity, and never says anything to support the trinity either.

If you're reading 1 John 5:7–8 in the King James Version, try any of the other versions and you'll see what I mean. Otherwise, everywhere throughout the New Testament, Jesus is represented as a separate person from God, the gateway to God, or as God's right-hand man, but never as a physical manifestation *of* God.

I've read Christian posts talking about how Jesus spoke to Adam and Eve in the Garden of Eden. No. The authors of the Tanakh (now known as the Old Testament) were not Christians; Judaism is a different religion. Jesus was not the god of Abraham. Moses didn't speak to Jesus; he spoke to Yehowa, Jehovah, the same god Jesus himself prayed to.

Bearing that in mind, it's funny when you compare the Old Testament with the New, because it appears that God and Jesus are arguing with each other over Jesus' divinity.

> *"Thou shalt have no other gods before me."*
>
> —Jehovah (God), 1st Commandment, Exodus

> *"No one comes to the Father but through Me."*
> —Jesus, John 14:6

> *"I am the Lord, and beside me there is no savior."*
>
> —Jehovah (God), Isaiah 43:11

> *"But from now on the Son of Man will be seated at the right hand of the power of God."*
>
> —Jesus, Luke 22:69

> *"I am Jehovah, and there is none else; besides me there is no God. I will gird thee, though thou hast not known me; that they may know from the rising of the sun, and from the west, that there is none besides me: I am Jehovah, and there is none else."*
>
> —Jehovah (God), Isaiah 45:5–7

> *"You have said so. But I tell you, From now on you will see the Son of Man seated at the right hand of Power and coming on the clouds of heaven."*
>
> —Jesus, Matthew 26:64

Please remember that Isaiah was Jewish, not Christian. He lived 700 years before Jesus' time, and Isaiah did not believe "the father" ever had a "son," nor did he believe that the prophesied "redeemer" would be the son of God either. So when Isaiah uses the name Jehovah, it is to describe the father only. The god of Abraham and of Isaac was known by several names—El, Allah, Abba, or YHWH (Yovah/Jehova)—but none of the authors of the Old Testament ever meant to refer to Jesus, because he didn't exist in their theology.

In the Old Testament, we had Jewish prophets pretending to speak for Yehowa, and in the New Testament, we have converted Christians pretending to speak for Jesus.

> *"But he, being full of the Holy Ghost, looked up steadfastly into Heaven, and saw the glory of God, and Jesus standing on the right hand of God."*
>
> —Stephen the Hellenist, Acts 7:55

Again we see Jesus and God as separate characters. Those trying to reconcile contradictions between the Old and New Testaments may have borrowed concepts of trinitarian gods and avatars from the Hindus. The Hindu Trimurti consists of Brahma the creator, Vishnu the sustainer, and Shiva the destroyer. Each of these characters are their own individual beings, and each have their own lives and are married to different wives, yet they act collectively as three aspects of a single unified godhead. This is the culture that invented the concept of the avatar, and that is how Lord Krishna describes himself in the Bhagavad Gita of the Mahabharata.

Consider the following quotations from Śrī Krishna from Rāja-Vidya-Guhya Yoga (Confidential Knowledge of the Ultimate Truth),

> Verses 6–7:
>
> *"Understand just as the mighty wind blowing everywhere is always situated in space, similarly all created beings are situated in Me. O Arjuna, all created beings enter into My nature at the end of a four billion, three hundred twenty million year cycle, and after another four billion, three hundred twenty million year cycle, I regenerate them all again."*
>
> Verses 11–13:
>
> *"Fools deride Me in My divine human form, unable to comprehend My supreme nature as the Ultimate Controller of all living entities. These bewildered fools of futile desires, futile endeavors, futile knowledge and futile understanding certainly assume the nature of the atheistic and the demoniac. But the great souls having taken refuge of the divine nature, O Arjuna, render devotional service unto*

Me with undeviated mind, knowing Me as the Imperishable origin of all creation."

And from Vibhūti-Vistara Yoga (Infinite Glories of the Ultimate Truth),

Verses 2–5:

"Neither the demigods nor the great sages understand My divine transcendental appearance; because I am the original source of the demigods and of the great sages in every respect. One who knows me as birthless, beginningless, and the supreme controller of all the worlds; he being undeluded among mortals is delivered from all sins. Spiritual intelligence, knowledge, freedom from false perception, compassion, truthfulness, control of the senses, control of the mind, happiness, unhappiness, birth, death, fear and fearlessness, nonviolence, equanimity, contentment, austerity, charity, fame, infamy; all these variegated diverse qualities of all living entities originate from Me alone."

Verse 41:

"Certainly wherever and whatever is majestic, beautiful or magnificent; you must certainly know that all these manifestations arise from but a fraction of My glory."

So Lord Krishna very plainly declares himself to be the supreme personality of the godhead, and the source of all the gods. He said that he himself not only created this world, but many worlds—the whole universe. At one point, he even makes reference to a multiverse. Such a futuristic concept isn't bad for scriptures that are so many centuries older than the Bible. The Bhagavad Gita has been contextually dated to at least 2,600 years old, and the preredacted origins of it are reputed to be much older. Most importantly, Krishna clearly and repeatedly declared that he was "God-made-flesh" and an avatar of Vishnu. He was quite explicit on that point.

But in the gospels, which are the only documents purporting to record the Christ's actual words, Jesus never implied any of that at all. The most compelling argument trinitarians have is one isolated comment that was poorly phrased. Jesus once said *"before Abraham was born, I am."* Somehow simply phrasing it "I am" instead of "I was" implies that Jesus is the same person as Jehova, God the father. By this logic, it could be argued that René Descartes declared himself to be God too. "I think, therefore I am … God." I understand there is some significance to that wording within the context of early Jewish culture, but I still have to compare that aberrant quip to the many unambiguous boasts of Lord Krishna. More importantly, Jesus also told the same group of people that they were gods too (John 10:34).

Throughout the Bible, regardless of whatever else he may claim about himself, Jesus always only ever described himself as separate from and subordinate to El/Allah/Abba/YHWH, and he said that the god of Abraham and bringer of the flood was someone else somewhere else, who knows things Jesus doesn't know, can do things Jesus can't do, and who did things Jesus didn't do. Jesus also spoke about God in third person and to God in second person, and God even talks about Jesus in third person too when he introduced his son to the Jews (Matthew 3:16–17). Then the Holy Ghost showed up and led Jesus to somewhere Jesus did not already know (Matthew 4:1). None of this could be if Jesus were an avatar or "God in the flesh" because then Jesus and God and the Holy Ghost would all still share the same knowledge, power, identity, and position in space and time. So it is pretty clear that Jesus did not believe himself to be the same god as the one he and the Jews both worshipped.

Another even more explicit example comes from Mark 10:18: "And Jesus said to him, 'Why do you call Me good? No one is good except God alone.'" Here we see that Jesus says he is not what God is, and that God is alone." Jesus is not good, Jesus is not God. This is a consistent theme.

Jesus never presented himself as God. He only did what the pharaoh Akhenaton did: promote himself as the sole prophet of the sun-god. In the context of the story, it is clear that Jesus meant to describe himself as one who had previously dwelt in Heaven since at least the time of creation, and that he had since been sent to the earth by his father. He said he was *"not of this world"* and that he would return to his father. As a magical anthropomorphic immortal, Jesus would be divine; he would be a god in that sense. But he never said nor in any way implied that he and his father were the same person. Nor could they be, according to anything Jesus ever said in the words attributed to him alone.

At one point, Jesus says he is "one" with God—but he clarifies that he is referring only to his purpose, and later in John 17 he prays to God rather than himself, pleading that his disciples could become "one" with God just as he is. This proves that Jesus is not "one" with God in the sense that Christians say he is.

Jesus encouraged his followers not to call any man "father," but to refer to the god whom Jesus himself worshipped as "father" (Matthew 23:9), while he referred to his disciples as his brothers of the same god and of the same father (John 20:17). Jesus was like a typical evangelical faith healer in some respects, including that he taught his disciples to perform his same routines, saying that they might do even greater things than he did.

There was one time when Jesus said, "You know neither me nor my father; if you knew me, you would know my father also." But he also says, "Not that anyone has seen the Father, except the One who is from God; He has seen the Father" and "You have neither heard His voice at any time nor seen His form." So knowing God is not the same thing as knowing, seeing, or hearing Jesus; they're different people.

Notice also that Jesus says these people have not heard the voice of God, yet there are at least a couple times when God speaks directly to those standing around Jesus (Matthew 17:5 and Luke 9:35). So either Jesus was wrong about that or these stories were compiled by people who weren't paying close attention to what they were saying.

Of course the story says Jesus was wrong about a few other things, too. He cursed a fruit tree because he didn't realize it had no figs left (Matthew 21:19). He thought mustard had the smallest of all seeds (Mark 4:31), but even first-century Palestinian farmers knew of a few other seeds that were smaller, such as the black orchid. Jesus also believed that all the kingdoms of the world could be seen at once from the top of a high mountain (Matthew 4:8), so he believed the world was flat, and he hoped that his "second coming" wouldn't happen in winter (Matthew 24:20). So he was only concerned with his own provincial position in that little corner of the world, and had no concept of the planet as a globe. He was racist against Canaanite "dogs" (as he called them in Matthew 15:21–28) and he boasted about how truthful he was when he lied about the father of the Pharisees being the devil and a liar (John 8:41-45). Such irony!

He said he came to divide families, separating husbands from their wives and parents from their children (Matthew 10:34–38), and he said that no one could follow him unless they hated their own kin (Luke 14:26). Jesus broke Jewish tradition by refusing to wash his hands at mealtime

(Mark 7, Matthew 15), because Jesus believed that diseases were caused by demons rather than germs. That's enough all by itself to call the whole thing into question. So it's not surprising that he couldn't perform his faith-healing shtick in his own home town because the people who knew him best thought he was full of shit (Matthew 13:54–58). He taught all the time, but only from the memorized manuscripts of men. He couldn't reveal anything that wasn't already known by the common folk of his day because that was the limit of his knowledge, and that's the problem with the entirety of scripture.

If the Bible is interpreted literally, then it is clear that its authors believed that the world was spread out like a map over a flat disc—not a sphere (*dur*), but a circle (*chug*) (Isaiah 40:22) divided into four quadrants (Isaiah 11:12), sometimes mistranslated as "corners." This disc-world stood on pillars (1 Samuel 2:8) like a table so that it would not move (Psalms 93:1 and 1 Chronicles 16:30). All of this was submerged in a watery abyss and covered by a giant transparent crystal dome, like a snow globe (Genesis 1:7). The sun, moon, and stars were contained within the expanse (Genesis 1:14) of this massive dome (Ezekiel 1:22). Fountains would allow water in from below the firmament, and windows in the expanse of it would allow rain in also (Genesis 7:11). This wasn't even an original idea; it was a common belief throughout many neighboring regions, but it was still wrong.

The biblical authors obviously knew nothing about the real state of this world nor the worlds beyond this one either. But we know what lies outside our atmosphere, and that proves that there is no water above where the firmament isn't, and no windows to let it drain in if there was either water or firmament there.

Some Persians at that time said that the god Mithras had the stars sewn into the lining of his cloak, which he would drape over the crystalline firmament to bring on the night. But we know that night is not a veil to be spread over the missing firmament like a curtain (Psalms 104:2) or a tent. We also know that the stars are not made to stand in the span of this expanse (Isaiah 48:13) because they are not "high" in the firmament (Job 22:12); there is no firmament, and they are so far beyond our puny world that "height" is meaningless and inapplicable. They are much too far away to be blown out of place by any storm (2 Esdras 15:34–35) and they couldn't be taken down by anything at all. We've also proven that the illusive heavenly firmament has no foundations either (2 Samuel 22:8), and neither does the earth (Job 38:4–6). There are no pillars holding Earth above the deep (1 Samuel 2:8) because there is no deep (Genesis 1:2). Outer space is not full of water!

We also now know what lies outside our gravitational field, and that proves that you can't have any passage of days and nights without a sun (Genesis 1:13–14) to measure them against an Earth that constantly moves (Psalm 104:5). We also know that the sun cannot be made to set at noon (Amos 8:9), and that neither the sun nor the moon can be stopped in the sky (Joshua 10:12–13).

We also now know what is beyond our solar system, which means we know the stars can't fall from the sky (Matthew 24:29). Even if they did, we still couldn't stomp on them (Daniel 8:10) because they're each thousands to millions of miles around, which makes it a bit silly to imagine a whole group of them having conscious minds (Judges 5:20) and ganging up in combat with a mere human being.

We even know now what lies beyond our galaxy. And that proves that nothing or no one could ever "seal up the stars" (Job 9:7). We also know that the earth with its fictitious firmament didn't predate the "lights in the heavens" by any amount of time (Genesis 1:17–19) and that the stars weren't "set" specifically to light the earth, because the earth is not at the center or the beginning (Genesis 1:1) of the universe in any respect. The way the Bible depicts Earth in relation to the rest of the universe is wrong, and has been known to be wrong for thousands of years.

Many creationists say that it is impossible to understand or believe the Bible unless it is read "in the spirit of the holy ghost." In other words, you must already assume its truth before you read it, and you have to read it through filters of faith, because it certainly isn't compelling on its own without those blinders on. If it doesn't make sense, then you've got to convince yourself that you must not understand it properly and that you've just got to try to make yourself believe it anyway somehow. That is precisely why creationist faith is deemed dogmatic, but that's also proof by admission that even a literal reading must be interpreted, so the Bible's very design is such that it cannot be either inerrant or "absolute truth."

THE 4TH FOUNDATIONAL FALSEHOOD OF CREATIONISM

"Belief Is Knowledge"

The next in our series of foundational falsehoods of creationism is a logical fallacy illustrative of the fundamental sophistry behind the creationism movement: the idea that really, really believing something is the same thing as knowing it. That is essentially the argument of presuppositional apologetics, a branch of religious belief wherein the believer asserts that he is right simply because he believes he is right.

Step 1. Assume the conclusion

Step 2. Assert your conviction

(Repeat)

I would say that the practice of apologetics amounts to the systematic evocation of excuses devised to rationalize away any arguments or evidence that might challenge a sacred assumption. The assumption is sacred because it is held *a priori*, meaning that those holding that position must never admit when they're wrong on any critical point. I have argued with many such people, and the same pattern consistently emerges: it seems that in their minds, they can't be wrong unless the whole universe is wrong. Somehow they believe that not even reality can be real if their god is not real, so the first thing they challenge is reality itself, or at least our perception of it, because the first thing they want to argue is how anyone could be sure they really know *anything*. Because if we don't really know what we know, then they can still make-believe something else.

I would also say that all faith-based beliefs stem from the logical fallacy of a question begging, circular argument routing back to an assumed conclusion. This is true of all forms of apologetics, but especially presuppositionalism. Their whole position can be summarized with the circular assertion "I know because I know," and further summarized with "'cuz I said so." There really is no more justification of their position than that.

Unlike other apologists, presuppositionalists say they can prove the existence of God. Here's how they do it: first they redefine "absolute truth" such that it only means that it is possible to know things; then they point out that logic is a concept, and as such is not made of material components. It doesn't matter if a thing is real outside of your own mind. Some apologists argue that if you can conceive of it, that makes it conceptual, and that means that you have to conceive of it for it to exist at all. The next assumption of course is that anything that can be conceived must have been conceived by God, and was thus imagined into existence.

Onto these two unrelated premises they somehow add the wholly unwarranted *non sequitur* that truth, knowledge, and logic cannot exist unless you assume their conclusion that God exists too. They don't give a reason for that; they just say, "The proof that God exists is that without Him, you couldn't prove anything." This exact argument (or rather, lack thereof) can be found on a website created by Sye Ten Bruggencate, a sort of vagabond king of presuppositional apologetics. His only claim to Internet fame is his inability to understand why his own reasoning is erroneous rather than everyone else's.

Knowledge is justified belief, meaning that it is demonstrable and measurable. As I explained in previous chapters, "truth" is whatever statement can be shown to be true; as such, since no one can show that there is any truth to anyone's religious beliefs, then those beliefs cannot count as knowledge. An assertion of unwavering conviction that is not based on factual evidence is irrational by definition and illogical as well. Thus, assuming the existence of a god should fly in the face of truth, knowledge, and logic.

If miraculous magic displaces the relevance of material evidence, then it would no longer be possible to prove anything if such a god *did* exist, because the only way to prove things is with evidence in accordance with natural laws. I'm using the word "prove" in the legal sense, where "proof" is defined as an overwhelming preponderance of evidence.

I have argued with Sye Ten Bruggencate on the Magic Sandwich Show and again on the Dogma Debate podcast, and on both occasions I showed he was making the same errors, which he simply will not acknowledge. Sye says that whether we read the Bible or not, the Bible still somehow forms the basis of all our knowledge claims. There is no concession given for people of other cultures who obviously know things but have never even seen a Bible, nor for those who have studied the Bible in-depth and know it far better than Sye does. Sye insists he cannot be wrong because "God has revealed it to me in such a way that I can know it for certain."

So the question I put to Sye was this: Sir Isaac Newton (possibly the most brilliant man who ever lived) was, embarrassingly enough, a deeply religious Christian and a creationist even by the modern definition. Newton declared that he had been specially chosen by God to receive a personal revelation leading to a greater understanding of the scriptures than that of any other man. That would mean that Sir Isaac Newton must have understood the scriptures better than Sye Ten Bruggencate, correct?

Sye makes a similar claim, albeit not quite as lofty and without near as much credence. He too is claiming an infallible knowledge of scripture. Both men said their beliefs are perfectly inerrant, being a personal revelation from God through scripture and revealed in such a way that they could each know it for certain. Of course there's no way to show that, and thus no way either man can honestly claim to actually "know" that either, but that's only part of the problem. The real trouble is that these men disagree on a critical tenet of their shared religion, and they can't both be right at the same time; that means that despite all their insistence of absolute certainty, one of them has to be wrong. The issue is that neither of them would admit that they even can be wrong—not about this, anyway.

Sir Isaac Newton said that Jesus is not the same person as God, and that this was revealed to him in divine revelation of scripture. As I explained in the previous chapter, this interpretation is recurrent because it is easily justified. Sye believes that Jesus is the essence of God made flesh, and he says that this was revealed to him in divine revelation of scripture too. Both of them state an unwavering conviction regarding their own supposedly flawless and divinely guided perspectives, but regardless how confident they both appear to be, one of them has to be mistaken, and both insist that's impossible because they both claim to know better.

Sye will not even consider that his belief could be in error. Instead, he deflects any possible blame away from himself and back onto God. If Sye is wrong, then God is wrong—and how could God be wrong? Newton apparently held the exact same position just as rigidly and for the same reasons. But the principle of noncontradiction demands that at least one (and probably both) of these men arrived at their respective conclusions on their own imperfect means, without the guidance of any deity. Logically that has to be the case, but such stoic believers will never admit that—even to themselves. They cannot accept that they might not really be telepathically guided to opposite points by the same psychic influence they both imagine. This shared "relationship" to their invisible friend means so much to them that neither one could bear the thought that God wasn't really on their side leading them on. Such a realization in the mind of a staunch believer could be so traumatic that it might potentially cause a psychotic episode. Such a person will immediately dismiss the other's position without a thought while refusing outright to thoughtfully reconsider their own position.

Sye's argument is that you can't know anything unless you know everything—that in order to have any knowledge at all, you have to have infinite knowledge, and only God has infinite knowledge. So before you can really know anything for sure, you either have to *be* God or you have to *know* God and have special revelation *from* God. Now every clause in this argument is logically invalid, but more importantly regarding the mirror positions of Sye and Newton, the mere fact that this disagreement exists is enough all by itself to refute Sye's presupposition. For Sye to admit the reason behind this logical impasse is to concede that the justification for his knowledge claim has already failed, regardless of who is right or what the details of the conflict might be. According to his own analogy, he can't claim to know anything because both he and Newton face the same conundrum. Either one could be wrong, even about the most important thing each man thinks he knows.

The point is that both men claim to know something that neither of them really knows. Philosopher Peter Boghossian famously defined faith as "pretending to know what you don't know," and that is exactly what is happening here.

So how could we determine which of these two men is correct, assuming either one even could be? Which one really knows what he's talking about, and which one merely believes with all his heart but is still wrong nonetheless? Sye said he determines that the same way he says atheists do: according to scripture. He will not admit that we *don't* do that, nor can he admit that Newton *did*. Newton and Sye both used the same book (and often the same verses) to prove opposite points.

Amusingly, Sye tried to dismiss Newton's revelation on the grounds that it was extrabiblical. On his own authority, Sye said that "such revelation doesn't happen anymore." While many creationists worship the Bible as God (being unable to distinguish doctrine from deity), Sye Ten Bruggencate takes bibliolatry to a new level: he holds the Bible *over* God! He says that any revelation that comes directly from God can be ignored and dismissed if it pertains to something that was not already in the Bible. So God himself could not reveal to Sye any new thing that wasn't already written by fallible human scribes back in or before the Iron Age. Sye would dismiss God's own voice as being extrabiblical, so if God wants to talk to Sye Ten Bruggencate, God had better limit his comments to citations of scripture.

Not even God could correct Sye's interpretation of that scripture either, and that is the real problem. Newton didn't claim extrabiblical revelation any more than Sye did. Both men claimed infallible knowledge based only on their different interpretations of the exact same source material, and both consider their erroneous reasoning and human biases to be divine revelations. Thus, it would not be possible to prove conclusively which of their interpretations is correct. The only way we'll ever know is if God shows up doing ventriloquism through a Jesus hand puppet, or if Jesus shows up conversing with a talking cloud as he did in Matthew 17:5, Mark 9:7, and Luke 9:35.

Once again, the real way to settle disputes of two men arguing the contradictions of mutually exclusive conclusions is with evidence. In this case, since both beliefs are mere interpretations of folklore, then the only evidence is the context of the mythology they're based on and which both men revere. We could default to Newton's brilliance, or we could list scriptural citations supporting his point, showing that Sye is wrong again even according to his own source. We could also list all the logical errors Sye always makes whenever he gets into these discussions, but it still comes down to interpretation rather than knowledge.

Like many apologists, Sye's belief is so unsound and indefensible that it requires an evocation of complete insanity to argue for it at all. One ploy is to say that we can't know who is right or wrong because we can't even know whether we ourselves are sane enough to make such judgments. Remember, creationists can't be right unless reality (or everything we know about it) is wrong.

Since science and faith are philosophically opposite in every aspect, then *they* can't know anything if *we* know anything. So another ploy is to imply that you don't know anything if you don't know everything, because the apologist can't claim to know better unless we don't know it either. Thus, whatever logical fact you try to point out will only prompt him to question how you could know that, or indeed how you could be sure you know anything.

"How do you know the laws of physics won't change five seconds from now?" Sye's question can be effectively paraphrased to inquire how we can know that reality is really real and will remain so. My answer is that reality is real by definition, that life is not an illusion—we're not all living in a dream, and that to imagine otherwise is only foolish nonsense. But foolish nonsense is what presuppositional apologetics depends on, and foolish nonsense is all it is.

Matt Slick proved my point for me when he and I debated on the Bible-Thumping Wingnut Show. I predicted at the onset that Slick couldn't address my evidence against the fables in Genesis and, having no other option, he would immediately have to argue that reality might not be real. A moment later, he told me I couldn't prove anything because I couldn't prove whether I existed a couple weeks ago, or because I might be a brain in a vat and that all my knowledge might be false memories planted to deceive me. Even when I told him in advance that he would do that, he still did it, and he denied having done it as soon as I called him on it. But that is the advantage of recorded conversations.

Apart from the double standard, the confirmation bias, straw man fallacies, and shifting of the burden of proof, I think the worst tactic of apologists is the game of equivocation and projection. It's an attempt to paint the illusion that science and religion are somehow comparable when they are not. The game is played by creationists pretending to be objective when everyone (including themselves) knows they're not, while projecting all of their own logical fallacies onto the science-minded, who of course do not share any of those flaws. That game typically has the creationist telling some or all of the following lies:

- Evolution is a religion;
- Science relies on faith, just like religion does;
- Science is biased just like religion is;
- There is no evidence for evolution, the Big Bang, abiogenesis, etc.;
- There is evidence for creation, Noah's flood, God, etc.;
- Religion is reasonable just like science is;
- Religion can be confirmed empirically and experimentally just like science;
- Creationism is scientific.

This is especially bad with presuppositionalists, who even go so far as to describe rationalism as irrational and who say that those who believe in reason are unreasonable, and even that we presuppose our opposition to presuppositions. This "I'm rubber, you're glue" form of playground posturing has to be the all-time favorite ploy among apologists, because they all use it every time.

But the most offensive tactic in this category is the assertion that atheists secretly believe in God. Worse than that, according to presupps like Sye Ten Bruggencate, atheists don't just *believe* God exists; we *know* he does. Why does he say that? Because of the following passage from the first chapter of the book of Romans:

> (18) For the wrath of God is revealed from heaven against all ungodliness
> and unrighteousness of men, who hinder the truth in unrighteousness;
> (19) because that which is known of God is manifest in them; for God
> manifested it unto them. (20) For the invisible things of him since the
> creation of the world are clearly seen, being perceived through the things
> that are made, [even] his everlasting power and divinity; that they may be
> without excuse: (21) because that, knowing God, they glorified him not as
> God, neither gave thanks; but became vain in their reasonings, and their
> senseless heart was darkened. (22) Professing themselves to be wise, they

> became fools, (23) and changed the glory of the incorruptible God for the likeness of an image of corruptible man, and of birds, and four-footed beasts, and creeping things. (24) Wherefore God gave them up in the lusts of their hearts unto uncleanness, that their bodies should be dishonored among themselves: (25) for that they exchanged the truth of God for a lie, and worshipped and served the creature rather than the Creator, who is blessed forever. Amen.

After this, the passage goes on to imply (in the modern interpretation) that all atheists are hateful, evil, twisted, perverted, arrogant, and full of gay pride. It also says we worship the creation rather than the creator. I take that to mean that we actually study the natural world with awe and wonder, and that we accept evidence while apologists deny it to believe something else on faith. Remember also that everywhere except the Bible, a fool is defined as one who too readily accepts improbable claims of credulous sources on insufficient evidence and is thus easily duped by a lie. Of course the Bible gives the opposite definition. So whenever someone says to me, "The fool says in his heart, 'There is no God,'" I like to reply with Jeremiah 8:8: "How can you say, 'We are wise, and the law of the LORD is with us'? Behold, the lying pen of the scribes has made it into a lie."

Somehow apologists interpret the above citation of Romans 1 from verse 18 onward to mean that everyone really knows that God exists, whether we admit it or not.

First off, even if we pretend that the Bible is the authority they imagine it to be, these comments still didn't apply to everyone in the world; they were directed only to a particular subset of Jews and Gentiles who were quarreling over who knew God better. Secondly, the passage actually requires that the reader simply assume the conclusion that "creation" requires a creator, but that's the fallacy of question begging. What if we call it "reality" instead? Otherwise the passage gives no explanation of how anyone (much less everyone) is supposed to "know" that God exists.

There are also many other verses which prove that this cannot be the correct interpretation, beginning with the 16th verse of this very chapter from Romans where it says that the gospel is the power of God for salvation "to everyone who believes," which implies that there are also those who do *not* believe. 2 Thessalonians 1:8 says the same thing. 1 Thessalonians 4:5 says the Gentiles don't know God. According to 1 Samuel 3:7, Samuel didn't know God, and Exodus 5:2 says Pharaoh didn't either. Jeremiah 9:3 has God himself complaining about people who don't know him. 1 Corinthians 8:6–7 also states unambiguously that there are people who do not know that there is a creator god, much less who he is. So the Bible clearly admits in several places that there are people who do not believe in the god of Abraham, either because they believe in other gods instead or because they believe in no gods at all.

This is interesting. Not because Sye Ten Bruggencate is wrong again, but because he gets his knowledge from scripture; because he says he knows God, and that God knows everything. Sye says everyone believes in God, but the Word of God says otherwise. So who is right this time—Sye or God? It seems that Sye Ten Bruggencate could be wrong about virtually everything he thinks he knows for the very reason that he thinks he knows things. He's certainly wrong about the things he boasts about knowing most proudly.

Hypothetically, what if God really did exist, and atheists knew it? Then they wouldn't even *be* atheists. Some apologists make the argument that (a) we only pretend not to believe in God because (b) we love sin, and (c) we don't want to be held accountable. So I guess we're supposed to think that if we pretend there is no god, then we can pretend to escape his judgment? I don't know how that's supposed to work. If we ignore him, will he go away? Can we disbelieve him out of existence? I don't think so: not if he was really real.

This is where we see the contrast between our two vastly different mindsets. Appalachian hillbilly snake handlers may read Mark 16:18 and believe they can "take up serpents" and the venom will not hurt them, but a fair number of them still die that way nonetheless. They think that if you believe hard enough, your wishes will come true—that faith will move mountains and all that, which obviously doesn't really happen. Atheists typically don't think like that. Mere beliefs do not change reality, but reality should change one's beliefs. So your typical atheist is smart enough to know that if there is a god and if we know that there is one, then we also know that pretending he isn't there isn't going to protect us from him.

Many atheists were schooled in scripture. That's the reason most of them cite when explaining why they became atheists in the first place. Such people realize that, according to the Bible, whether you're damned or saved has nothing to do with how good or bad you were. That's not what you're being judged on. You can't get into Heaven if you don't believe in the contradictory nonsense accounts of Jesus' impossible miracles, and you're supposed to believe all this on faith, regardless of evidence, which is the only good reason to believe anything. So if you're an amiable, generous, compassionate humanitarian, you're still damned if you don't believe impossible nonsense for no good reason. You're not getting into Heaven, and it doesn't matter how wonderful you were in life.

If you're a greedy, abrasive, apathetic, cruel, and callous criminal, that won't matter either as long as you believe in Jesus. All sins will be forgiven if you but believe. You don't have to keep any of the hundreds of commandments in the Old Testament. Jesus says you're supposed to obey them all; keep the feast of the leavened bread, sacrifice the first of your harvest to God, kill everyone who believes differently than you do, don't make any artistic renderings of living things, and so on. According to Jesus, if you disobey even one of these commandments, you will be called "least" in the Kingdom of Heaven. But who cares what they call you as long you made it into Heaven? Sin doesn't lead to damnation; only disbelief does.

It is actually fundamentalist Christians who use their belief in God to avoid being held accountable for anything. If they do evil, they can quite literally say the devil made them do it; atheists are responsible for their own actions. Christians don't have to be responsible for damaging the environment either because they believe that these are the last days, and Jesus will destroy the world soon anyway. But while they're dominating and subduing the earth, secular environmentalists are trying to be stewards of the world for our heirs.

The Christian god doesn't care about morality; gullibility is the sole criteria for redemption. So if you "love sin," become a Christian. Then you can commit all the sin you want and you'll still be forgiven for it.

According to Deuteronomy 17:2,7, Isaiah 64:6, Mark 3:28–29, Revelation 21:8, and Ecclesiasticus (Sirach) 2:13, it doesn't matter how good you were in life, because you still can't be saved if you don't believe. According to Mark 16:16, John 3:18,36, 5:24, 6:40,47, 11:25–26, Acts 10:43, and Romans 4:5 and 10:4, believers can still get into Heaven regardless of their works in life. So if you "love sin," be a Christian. There are many Christians who believe "once saved, always saved."

According to Matthew 12:22–32, the only unforgivable sin is blaspheming the Holy Spirit. Essentially what this means is to deny some aspect of God through mockery or apostasy. Here we see that the surest way to piss God off is to not believe in him. So if there really was a god, and we were sure of that, we *damned* sure wouldn't call ourselves atheist, because that's the only way to be sure we'd be damned. We're usually smart enough to figure that much out.

Anyone who knows the risk and is still comfortable calling himself "atheist" is pretty confident that the chances of divinity are slim to none, and even if some god did exist, it wouldn't be the god of Abraham. So when someone tells you they don't believe in God, it usually means they really *really* don't believe. It is *never* the case that they're just pretending. That would be a stupid thing to do, and we're just not that dumb.

Some of the most brilliant minds of the modern age are atheist. They may not always identify as such, but an atheist is simply one who does not believe in any god. A lot of people say they're not atheist; they're agnostic, because they don't know whether God exists or not. A-gnosis means "without knowledge" just as a-theism means "without theism." Theism is a religious belief in a deity. You either have theism or you don't; you're either convinced or you're not. But you wouldn't "know" in any case. The agnostic position is that it is impossible for mere mortal men to have certain knowledge of supernatural things. This is because it is impossible to demonstrate that knowledge or test it. If we can't confirm your accuracy to any degree at all by any means whatsoever, then it is a fact that you cannot actually know what you think you know; you only *believe* that you know that.

You can be a gnostic theist: "I know Krishna is the supreme personality of the god-head." That cannot be a completely honest position, but it is a common one. Or you can be an agnostic theist: "Well I believe there's a god, but who is to say whether it is Vishnu or Allah, or which religion is the right one?" You could also be an agnostic atheist, which most atheists are: "I don't know if there's a god or not, but I don't really believe there is one." There are relatively few gnostic atheists, and I am one of those. I know God doesn't exist the same way I know there are no leprechauns, Sasquatch, or extraterrestrial aliens anally probing drunken bumpkins in the rural areas behind the trailer park. Since knowledge is demonstrable, I can show that there is no evidence of such things, but more than that, I can also show substantial evidence against each one. Conversely, those who claim to "know" the opposite are not able to demonstrate their knowledge the way I can. I know God doesn't exist the same way I know that Superman, Bugs Bunny, and Paul Bunyan don't exist. They're each impossible fictional characters with no evidence indicating otherwise. You don't have to prove a negative to know the Bible god isn't real, just like you don't have to visit every galaxy far, far away to know there was never a Darth Vader either. The principle is the same.

Belief is a conviction. A lack of belief is a lack of conviction. So if you're not convinced that an actual deity really exists, then you're atheist: you lack theism. You don't have to like the label; everyone who is not convinced that a god exists is an atheist, regardless of whether they're comfortable admitting that or not. A lot of people say, "Just because I don't believe in gods doesn't make me an atheist." Well, yeah—it kinda does. That is the sole criteria, and it applies in all cases.

A lot of people think you can't be atheist unless you know for certain that no god could possibly exist, but that definition was devised by the opposition in order to suppress our expression. Most atheists don't fit that mold. One of my best friends held that opinion. When I told him he was atheist, he objected: "I'm not convinced there's a god, but I'm not convinced there's *not* one either!" "Sorry man," I told him, "but since your worldview doesn't include any god, and you have no theism, you're still an atheist by definition."

Desperately trying to shirk the label, my friend then said, "Well, I believe there's *something*." But I said, "Okay, so what? We all believe there's 'something,' but is that something a *god?*" You have to know what a god is before you can tell whether you believe in that or not.

A lot of people define "deity" such that it only applies to their version of God and excludes all others by definitional fiat. But we cannot fairly dismiss all the hundreds of gods who were worshipped by millions of people for thousands of years. However we define what a god is, that definition must include every entity already universally accepted as a deity by those who worship it. My research in this area would have me submit that all gods are magical anthropomorphic immortals, because that description does seem to apply to all of them. They all have miraculous powers and human characteristics, and even if you can kill the body, they still exist and can still return in some other form, or the same form, or be literally born again as it was with Dionysus II.

A survey of the American Association for the Advancement of Science was conducted by the Pew Research Center for the People and the Press in May and June 2009. They determined that 12% of the general public and 18% percent of scientists did not believe in a god, but did believe in some other form of higher power that could not be accurately defined as a god. Examples of this could include the pantheist biosphere of Gaia or "the Force" from *Star Wars*, among other things. The majority of scientists polled (41%) did not believe in either a god or any other sort of "universal spirit" either. So it seems that the more you know, the less you believe.

Sye Ten Bruggencate likes to say that those who identify as rational nonbelievers use their reason to justify their reasoning. He uses that phrase to pretend that we rely on circular arguments just like he does, and that this should indicate that we're crazy. However, he openly admits that his own argument is circular and that it is bolstered by confirmation bias, prohibiting him from admitting that he could be crazy.

Now as I said, we may assume that reality is real, because we kind of have to. It's an inescapable conclusion, and even the presupps know that. Beyond that, we obviously don't rely on our reasoning to justify our reasoning; instead, we use the reasoning of others to confirm or correct our own perception by relying on objective analysis. We can test ourselves with a consensus of independent observers weighing our suspicions against measurable evidence and well-defined criteria, people who consistently demonstrate the ability to communicate rationally and evaluate properly. These are things that Sye simply refuses to do and refuses to acknowledge that others do.

At a minimum, the arm-chair psychiatrist should accept the colloquial definition of a sane person as one who is rational, meaning that one has reason, can reason, and is amenable to reason, that one is reasonable and can be reasoned with. None of this applies to apologists arguing the beliefs of their religion, because religions are belief systems; they have required beliefs that are unsupported but still must be held even when they're wrong, and they have prohibited beliefs that must be execrated even after they've been vindicated. And they all claim direct revelation of supernatural knowledge—which is of course impossible. As Thomas Paine wrote in "Of the Religion of Deism Compared with the Christian Religion and the Superiority of the Former over the Latter,"

> The Persian shows the Zend-Avesta of Zoroaster, the lawgiver of Persia, and calls it the divine law; the Bramin shows the Shaster, revealed, he says, by God to Brama, and given to him out of a cloud; the Jew shows what

> he calls the law of Moses, given, he says, by God, on the Mount Sinai; the Christian shows a collection of books and epistles, written by nobody knows who, and called the New Testament; and the Mahometan shows the Koran, given, he says, by God to Mahomet: each of these calls itself revealed religion, and the only true Word of God, and this the followers of each profess to believe from the habit of education, and each believes the others are imposed upon.

Every religion claims to believe as they do because of reason, education, or intelligence given by their god in revelation. But whether they admit it or not, all of them are assuming their preferred conclusions on faith, and this would still be true even if all of their gods exist. Believe as hard as you want to, but convincing yourself however firmly still can't change the reality of things. Seeing is believing, but seeing isn't knowing. Believing isn't knowing. Subjective convictions are meaningless in science, and eyewitness testimony is the least reliable form of evidence.

For example, if I go into my front yard and I see a large sauropod walking down the middle of my street, I will of course be quite convinced of what I see. I may be even more satisfied when I follow the thing and find that I can touch it, maybe even ride it if I want to. When I gather sense enough to run back for my camera, I may not be able to find the beast again because I don't know which way it went. But that doesn't matter because I saw it—I heard it, felt it, smelled it, and I remember all that clearly with a sober and rational mind. But somehow I'm the only one in the whole neighborhood who ever saw it, and of course no one believes me.

Some other guy says he saw a dinosaur too, but his description is completely different, such that we can't both be talking about the same thing. So it doesn't matter how convinced I am that it really happened; I have to accept that it might *not* have. When days go by and there are still no tracks, no excrement, no destruction, no sign of the beast at all, no other witnesses whose testimony lends credence to mine, and no explanation for how a twenty-meter-long dinosaur could just disappear in the suburbs of a major metropolis, much less how it could have appeared there in the first place, then it becomes much easier to explain how there could be only two witnesses who can't agree on what they think they saw than it is to explain all the impossibilities against that dinosaur ever really being there.

Carl Sagan famously said, "Positive claims require positive evidence;

extraordinary claims require extraordinary evidence." Christopher Hitchens added his razor that "what can be asserted without evidence can be dismissed without evidence." These two concordant comments form the basis of empirical rational science.

My testimony is meaningless without evidence to back it up, and the more bizarre the claim, the more compelling evidence I'd need—especially since what I propose in this example isn't just extraordinary, it's impossible. But since there's not one fact I can show that anyone can measure or otherwise confirm, then my perspective is still subjective, and thus uncertain. Eventually, even I the eyewitness would have to admit that although I clearly remember seeing it, I still don't know if it was ever really there regardless of whether I still believe that it was.

I have actually faced a situation like this. As a young man experimenting with meditation and neopagan spiritualism, there were several instances that might be considered visions, manifestations, or even encounters with supernatural entities. I find them all somewhat embarrassing now and will not discuss them in this book. Suffice to say there were distinct "vibrations" from a transcendental trance, a visit from an astral projectionist, a flurry of gargoylian shadow demons, and a human-shaped apparition of warped air that looked like heat distortion—all conjured during focused concentration.

At that time, these experiences seemed deeply compelling and shaped my worldview. For decades, these memories were all I had to convince me that there was a supernatural aspect to existence. Eventually, however, I had to accept that these things probably didn't really happen the way I remembered them simply because they couldn't have. Once I realized that and could accept it, I began to see some alternative explanations.

On two of these occasions the experience was not so easily dismissed, because it was shared. Normally that would lend more credence to the event; however, twenty years later I happened across each of the other parties involved, and found that they didn't remember these things the same way I did, nor did either one want to talk about it. It's like when someone recounts the silly things you said as a child, or what you did when you were drunk. Although my friends were just as spiritually inclined as I was back then, they both later faced the same conundrum I did. Despite our shared spiritual experience, we all became unbelieving rationalists, independently and each unbeknownst to the others.

I see this issue being divided between those with a deep-seated need to believe and those who have instead only a desire to understand. It's fantasy versus reality, science versus faith, skepticism versus credulity. Whichever side of that issue you're on determines your value of truth. Those who favor faith often hint that whether you believe something matters more than whether it is true. This is a subtle but frequently repeated theme among the religious, and is further illustrated by the words attributed to sixteenth-century Reformed Protestant ex-father Peter Martyr Vermigli. As Andrew Dickson White quotes him in *A History of the Warfare of Science with Theology in Christendom*,

> So important is it to comprehend the work of creation that we see the creed of the Church take this as its starting point. Were this Article taken away, there would be no original sin; the promise of Christ would become void, and all the vital force of our religion would be destroyed.

This sentiment heralds the very beginning of the modern creationist movement. Following this, Ken Ham, president of Answers in Genesis, reinforces this philosophy in his seminars:

> The message of Jesus comes from this book, does it not? In young people's minds today, if that book has been undermined, if they have been caused to doubt that it's really trustworthy, then why will they listen when you talk about the message of Jesus from the same book?

In an attempt to defend the newest incarnation of creationism, intelligent design proponent William Dembski parroted the same notion during his debate with Christopher Hitchens:

> If we don't see ourselves as made in God's image, as exceptional, not in a prideful sense, but exceptional in the sense that we are made in God's image, that humans have a special place in the scheme of things, then I think you do very easily run into eugenics, euthanasia, abortion, and things like that.

In each of these cases, we are encouraged to believe for reasons that have nothing to do with whether it is true. On this topic, Nobel Laureate Bertrand Russell said:

> There can't be a practical reason for believing what isn't true; at least I rule it out as impossible. Either a thing is true or it isn't. If it is true, you should believe it, and if it isn't, you shouldn't. And if you can't find out whether it's true or whether it isn't, you should suspend judgment. It seems to me a fundamental dishonesty and fundamental treachery to intellectual integrity to hold a belief because you think it's useful, and not because you think it's true.

This seems to me to be completely reasonable. At the other end of the scale, we have William Jennings Bryan, the prosecuting attorney at the Scopes "Monkey Trial," when it was illegal to teach evolution in school, he said, "If we have to give up either religion or education, we should give up education." I find this sentiment breathtakingly myopic and moronic. But it gets worse, because Bryan is also known for another comment that is so logically unsound that it should call into question his very sanity and competence: "If the Bible had said that Jonah swallowed the whale, I would believe it."

Whether Bryan was clinically insane or not, it is clear that he has rejected his reason. Even worse, Bryan wasn't just a lawyer; a century ago, he was a U.S. presidential candidate. Think of the damage that a world leader can do with that mind-set. This is the same philosophy that also believes that the ends justify the means; thus, it is even permissible to lie as long as you're lying in order to promote your faith. Conservative Christian Scott Sloan interviewed me on his talk radio show wherein he admitted on the air that "there's nothing wrong with lying if no one knows you're lying." Many other admissions by other religious people make clear that belief matters more to the faithful than does truth, and this is explicitly expressed by the second-century apologist Tertullian:

> We want no curious disputation after possessing Christ Jesus, no inquiring after enjoying the gospel! With our faith, we desire no further belief.
>
> And the Son of God died; it is by all means to be believed, because it is absurd. And he was buried and rose again; the fact is certain because it is impossible.
>
> After Jesus Christ we have no need of speculation, after the Gospel no need of research. When we come to believe, we have no desire to believe

> anything else; for we begin by believing that there is nothing else which we have to believe.

I have the opposite philosophy. I find the pursuit of curiosity richly fulfilling. I consider it unconscionable to avoid knowledge in order to preserve a preferred belief against any inconvenient truth. Truth won't always set you free and it will often piss you off, but that is still better than clinging to a lie. The placebo effect is trivial and notwithstanding. Otherwise only accurate information has practical application. So I believe it is better to understand things as they really are, to improve that understanding regardless of my own desires or biases, and not to be fooled into believing any improbable thing on insufficient evidence.

Those who embrace faith instead often say that they are happy because of it, but that is no more relevant than the fact that drunks often think they're happier than sober people. This is where we get the colloquialism, "If this is wrong, I don't want to be right."

We are often culturally conditioned toward the notion that it is better to hold onto a preferred belief than to risk discovering some ugly truth. This is when we're told that ignorance is bliss. A good example of this mind-set was depicted in the 1950 film classic *Harvey*. James Stewart plays the part of Elwood Dowd, a warm, polite, and charming fellow who is always accompanied by a six-foot-tall invisible magic rabbit named Harvey. Everyone else in the film is typically depicted as judgmental cynics with some obvious void in their lives.

The irrational belief seems to be the cause of Dowd's happiness. Dowd explains it thusly: "In this world … you must be oh-so-smart or oh-so-pleasant. For years I was smart; I recommend pleasant." The point of transition is implied to be the moment when Elwood met Harvey. The problem is the implication that those who are "smart" can't also be pleasant, because "smart" people don't believe, and belief is what makes one "pleasant." This point is hammered home when Dowd's family tries to have him committed. They are warned that patients who have these delusions go into the hospital as gentle, kind, and endearing, but when they are cured, they come out bitter, harsh, and cynical.

At this point, the film not only endorses mental illness, but it also again implies that believing in the invisible friend makes one nice, and not believing leaves one empty and bitter. Then just to relieve the worried minds of the audience, at the end of the movie, we find out the rabbit is real. Of course.

However, if Elwood Dowd was a real person, he might be really sweet sometimes, but he would likely also have bouts of hatred and paranoia, because that's how such people typically are. Whether they're delusional or not, when irrational beliefs are seriously challenged, such people often lash out in defensive rage. That is the reality of these conditions.

This chapter is an attempt to undo some of that mental conditioning, to explain the real importance of critical thinking and the value of empirical rational analysis. There are certain rules of logic that science has to adhere to, and there are good reasons for that; faith by contrast ignores all of that outright, preferring to believe whatever makes one happy. I want people to understand that accuracy and accountability actually matter, not just in academics but also as a point of integrity and honor and as a general rule in life. That's why these rules apply, and I have prepared the following parable to help illustrate that.

When I was a geoscience student at the University of Texas, I was pursuing an interest in paleontology, the study of ancient life forms. We're always finding new things in the fossil record. That record is already much more rich than any layperson would ever suspect, and some of the many things we've found were pretty weird, so all kinds of things might be there.

When I was a kid, I loved Godzilla, and I remember imagining how people would react if they found something like it in the fossil record. Even if it were only one-fifth of its depicted size, it would still be cool.

I drew Godzilla more realistically than we see in the movies, and I envisioned it being closer to a lizard than a dinosaur. It's funny how Toho Pictures made the original 1954 Gojira look like a lizard but called it a dinosaur, then in 1998 Tri-Star made the monster look like a dinosaur but called it a lizard. What do moviemakers know about taxonomy? Of course I liked the original, and that is the structure of a lizard regardless of what the scriptwriter calls it. If a fossil matching that description were discovered, I would call it *Godzillasaurus dios*. (The name *Gojirasaurus* was already taken.)

Is it possible that such a thing once existed? A bipedal Lepidosaur with this sort of look and these dorsal spines? Well, I don't know how big it could be (it couldn't possibly be *that* big!), but otherwise to be philosophically correct, I would have to say yes, it is technically "possible" this exact animal actually could have existed, and I concede that it is even conceivable that we could find it in the fossil record someday.

But let's forget what is possible and concentrate on what is probable. Is there any reason to believe that an ancient reptile with these specific traits ever existed in real life? No, nothing at all. I mean, there were several old folklorish movies about it, and there are a heckuva lot of Kaiju fans who would love it if this thing were really real once. But apart from some fanatic devotees and their beloved fiction, what evidence is there for Godzilla? Not one thing that could be verified by anyone. Consequently, there is every indication that the king of the monsters is just a made-up character.

If I found a four-toed footprint the size of a whole T-Rex, that at least would be something. But it still wouldn't be enough to justify this particular animal, would it? I would need volumes more evidence than that! I mean, how can I claim to know all these details about something I can't even show was ever real? Especially if I have no reason to imagine such a thing in the first place.

Still, I live in a country where I have a constitutionally guaranteed freedom to believe whatever I want, and I'd really like to believe that something like that existed once. No one can conclusively prove that no extinct reptile could have looked like that, right? If you can't prove something doesn't exist, would that mean that it does? Should it even imply that it could? Even if you explore all the most inaccessible depths of the world's oceans, you still couldn't prove it was never there! Maybe it just wasn't in that spot at the same time you were. We'll never discover every species that ever lived, and absence of evidence is not evidence of absence, right? I don't even need evidence if I "know it in my heart" that he's real, right? So don't I still get to believe in Godzilla if I want to?

What if I then went on to list all sorts of other details I supposedly knew about Godzillasaurus, such as what color it was, or what its reproductive peculiarities were, or the unique way it would respond to certain stimuli? And I say all this as if I had the facts and test results necessary to prove each point, when I really don't have any indication that anything like this ever even lived at all.

What if I still didn't stop there? What if I didn't just say that Godzilla *could* have existed? Given the utter lack of evidence, just that comment alone might have cost me my credibility as a scientist. Even if I said he "probably" existed, my reputation would be ruined because I can't substantiate that claim. Scientists can't say that anything is probable until they have solid data indicating that probability. Scientists don't get to say that anything is possible either—there first has to be some precedent or parallel showing that it is possible, or there have to be strong indications that that is the case regardless of whether we can yet explain it. For example, it is not possible for any lizard to be fifty meters tall. That's how tall Gojira is in *Godzilla*, the original Japanese movie, before they reedited the movie and added scenes to it in an adaptation titled *Godzilla, King of the Monsters!* In the Americanized version, actor Raymond Burr, as reporter Steve Martin, says, "George, here in Tokyo, time has been turned back two million years. This is my report as it happens. A prehistoric monster the Japanese call Godzilla has just walked out of Tokyo Bay. He's as tall as a thirty-story building."

Regardless, there would have to be myriad unprecedented structural modifications throughout the animal's body to accommodate weight distribution, oxygen intake, and so on at such colossal size. But if we found a set of footprints that were each the size of a city bus, then we'd have to accept that those modifications must exist or the monster still couldn't.

But if I were speaking as a paleontologist, let's say I went several steps beyond too far, and stated flat out that Godzilla *did* exist—not that I think he existed, or that I "believe" that he did, but that I *knew* he did. What happens to my credibility then? Can one even say something like that and still be trusted anymore? If I have no positively indicative evidence at all to back me up and thus can't prove that I'm right about anything I profess to know, then if I go ahead anyway and confidently posit that Godzillasaurus did in fact roam the Japanese islands two million years ago (like the movie said), would that be an honest claim?

Normally, anyone disreputable enough to flatly affirm such positive proclamations without adequate support would lose the respect of his peers and be accused of outright fraud. Anyone but a religious advocate, that is. When allegedly holy men do the exact same thing, then it's not called fraud anymore—it's called revealed truth instead. That's quite a double standard, isn't it?

Like when some minister gets on stage at one of those stadium-sized churches to state as fact who and what God is, and what he wants, hates, needs, won't tolerate, or will do for whom, how, and under what conditions; they don't have any data to show they're correct about any of it. Yet they speak so matter-of-factly. Even when they contradict each other, they're all still completely confident in their own empty assertions!

So why do none of the tens of thousands of head-bobbing, mouth-breathing, glassy-eyed wanna-believers in their audience have the presence of mind to ask, "How do you *know* that?" Well, for all those who never asked the question, here's the answer: they *don't* know that! There's no way anyone *could* know these things. They're making it up as they go along. These sermons are the best possible example of blind speculation, asserted as though it were truth and sold for tithe.

It is dishonest to assert as fact that which is not evidently true, yet that is what all religions do. So if anyone or everyone else would be called liars for claiming such things without any evidentiary basis, then why make exceptions for evangelists? These charlatans are obviously liars too! The clergy are in the same category of questionable credibility as are commissioned salespeople, politicians, and military recruiters.

One famous example is Marjo Gortner, who began preaching his evangelical ministry as a preschool child. He grew up in the church and became a traveling revivalist. Then as a young man in the early 1970s, he did something unprecedented: he invited a documentary crew to film one of his tours. They showed his emotive preaching of the gospel, speaking in tongues and working miracles, and his confident conversations with supportive believers, and finally they provided an accounting of all the money he was taking in from each of these revival meetings. Then he had a confession of conscience for the film crew. Marjo admitted on camera that he never believed in God, not even when he was a little boy. He had been scamming people his whole life and was tired of it. With that, he quit the ministry and moved to Hollywood to become an actor.

Since then, so many of the televangelists and other clerical leaders have been repeatedly exposed for corruption, drug use, sex scandals, and other abuses that fraudulence in some aspects of the clergy is considered commonplace, even typical. Consequently, religion is increasingly seen as a matter of mental conditioning, emotional manipulation, and inculcation—a mild form of mind control via the autohypnotic power of pretend. It is literally make-believe.

An excellent anecdotal example was when I was nineteen years old, on the day that I thought I was "reborn in Christ." I felt an overwhelming euphoria, but I still had serious doubts. How could I know whether this sensation was demonic or divine? A good friend of mine at that time (who later became a Southern Baptist minister) could not tell me how to be sure whether anything I believed was actually true; instead he told me how to "reaffirm" my faith. It was a procedure he still refers to by the colorful rhyme, "Fake it 'til you make it." He grabbed me by both shoulders with a beaming smile and said, "Just keep telling yourself it's Jesus until you believe it!" It was such an admission of the intellectual dishonesty of religious belief that it was also my last moment as a Christian.

Homicidal cult leader Charles Manson was one of many who reportedly said that if you repeat a lie often enough, people will believe it, and eventually you will even believe it yourself. But Manson added an important addendum when he said that they may not believe it completely, but that they'll still have to draw on it if that's all they know because they've never been told anything else. That is an important point to remember when you think about indoctrination. You could raise a community of children to believe in Cthulhu if you always insist that he's true. If you make them worship him regularly and pray to him in fear begging for signs or impressions revealing his existence to them, then at least some of those children will eventually claim to have experienced that god, despite the fact that he only ever existed in fiction.

Occultists, transcendentalists, and faith healers of every religion know the autodeceptive power of faith. It doesn't matter which gods or spirits they pray to; no matter which devotion one practices, if the ambience of the ritual is right, then faith can prepare the mind and psych the senses into perceiving or experiencing whatever the subjects want to believe. Seemingly miraculous feats and visions occur in every faith because faith itself is the cause of them, rather than whatever devotees may have faith in. That has to be the case, because faith is the only common bond between all religious beliefs.

Believers often say they "know for a fact" that their beliefs are the truth. They testify to things they don't know anything about. They pretend to witness things they've never really seen, and they like to use other confident-sounding terms like "conclusively proven" when they're really only talking about baseless assumptions, and vice versa. They often claim absolute truth when they're really telling bald-faced lies, and all too often, they will continue to repeat and appeal to arguments they know have already been proven wrong.

But if you believe in truth at all, then you should make sure that the things you say actually are true, that they are defensibly accurate, and academically correct. And if they're not correct, you should correct them! You wouldn't claim to know anything you couldn't prove that you knew, and you wouldn't talk about anything being proven at all, unless you're clearly using that term in the sense that a court of law would use. Scientists must choose their words very carefully, because science is brutal in peer review, and no scientist would ever get away with any of the wild raving propaganda that religious zealots or the news media use. That's why they say the devil is in the details!

The truth is what the facts are. But in a philosophical sense, it's more than just that. It implies something that is completely true, the whole truth, and nothing but the truth. So every word of it better be accurate, or it isn't truth at all—and depending on the topic, such a concept is likely beyond human comprehension anyway. In that sense, truth may be pursued but never possessed. That's why we should trust those who seek the truth and doubt those who claim to have it. In any case, truth doesn't exist on its own—it only exists as a declaration or observation. Reality is not truth; truth is the statement which conforms to reality. That means it has to be tested before we can call it "truth."

A fact is a unit of data that is either not in dispute or is indisputable in that it is objectively verifiable, and obviously beliefs based on the conflicting faiths of different religions cannot qualify as that. Belief may be either rational, or assumed on faith. But in either case, it doesn't matter how convinced you are; belief does not equal knowledge. The difference is that knowledge can always be tested for accuracy where mere beliefs often cannot be. No matter how positively you think you know it, if you can't show it, then you don't know it, and you shouldn't say that you do. Nor would you, if you really cared about the truth. Knowledge is demonstrable, measurable. But faith is often a matter of pretending to know what you know you really don't know, that no one even can know, and which you merely believe, often for no good reason at all.

THE 5TH FOUNDATIONAL FALSEHOOD OF CREATIONISM

"Evolutionism" Part I: "Evolution Is the Religion of Atheism"

"As the government of the United States of America is not in any sense founded on the Christian religion, as it has in itself no character of enmity against the laws, religion or tranquility of Musselmen [Muslims] and as the said States have never entered into any war or act of hostility against any Mahometan [Islamic] nation, it is declared by the parties that no pretext arising from religious opinions shall ever produce an interruption of the harmony existing between the two countries."

—Treaty of Tripoli, approved by President John Adams in 1797 and ratified by the U.S. Senate by *unanimous* vote

I have frequently heard the Religious Right assert that the U.S. government was founded on Judeo-Christian principles, exemplified by the Ten Commandments. However, only three of the commandments they're talking about ever had any parallel in American law, the same ones that were already represented in everyone else's laws too—since the dawn of civilization. The other popularly cited "commandments" are unenforceable or even immoral as legislation.

The traditionally accepted commandments read as follows:

1. *"I am the LORD thy God, which have brought thee out of the land of Egypt, out of the house of bondage. Thou shalt have no other gods before me."* Yet Jesus deifies himself and positions himself before the Hebrew god as the gateway to the father-god, YHWH, which is the deity speaking here. Belief in such things was a crime punishable by death in the Old Testament. The 1st Amendment contradicts the 1st Commandment, because it ensures religious freedom instead of mandating one belief for all under penalty of death.
2. *"Thou shalt not make unto thee any graven image, or any likeness of anything that is in heaven above, or that is in the earth beneath, or that is in the water under the earth: Thou shalt not bow down thyself to them, nor serve them: for I the LORD thy God am a jealous God, visiting the iniquity of the fathers upon the children unto the third and fourth generation of them that hate me; And showing mercy unto thousands of them that love me, and keep my commandments."* This commandment (also opposed by the 1st Amendment) prohibits all artistic renderings of any recognizable thing in any medium. There was never any such law in the history of any government that I am aware of, but certainly not within the United States.
3. *"Thou shalt not take the name of the LORD thy God in vain; for the LORD will not hold him guiltless that taketh his name in vain."* There are a few countries with blasphemy laws, but the United States is not one of them, and we have never had such a law since this nation was founded.
4. *"Remember the Sabbath day, to keep it holy. Six days shalt thou labour, and do all thy work: But the seventh day is the Sabbath of the LORD thy God: in it thou shalt not do any work, thou, nor thy son, nor thy daughter, thy manservant, nor thy maidservant, nor thy cattle, nor thy stranger that is within thy gates: For in six days the LORD made heaven and earth, the sea, and all that in them is, and rested the seventh day: wherefore the LORD blessed the Sabbath day, and hallowed it."* Working on weekends was another capital crime in the Old Testament, punishable by death, but it was never even a misdemeanor in this country.
5. *"Honour thy father and thy mother: that thy days may be long upon the*

land which the LORD thy God giveth thee." The Old Testament encouraged parents to murder disobedient children, and it even permitted them to be beaten, molested, or sold as slaves (the Bible describes and condones both "bondservants" and slaves). However, the U.S. federal government has a number of protections for abused or neglected children. Here it is even possible for kids to sue their parents.

6. "*Thou shalt not kill.*" Killing people is illegal except for executions and abortions and acts of war. Euthanasia in the form of assisted suicides is today legal in some parts of the United States, including in California, Oregon, Vermont, and Washington, and may soon be legal in other states too.
7. "*Thou shalt not commit adultery.*" This too would be a death sentence in the book of Exodus, but there is no such law in any U.S. state. Divorce is also illegal in a few places in the Bible, but not in this country. According to American law, rape is a felony, as is slavery, yet both of those crimes are excused by other commandments immediately following the usual "ten" found in Exodus 20.
8. "*Thou shalt not steal.*" This is the second of only three commandments that have any parallel in American law, and this was already established by polytheist cultures centuries before estimates for the time of the Ten Commandments.
9. "*Thou shalt not bear false witness against thy neighbour.*" While perjury, slander, and libel are crimes in the United States, there are requirements for each that are not specified here. Otherwise, lying is often perfectly legal, and religion depends on that. As long as you're not under oath, slander and libel each might only result in a fine from a lawsuit.
10. "*Thou shalt not covet thy neighbour's house, thou shalt not covet thy neighbour's wife, nor his manservant, nor his maidservant, nor his ox, nor his ass, nor anything that is thy neighbour's.*" This is a thought crime. Not only is it in violation of the most basic of all American principles, it could also be argued that coveting thy neighbor's goods is the primary motivation behind our capitalist economy.

There are actually eleven commandments given in Exodus 20, the last one being, "*Thou shalt create no other gods besides me, neither of silver nor gold."* That one is not illegal in these United States either. Contrary to popular belief, the Bible doesn't describe that first set as the Ten Commandments; it was actually a much longer list, which continues through Exodus 23. It included dozens more commandments that most Christians never knew about, mostly ethno-specific traditions pertaining only to Israel and allowing special privileges for Jews. It is all written in the regional currency of sheep and sheckles. There is no timeless wisdom or moral guidance to be found in any of those commandments.

Finally, Moses comes down from the mountain in Exodus 24, and recounts to the people *"all of the commandments and all of the words of the Lord,"* so we know that list did not stop at ten. But as the story goes, Moses smashed the original tablets and had to go back up the mountain for the second edition. God promised that the next set would be the same as the first, and for the most part they are. But they are only a sampling of the previous volume that took four whole chapters to list. It was only the second set that is finally described in Exodus 34 as "the Ten Commandments," and they are a very different collection that lacks any moral value. Only three of them match the popular favorites. None of them have any correlation with U.S. law whatsoever, nor with the laws of any other nation either—and with good reason. They are barbaric provincial practices that include a mandate of human sacrifice of all firstborn male children (Exodus 22:29 and 34:19). Exodus 34:20 allows that firstborn sons may be redeemed if their parents own a scapegoat, but Leviticus 27:28–29 says they may *not* be redeemed, and must surely be put to death. Ezekiel 20:26 confirms that this horrific injustice was repeatedly carried out.

For this and many other reasons that I will not explore here, anyone attempting to use God's commandments as a moral guide would be a criminal in every nation on this planet, and that is also why *this* nation was definitely not founded on *any* of the damn commandments. As Madalyn Murray O'Hair stated in "Consider the Atheist":

> These men devised a government which was contractual with the people, and which established certain unique features, all of which you know: Separation of powers, a grant of specific authority to government bodies with a large residue of the power retained by the people, a curious rotating system of in-office placement which would bring about an opportunity for removal from office on a constant and continuing basis.

> But none of that was any good, because as they played with these ideas, they came to realize that there was a certain decision which had to be made in respect to religion (theism) before they could begin. So as an underlying platform, they programmed in complete and absolute separation of church and state, a phenomena that had never occurred anywhere else in the history of man. They disestablished the church, and as we review it, we can see it was because—actually—they hated Christianity.
>
> Indeed a striking feature of the times was the evasion of the use of the word, "Christian" in respect to the religion embodied in official national declarations. The Constitution of the United States does not contain one word of reference to this religion, which was then the dominant one of each state of the union. Article 2, Section 8 was written as an escape clause for those who did not care to be identified with the oath-taking of Christianity. The Declaration of Independence refers in ordinary Deistic manner to Nature and Nature's God, Creator and Divine Providence, and concentrated reading of any documents of this entire era brings one to an immediate realization of to what extent the writers would go in order to avoid the mention of the word "Christian."
>
> And our national leaders were for the most part Deists, and included in their ranks were the first four presidents of the United States, George Washington, John Adams, Thomas Jefferson, and James Madison. And they include Thomas Paine, Benjamin Franklin, George Mason, and Ethan Allen.
>
> Thomas Paine's writings on this matter of Deism vs. the Hated Christian Religion were so widely known and accepted by the leaders of the Revolutionary movement that they were cited in the arguments before the Supreme Court of that day. Indeed at the Constitutional Convention, when Benjamin Franklin—being aged—attempted to have deliberation open with a prayer, he was defeated by a parliamentarian move.

On Monday, May 5, 2014, the United States Supreme Court decided 5–4 to contradict the Constitution as well as the Gospel of Matthew 6:5 by permitting overtly Christian prayers as part of official government assembly. The 1st Amendment promises that "Congress shall make no law respecting an establishment of religion, or prohibiting the free exercise thereof," but the Supreme Court's new ruling not only effectively establishes the state religion, but it also requires that non-Christians bow their head to someone else's god as a prerequisite to official government business.

In that same week, the Massachusetts Supreme Judicial Court also overturned earlier rulings, such that the state may now force non-Christian students to recite the Pledge of Allegiance, including the phrase "under God." This phrase was not a part of the original pledge, but was added several decades later in 1954. The same thing happened to American currency. The national motto found on our dollar bill was originally *E pluribus unum*—Out of many, one. It was changed to "In God We Trust" in 1957.

As if it was not already bad enough just to force someone to swear an allegiance (which the state should rightly earn), the government can now also force nonbelievers to acknowledge an imaginary being and to pretend as though it were really real. That is an enforcement of established religion. These new rulings were seen by many as the destruction of the "wall of separation" between church and state that had been ordered by the founding fathers.

U.S. politics have flip-flopped since our revolutionary beginnings. The United States was originally famous as a haven for immigrants with any religion or none, precisely because it wasn't founded on any one. Now the Religious Right is trying to say that America was founded as a Christian country. But the fact is that while we have become the most religious of any of the predominantly Christian First World nations (due to repeated surges in rural revivalism), the United States in its infancy was once the most secular government in history.

The original colonies were primarily peopled by refugees fleeing religious persecution in other countries. But almost upon arrival, the Puritans only continued that practice against native Shamans, then against Quakers, and even against each other over religious differences. As noted in the first volume of the textbook *The American Promise*,

> New England communities treated Quakers with ruthless severity. Some Quakers were branded on the face "with a red-hot iron with [an] H. for heresy." Several Quaker women were stripped to the waist, tied to the back of a cart, and whipped as they were paraded through towns. When Quakers refused to leave Massachusetts, the Boston magistrates sentenced two men and a woman to be hanged in 1659. The three Quakers walked to the gallows "hand in hand, all three of them, as to a Wedding-day, with great cheerfulness of Heart," one colonist observed. (2nd ed., p. 87)

In 1661, a Boston statute ordered that Quakers, both men and women, "be stripped naked from the middle upwards, and tied to a cart's tail and whipped through the town," and also to "be branded with the letter R on their left shoulder." Dutch Anabaptist Mennonites were murdered by Catholic authorities in 1554. Catholics to the South were even worse! Their list of atrocities directly linked to religious oppression alone is staggering. Witnesses wrote that Catholic conquerors delighted in conjuring new cruelties to practice on captured tribal people.

During the Valladolid debate of 1550–1551, friar Bartolomé de Las Casas argued that the native inhabitants of Greater Antilles were "fully human." He was alone in that assessment, as all the other Catholic missionaries poised against him.

> *"It is said they are like talking animals"*—Gil Gregario
>
> *"thoughtless natural slaves"*—Juan Ginés de Sepúlveda
>
> *"soulless animals without reason"*—Domingo de Betanzos
>
> *"soulless parrots in human guise"*—Garcia de Loaysa

(A few centuries later, creationists would deny that such attitudes came from pious Christians, but say instead that they were caused by belief in "Darwinism.")

Nearly everything Americans were taught about the origin of our Thanksgiving tradition is a lie. The first Thanksgiving was held by Governor John Winthrop to celebrate the massacre of over 700 Native American men, women, and children in the Pequot War of 1637. This makes the otherwise charming modern myth of the first Thanksgiving chilling and offensive.

The Salem witch trials of the 1690s are another among many horrific examples showing precisely why there need be a wall of separation between church and state. Benjamin Franklin, the towering figure of the American Enlightenment, wrote eloquently on the matter in his 1772 Letter to the London Packet:

> If we look back into history for the character of present sects in Christianity, we shall find few that have not in their turns been persecutors, and complainers of persecution. The primitive Christians thought persecution extremely wrong in the Pagans, but practised it on one another. The first Protestants of the Church of England, blamed persecution in the Roman church, but practised it against the Puritans: these found it wrong in

> the Bishops, but fell into the same practice themselves both here and in New England. To account for this we should remember, that the doctrine of toleration was not then known, or had not prevailed in the world. Persecution was therefore not so much the fault of the sect as of the times. It was not in those days deemed wrong in itself. The general opinion was only, that those who are in error ought not to persecute the truth: But the possessors of truth were in the right to persecute error, in order to destroy it. Thus every sect believing itself possessed of all truth, and that every tenet differing from theirs was error, conceived that when the power was in their hands, persecution was a duty required of them by that God whom they supposed to be offended with heresy.

The American Revolutionary War patriot Ethan Allen, in his 1784 book *Reason: The Only Oracle of Man*, wrote candidly about his views toward religion:

> In the circle of my acquaintance, (which has not been small,) I have generally been denominated a Deist, the reality of which I never disputed, being conscious I am no Christian, except mere infant baptism make me one; and as to being a Deist, I know not, strictly speaking, whether I am one or not, for I have never read their writings; mine will therefore determine the matter; for I have not in the least disguised my sentiments, but have written freely without any conscious knowledge of prejudice for, or against any man, sectary or party whatever; but wish that good sense, truth and virtue may be promoted and flourish in the world, to the detection of delusion, superstition, and false religion; and therefore my errors in the succeeding treatise, which may be rationally pointed out, will be readily rescinded.

Similarly, Thomas Paine, arguably the most influential political activist in the Revolutionary War period, made explicit his appraisal of religion in his 1794 work *The Age of Reason*:

> I do not believe in the creed professed by the Jewish church, by the Roman church, by the Greek church, by the Turkish church, by the Protestant church, nor by any church that I know of. My own mind is my own church. All national institutions of churches, whether Jewish, Christian or Turkish, appear to me no other than human inventions, set up to terrify and enslave mankind, and monopolize power and profit.

The founding fathers were largely Deists, the least devout form of theism. They were brilliant men who knew better than to let religion rule over law because theocracy has in all instances almost automatically violated human rights, and it inevitably always does. Consider the following words by the first six U.S. presidents:

> *"Of all the animosities which have existed among mankind, those which are caused by a difference of sentiments in religion appear to be the most inveterate and distressing, and ought most to be deprecated."*
>
> —1st U.S. President George Washington (1732–1799), Deist

> *"The priesthood have, in all ancient nations, nearly monopolized learning. And ever since the Reformation, when or where has existed a Protestant or dissenting sect who would tolerate A FREE INQUIRY? The blackest billingsgate, the most ungentlemanly insolence, the most yahooish brutality, is patiently endured, countenanced, propagated, and applauded. But touch a solemn truth in collision with the dogma of a sect, though capable of the clearest proof, and you will find you have disturbed a nest, and the hornets will swarm about your eyes and hand and fly into your face and eyes."*
>
> —2nd U.S. President John Adams (1735–1826), Unitarian

> *"Believing with you that religion is a matter which lies solely between man & his god, . . . that the legitimate powers of government reach actions only, and not opinions, . . . legislature should make no law respecting an establishment of religion, or prohibiting the free exercise thereof, thus building a wall of separation between church and state."*
>
> —3rd U.S. President Thomas Jefferson (1743–1826), Deist

> *"... Experience witnesseth that ecclesiastical establishments, instead of maintaining the purity and efficacy of Religion, have had a contrary operation. During almost fifteen centuries has the legal establishment of Christianity been on trial. What have been its fruits? More or less in all places, pride and indolence in the Clergy, ignorance and servility in the laity, in both, superstition, bigotry and persecution."*
>
> —4th U.S. President James Madison (1743–1826), Deist

"It is only when the people become ignorant and corrupt, when they degenerate into a populace, that they are incapable of exercising the sovereignty. Usurpation is then an easy attainment, and an usurper soon found. The people themselves become the willing instruments of their own debasement and ruin. Let us, then, look to the great cause, and endeavor to preserve it in full force. Let us by all wise and constitutional measures promote intelligence among the people as the best means of preserving our liberties."

—5th U.S. President James Monroe, (1758–1831), Episcopalian

"When I observe into what inconsistent absurdities those persons run who make speculative, metaphysical religion a matter of importance, I am fully determined never to puzzle myself in the mazes of religious discussion [and] to content myself with practicing the dictates of God and reason so far as I can judge for myself."

—6th U.S. President John Quincy Adams (1767–1848), Unitarian

Consequently, the largely non-Christian and irreligious framers of the U.S. Constitution produced the first government ever to grant all its citizens the right to religious freedom, and they did so by forbidding the government from sponsoring or promoting one religion over any other. This meant that whichever religion held dominance in that place at that moment should not be legally able to persecute any other, as had historically always happened before every single time. History continuously demonstrates that it is not possible to have freedom *of* religion without freedom *from* religion. A secular government is necessary to that end, and religious extremists themselves will readily (albeit unwittingly) explain why.

For example, Abu Mohammed, a senior member of the radical pan-Islamic organization Hizb ut-Tahrir (Party of Liberation), is quoted in a May 10, 2006 *Christian Science Monitor* article as saying,

> Islam obliges Muslims to possess power so that they can intimidate—I would not say terrorize—the enemies of Islam. In the beginning, the Caliphate would strengthen itself internally and it wouldn't initiate jihad. But after that we would carry Islam as an intellectual call to all the world, and we will make people bordering the Caliphate believe in Islam. Or if they refuse then we'll ask them to be ruled by Islam. And if after all

> discussions and negotiations they still refuse, then the last resort will be a jihad to spread the spirit of Islam and the rule of Islam. This is done in the interests of all people to get them out of darkness and into light.

Whatever laws we make mixing religion with government for the sake of appeasing our (currently) predominantly Christian constituency will only pave the way for the next religious majority, who will take advantage of whatever permissions we've already established and which we thought were harmless enough at the time. Christians very often have no concept of the value of a secular government, until or unless they understand that they won't always hold the majority. Once they realize that their religion is in a general state of decline even in this country and that the fastest growing religion globally is Islam, then the math isn't very hard to work out.

At this moment, the combined denominations of Christianity collectively account for roughly one-third of the global population, where roughly a quarter of the world is Muslim. Protestant Christians also often dismiss Catholics as a non-Christian religion. But Catholicism makes up slightly more than half of the whole of Christianity. If Catholics are not a part of that whole, then Islam is already the dominant religion on Earth, and all the various Protestant and Orthodox denominations and every other sort of Christian combined only amount to the fourth largest religion, after Islam, Hinduism, and the irreligious "Nones"—those who reject religion, or who claim "None" on forms that ask about religious affiliation. There are surely many nondenominational Christians as well as neopagans among the so-called Nones, but there are a lot of disenfranchised skeptics and free-thinking types there too.

At this rate of comparative growth and decline, even if Catholics, Orthodox, and Protestants stand together against the rise of Islam, then the time of Christian dominion is nearing an end, even in the West. Ironically, that makes Christian fundamentalists right about only one thing—these are their "end times." Islam is projected to eclipse Christianity over the next few decades.

As the Treaty of Tripoli already indicates, it is not possible to fight religion with religion, but it is possible to fight faith with reason. That is one reason why the fastest growing perspective with regard to religion is represented by those irreligious Nones. The growth of the Nones is outpacing the growth of Islam in much of the world.

Dominionist fundamentalists from all three of the major Abrahamic religions have historically hated democracy and any republic that does not recognize their version of their god as the source of the state rule. This is why creationists campaign continuously to overturn our secular system, trying to undermine science education and teach their particular Bible-based religious beliefs in its place by pretending that creationism is science too. They know it isn't, but they say it is anyway.

Creationism moved from farcical antiscience nonsense to a powerful political movement as a calculated tactic referred to as the wedge strategy. The plan initially concocted by Christian Reconstructionist Rousas J. Rushdoony of the Chalcedon Institute was that the United States should be a "theonomy," a government that enforces the laws of the Old Testament, including the barbaric book of Leviticus. Any Bible scholar with a sense of natural morality and knowledge of political science should shudder at that thought.

The Christian Reconstruction movement reached a peak in 1990, but it strongly influenced the next generation of Right Wing evangelicals, religious libertarians, presuppositionalists, and fundamentalist homeschoolers, all of whom like the idea of America being a Christian nation that teaches "God's word" and enforces "God's laws."

The means to achieve this goal was outlined by attorney Philip Johnson of the the Discovery Institute, a "think tank" that focuses on the promotion of intelligent-design creationism. He explained it in a memo called the "Wedge Document," which revealed the institute's insidious agenda: "Discovery Institute's Center for the Renewal of Science and Culture seeks nothing less than the overthrow of materialism and its cultural legacies."

Remember that "materialism" here essentially means "not believing in magic." It pertains to philosophical naturalism (the belief that there is nothing beyond the natural) but refers to methodological naturalism, which is the primary principle of scientific methodology, wherein everything must have a natural explanation according to the laws of physics and so on. Creation scientists applying for a position at "Patriot University" might as well be asked the same question Ghostbusters secretary Janine Melnitz asked of applicant Winston Zedmore:

> Do you believe in UFOs, astral projections, mental telepathy, ESP, clairvoyance, spirit photography, telekinetic movement, full-trance mediums, the Loch Ness monster, and the theory of Atlantis?

The "Wedge Document" was meant as an internal memo, but it was accidentally leaked to the outside world. Rob Boston of *Church and State* magazine summarized it thusly:

> The objective is to convince people that Darwinism is inherently atheistic, thus shifting the debate from creationism vs. evolution to the existence of God vs. the non-existence of God. From there people are introduced to "the truth" of the Bible and then "the question of sin" and finally "introduced to Jesus."

So the planned Christianization of American politics had to begin with the denigration of science in order to promote the fables of the Bible the only way they can be: through legislation. That meant that religious activists had to infiltrate various levels of government in order to place senators, judges, and lobbyists at necessary positions, which they have obviously done. As the Southern Poverty Law Center notes,

> Although most fundamentalist leaders now deny holding Reconstructionist beliefs, several—including Beverly and Tim LaHaye (see Concerned Women for America), Donald Wildmon (see American Family Association) and D. James Kennedy (see Coral Ridge Ministries)—did serve alongside Rushdoony and other Chalcedon associates on the Coalition for Revival, a group formed in 1981 to "reclaim America."

So far they've still been beaten in most courts, because they're not trying to teach "better" science, nor any "alternative" science. Many of them don't want students to learn science at all. They want to impose their religion instead. They don't want to educate; they want to indoctrinate. They want the government to support them in public schools because they're intolerant of other views, and they want to condition everyone else's children to believe as they do. But America's pioneer concept of separating church from state still prohibits this. This was supported by the U.S. Supreme Court decision on the case of *Edwards v. Aguillard*.

> We agree with the lower courts that these affidavits [pleading to teach creationism in public schools] do not raise a genuine issue of material fact.

With that decision in 1987, creation "science" was outlawed in public schools as being exclusively religious and not scientific in any sense, and was thus deemed unconstitutional. In 2005, a U.S. federal court similarly ruled in *Kitzmiller v. Dover* that the teaching of intelligent design in public schools was unconstitutional. Since even intelligent design creationism cannot compete with evolution as a science, creationists try instead to evoke the same constitutional prohibitions against them and turn them against science by asserting that evolution is a religion too. They know it isn't, but they say it is anyway. As creationist charlatan "Dr." Kent Hovind argues, "Evolutionism is a religious worldview not supported by science, scripture, popular opinion, or common sense. The exclusive teaching of this dangerous, mind-altering philosophy in tax-supported schools, parks, museums, etc. is a clear violation of the First Amendment."

And so we have the fifth in our series of fundamental falsehoods of creationism: the persistent insistence that perspectives opposed to faith and religion somehow still require faith as religion.

A religion is not just any ol' thing you happen to believe, and it's not just anything you believe strongly either. You can only hold one religion at a time, and not everything you're passionate about qualifies as your religion.

Every concept that is commonly accepted as a religion by both its adherents and its critics is a doctrine of ritual traditions, ceremonies, mythology, and the associated dogma of faith-based belief systems that all include the idea that some essence of "self" (be it a soul or consciousness, or memories, etc.) may in some way transcend the death of the physical body and continue on in some other form. This applies to every religion and only to religion, but it doesn't apply to evolution or atheists either, unless they happen to be Druids or Shamans or one of those other religions that don't happen to include gods.

Several people have tried to tell me that my definition of religion is invalid, because there are exceptions to the rule. If there were, they'd have a point. But so far I haven't found any exceptions. There are some beliefs that don't qualify as religions but are also not typically considered such, Taoism being one example. The core of the *Tao Te Ching* is that "nature acts without intent, and therefore cannot be said to be benevolent or malevolent to anything." But Taoism is typically intertwined with Confucianism and integrated with traditional Chinese spiritualism, so it becomes a component of religion without being a religion by itself. It is still an independent concept.

I've been told that the Pharisees and the Sadducees didn't believe in any supernatural afterlife, but according to the Jewish Virtual Library, they did, and they were the spiritual fathers of modern Judaism. Also, Judaism holds that there is a land of the dead, called Sheol, which is essentially the same as the Greek Hades. So there is still an afterlife of sorts, even if the dead are unconscious while they are there.

I went into a Theravada Buddhist temple, armed with my new definition and intending to test it. Their teacher said that Buddhism doesn't count as a religion because (1) Buddha was not a god, (2) they don't believe in any concept of self, and (3) they don't believe in any sort of afterlife, not even reincarnation. They said you can't be born again, because you've never been born before. That's when I pointed out that everyone in that room had a birth certificate.

Moments later, this same teacher said that he prays to Buddha, and that Buddha hears him and can even answer prayers by some miraculous means. So I pointed out that by his own description, Buddha is a magical anthropomorphic immortal who answers prayers and performs miracles. So he *is* a god after all!

The teacher was losing his patience with me, but he continued. Moments later he said that after you die, you may be a ghost for a while, before you come to live in another body. So I jumped in again, saying, "Whaddya mean, after *you* die? You mean 'you,' as in yourself? The self you don't believe in will survive beyond death to live again in some other form? That's the very definition of religion!"

After that, they didn't want to talk to me anymore.

Some Buddhists believe in a god and some don't. Some traditional Chinese beliefs are the same way. Not all religions have creative deities, but every religion must propose something paranormal (beyond tangible existence) that they believe we'll experience after we die. You can't posit something like that without faith, and if you don't have faith, you can't have religion.

And when creationists complain about atheists, they're not talking about Buddhists or Shaman. They're referring to material empirical rationalists—people they know don't have any faith in anything supernatural at all, which only makes their lie that much more brazen. Creationists know science isn't based on faith, but they say it is anyway.

I myself identify as an apistevist (one who rejects faith), because faith is an assertion of unreasonable conviction that is assumed without reason and defended against all reason. Even creationist Kirk Cameron of Living Waters Ministries understands this, if in word rather than in practice. For example, in attacking some of Richard Dawkins' views toward Darwinian theory, he has stated: "That is unreasonable, and it's unscientific. That is the definition of blind faith: 'I believe something even though there is no evidence to support it.'"

Faith is neither virtuous nor moral; it's a matter of self-deception and manipulation of the masses. It's fundamentally fallacious and inherently dishonest, but I haven't described the worst of it yet. Faith takes away reason and accountability. It means never having to accept responsibility for your own mistakes, because it means never having to admit when you're wrong. Instead, you label your lies as Truth with a capital T and hope that a show of confidence is convincing. Faith is not an allegiance to truth; in fact, faith is the most dishonest and unreasonable position it is possible to have. It is the most counterproductive, too. Faith has always only ever served to impede, retard, or reverse progress in whatever sociopolitical, medical, educational, economic, or environmental application it has ever touched.

According to a consensus of every authoritative or definitive source available anywhere—including dictionaries, scriptures, hymns, sermons of theologians, past and present—faith in the context of religion can be accurately defined as a stoic, unwavering conviction—a positive belief which is not dependant on evidence, and will not change because of evidence. Believers usually want to argue this point trying to conceal the fact. So to prove it here, I'll cite several dictionaries just to establish consensus:

Faith:

"*Confident* belief in the truth, value, or trustworthiness of a person, idea, or thing, *that does not rest on logical proof or material evidence.*"

—*Dictionary.com*

"1. *Complete* trust or confidence.
2. Strong belief in a religion.
3. A system of religious belief."

—*AskOxford*

"Belief; the assent of the mind to the truth of what is declared by another, *resting solely and implicitly on his authority* and veracity; *reliance on testimony*."

—*Accurate and Reliable Dictionary*

"A *firm* belief in something for which *there is no proof*."

—*Merriam-Webster Online Dictionary*

"Belief in, devotion to, or trust in somebody or something, *especially without logical proof*."

—*Encarta*

"For quite a lot of people, faith or the lack thereof, is an important part of their identities. E.g. a person will identify him or herself as a Muslim or a skeptic. Many religious rationalists, as well as *non-religious people, criticize implicit faith as being irrational. In this view, belief should be restricted to what is directly supportable by logic or evidence*."

—*Wikipedia* [All emphasis added]

Dictionaries give common usage regardless of whether it is appropriate or accurate. I wouldn't trust a common dictionary for scientific terms, because then I can demonstrate where dictionaries are sometimes wrong. (None of them properly define the word "animal," for example.) I wouldn't limit myself to dictionaries for this purpose either, because everyone wants to contest me. Another reason is that there are two different contexts in the dictionary; one of them exists only in the dictionary and in common vernacular, but it does not relate to religion, and does not derive from any of the writings of religion. I'm only talking about faith in the religious context.

Defenders of the faith want to pretend that "faith" is a synonym of "trust," as if the focus could be shown to be worthy of that trust. Or they deliberately use the wrong context, pretending that we must have "faith" that an airplane will land safely before we get on it. That is quite a bit different than the religious context. They won't admit what faith really is until they try to project their own faults onto nonbelievers in their frequent attempts at false equivalence—at which point they'll either say that I believe on faith in lieu of evidence just like they do (ignoring all the evidence I present) or they'll say that they have evidence just like I do, though they can never show it.

If faith is defined as an unsupported conviction, then they have it and I don't. If faith is defined as a secure confidence in the truth, value, or reliability of a given position, then I have it and they don't, according to the behaviors I typically see when debating such people—like when they ignore all my questions and won't acknowledge my answers either. But we are definitely talking about a religious context here, not my estimation of evident probabilities when boarding an airplane.

If I were arguing scientific terms, I would have to cite peer-reviewed studies. Since faith is a religious term, I'll have to turn to religious authorities, beginning with the most familiar scriptures in Western society.

- John 20:29: *"blessed are they who have not seen but yet believe."*
- Romans 1:20: *"the invisible things of him from the creation of the world are clearly seen, being understood from the things that are made."*
- Romans 14:22: *"The faith which you have, have as your own conviction"*
- 2 Corinthians 4:18: *"We look not at things seen, but at things not seen."*
- 2 Corinthians 5:7: *"for we walk by faith, not by sight."*
- Hebrews 11:1: *"Now faith is the assurance of things hoped for, the conviction of things not seen."*

Here we have things hoped for but not seen, looking at things that are not seen, not seeing what *is* seen, and, in Romans 1:20, the most common combination of logical fallacies: the circular argument routing back to an assumed conclusion.

Note that we are expected to see what is not there. Not only that, but we are blessed if we make ourselves see what cannot be seen. This is not a reasonable request, and these are not reasoned responses. As Dan Barker writes in *Losing Faith in Faith: From Preacher to Atheist*,

> Faith is the acceptance of the truth of a statement in spite of insufficient evidence. . . . Faith is a cop-out. If the only way you can accept an assertion is by faith, then you are conceding that it can't be taken on its own merits.

Faith is the very opposite of reason, and where faith is encouraged, reason is discouraged. We are expected to believe without reason; in fact, we are blessed if we readily believe the most outrageous illogical, inconsistent, and contradictory claims from even the most credulous and questionable people without any evidence at all, according to the sermons of theologians past and present. Consider the words of Martin Luther, founder of Protestant Christianity, in the following excerpts from his Commentary on the Epistle to the Galatians:

> What makes matters worse is that one-half of ourselves, *our own reason, stands against us*. . . . To turn one's eyes away from Jesus means to turn them to the Law. . . . When the conscience is disturbed, *do not seek advice from reason or from the Law*, but rest your conscience in the grace of God and in His Word, and *proceed as if you had never heard of the Law*. . . . The person who can rightly divide Law and Gospel has reason to thank God. He is a true theologian. I must confess that in times of temptation I do not always know how to do it. To divide Law and Gospel means to place the Gospel in heaven, and to keep the Law on earth; to call the righteousness of the Gospel heavenly, and the righteousness of the Law earthly; to put as much difference between the righteousness of the Gospel and that of the Law, as there is difference between day and night. *If it is a question of faith or conscience, ignore the Law entirely*. . . . We have two propositions: To live unto the Law, is to die unto God. To die unto the Law, is to live unto God. *These two propositions go against reason*. . . . When we pay attention to reason, God seems to propose impossible matters in the Christian Creed. To reason it seems absurd that Christ should offer His body and blood in the Lord's Supper; that Baptism should be the washing of regeneration; that the dead shall rise; that Christ the Son of God was conceived in the womb of the Virgin Mary, etc. Reason shouts that all this is preposterous. *Are you surprised that reason thinks little of faith?* Reason thinks it ludicrous that faith should be the foremost service any person can render unto God. . . . *Let your faith supplant reason*. Abraham mastered reason by faith in the Word of God. Not as though reason ever yields meekly. It put up a fight against the faith of Abraham. Reason protested that it was absurd to think that Sarah, who was ninety years old and barren by nature, should give

> birth to a son. But *faith won the victory and routed reason, that ugly beast and enemy of God. Everyone who by faith slays reason, the world's biggest monster, renders God a real service,* a better service than the religions of all races and all the drudgery of meritorious monks can render. *Do not consult that Quackdoctor, Reason.* Believe in Christ.[Emphasis added]

As you can see, where faith is encouraged reason is discouraged. They're opposites. This is not just my interpretation, but the common understanding of scholars and philosophers. Friedrich Nietzsche said, "Faith means not wanting to know what is true." Or, to put it another way, as Mark Twain did, "Faith is believing what you know ain't so."

Faith requires that we literally make-believe; that we presume, presuppose, and pretend; that we ignore what we really do see, and imagine something is there when it apparently isn't. It means that we lie to ourselves and fool ourselves. Worse than that, faith requires that we believe the unbelievable. This is reflected in the hymns of Michael Card, especially the appropriately titled "That's What Faith Must Be":

> *To hear with my heart,*
> *to see with my soul,*
> *to be guided by a hand*
> *I cannot see,*
> *that's what faith must be.*
>
> *So we follow God's own Fool,*
> *for only the foolish can tell.*
> *Believe the unbelievable,*
> *and come be a fool as well.*

This isn't just willful ignorance; this is dementia, a deliberately induced delusion. According to the National Alliance on Mental Illness, a delusion is a persistent false belief that is maintained despite indisputable evidence to the contrary, to falsely claim something even when there is evidence otherwise. What makes these beliefs delusional is that they don't change when the person is presented with conflicting information—the beliefs remain fixed even when the facts contradict them.

There was a creationist propaganda documentary called *Questioning Darwin* that exclusively interviewed believers and was intended to promote creationism; however, every nonbeliever and every believer who considers themselves reasonable or moderate should watch that documentary, because it is a psychological and philosophical freak show. One of the people interviewed said that any scientific fact that contradicts the Bible must be false, because the Bible can't be wrong. This prompts me to ask, what do educated expert specialists really know compared to ignorant superstitious primitives? Another person in that same show admitted that if the Bible said that 2+2=5, he wouldn't question it; he would find some way to believe it. And of course, he's not the only one like that.

So it doesn't matter whether you prove that it's wrong, nor how easily or obviously or eloquently you do that; once people have been thoroughly indoctrinated, their logical centers selectively shut down, and it's almost impossible to reason with them ever again where that subject is concerned. Fantasy is adopted as reality, and truth is dismissed as irrelevant.

Faith is often a belief in things that are impossible according to everything we know about anything at all. The belief is sacred, meaning that it is never to be questioned or critically examined but must be believed no matter what. You just gotta be-LEEVE! Skeptical inquiry is strictly forbidden, and apologetics exists only to obligately rationalize away any criticisms so that they may be dismissed without consideration.

In other words, faith assumes its own conclusions, believes impossible nonsense without reason, and defends those beliefs against all reason to the contrary. It can't help but be wrong to some degree to start with, and any errors will never even be acknowledged, much less sought out or corrected, so that situation can never improve. However wrong it already is is however wrong it will forever be. Faith offers no way to discover the real truth about anything, but it's a great way to stay wrong forever and never admit it—even to yourself.

Science is completely opposite in every respect. Religious apologists are like litigators who must keep defending their clients and pretending they're innocent, even when they know they're guilty. But scientists are investigators. They're often forced to reconsider their own initial perceptions, and are free to follow the facts wherever they may lead.

This is probably the most important distinction between faith and science. Religious dissention is punished as heresy, considered a deadly serious offense. And when they can't suppress dissention and enforce conformity, the result is another in an endless series of heretical cults.

Science acts exactly opposite; yes, there is a status quo, but the best way to become famous as a scientist is to challenge that, and you may be richly rewarded if you do, assuming you live long enough. Deep-seated positions are harder to dislodge, so minds don't always change quickly.

It's not enough that you *believe* differently; you have to show that you *know* better. There might be some crackpot fringe groups prone to human error, but otherwise, pioneer concepts in science will either be confirmed or disproved as soon as can be—usually before there are more than two divisions to compare. Because in science, it doesn't matter *what* you believe so much as *why* you believe it: science is all about what the data shows, and unsupported notions should be readily rejected. As Carl Sagan said in episode thirteen of *Cosmos*, "Who Speaks for Earth?"

> Science . . . Its only sacred truth is that there are no sacred truths. All assumptions must be critically examined. Arguments from authority are worthless. Whatever is inconsistent with the facts, no matter how fond of it we are, must be discarded or revised.

Rather than any need to believe, science is driven by a desire to understand—and the only way to improve your understanding of anything is to seek out errors in your current perception and correct them. You can't do that if you claim your initial assumptions are already infallible, and you can't even begin to seek the truth if you won't admit that you might not already know it, or that you don't know it all perfectly. You can't know when you're right if you can't admit when you're wrong.

The clearest way to know that evolution is not religion is that all postulations have to be based on evidence. But more than that, a scientific explanation requires that you devise a hypothesis and then determine whether the evidence is concordant with it or contradicts it. In this case (as in all others), we find that the evidence is exclusively concordant with evolution. It's the same in every science: creationism conversely has never been supported or indicated by any evidence in any field of study ever, and evolution has never been contradicted the way creationism consistently always has been.

But the important point is that there has to be a way to test it. Religious people will not subject their beliefs to that, because the whole point of faith is to convince yourself that it's true regardless of whether it is or not.

Religion can't seriously consider the question of whether you're wrong or how you could determine that; that would mean a denial of faith, which is exactly what science must do. There has to be some way to test your beliefs to determine whether they actually have merit.

That hypothesis should have at least two perspectives to it: (1) what if I'm right? and (2) what if I'm wrong? Regardless of how accurate my understanding is, if my primary suspicion is correct and the situation is as I describe it, then there are certain other conditions or elements that would have to be concordant with that. How can I test to see if they are? Conversely, what if I'm wrong? If I were to perform this specific test, then I should expect this specific sort of result. But if that doesn't happen or if I get some other unexpected result instead, then that would falsify my hypothesis. That would prove me wrong, and I would have to change my mind.

This is what science always has to do, but religion will not do that. While scientists themselves may be religious individuals of many different faiths, their methodology was designed to be the antithesis of faith—it requires that all assumptions be questioned, that all proposed explanations be based on demonstrable evidence, and that all hypotheses be testable and potentially falsifiable. Blaming magic is never acceptable because miracles aren't explanations of any kind, and there has never been a single instance in history when assuming the supernatural has ever improved our understanding of anything. In fact, such excuses have only ever impeded our attempts at discovery. Hippocrates, the father of medicine, understood this as far back as 400 BCE, when he wrote in On the Sacred Disease,

> It is thus with regard to the disease [epilepsy] called Sacred: it appears to me to be nowise more divine, nor more sacred than other diseases, but has a natural cause from the originates like other affections. Men regard its nature and cause as divine from ignorance and wonder, because it is not at all like to other diseases. And this notion of its divinity is kept up by inability to comprehend it.
>
> . . . Men think epilepsy divine, merely because they do not understand it. But if they called everything divine which they do not understand, why, there would be no end of divine things.

Hippocrates said diseases can only be cured if their cause is understood. This is one of many reasons why science depends on methodological naturalism: unlike religion, science demands some way to determine whose explanations are the more accurate, and which changes would actually be corrections. Science is a self-correcting process that changes constantly because it's always improving. Sure, science has been wrong before and is almost certainly wrong about something now, but that doesn't mean that everything we know today might turn out to be wrong tomorrow. Believers often complain that science changes, and that it's been wrong before. That's true. But it's like a game of Twenty Questions; with every question asked, your position may change. But the change is always an improvement. The more information you have that can be confirmed correct, the more fact is winnowed from falsity, the more accurate your perspective becomes. It has to. The laws of logic demand it.

Only accurate information has practical application, so it doesn't matter what you wanna believe. All that matters is why we should believe it too, and how accurate your perception can be shown to be. You can't just make up stuff in science like you can in religion because you have to substantiate everything, and you have to be able to defend it—even against peers who may not want to believe as you do. Be prepared to convince them anyway. It's possible to do that in science because science is based on evidence and reason. That means you must be ready to reject or correct whatever you hold true should you discover evidence against it.

All this stands completely counter to faith, and religious assumptions cannot withstand any of these rigors. But evolution can, and does, and has. For more than 150 years so far, the greatest minds of the modern age have collectively failed to controvert the essence of evolution. It is a study that neither requires nor desires faith, and doesn't even permit it. Nor is it needed, because evolution is easily indicated and evidenced, measured and tested myriad ways even against the harshest scrutiny. And remember that evidence must be objective, meaning that it can still be verified whether you want to believe in it or not. Evolution has all that in spades, and is a unifying theory that has enhanced our knowledge of many different aspects of biology enormously, and that's why so many religious as well as nonreligious scientists endorse it.

Believe what you want about the supernatural realm, but those beliefs have to rely on faith because there's no way to know whether any of it is true or not—because science can't look at the *meta*physical. We can't even see if there is a supernatural to look at. Believers can't even show there's a "there" there. So science can only help us understand the material world we can actually study, and which we know is really real. Thus, "evolutionists" may still believe in any religion they like. Lest we forget, when creationists complain about evolution, they're really complaining about science in general, both in principle and practice.

Most Christians accept evolution, and some atheists do not. One notable exception in the latter category is Dr. Periannan Senapathy, a molecular biologist and genome researcher. He is an atheist whose name is listed on RationalWiki between two people I know, Eugenie Scott and Michael Shermer. Senapathy's book *The Independent Birth of Organisms* actually posits what was the absolute worst description of evolution I had ever heard, even by the standard of oft-repeated creationist arguments: that "one group of molecules gathers together to become a man, and another group of molecules gathers together to become a fish." Of course, that bewildering absurdity is not evolution and Senapathy knows that, but that is exactly what he posits as his alternative to evolution.

Senapathy argues against common ancestry with the inane proposition that all different species of animals emerged from primordial ooze in more or less their current condition as crabs, frogs, rabbits, and so on. Not only is this argument bizarre in that it calls for millions of individually impossible abiogenesis events as already multicellular organisms, but this claim is also made in complete ignorance of the taxonomic hierarchies established by morphology and confirmed by phylogenetics. Senapathy's position would not allow crabs to be related to lobsters or shrimp, or even other crabs. Neither would rabbits be related to hares or rodents; each of them would be incidental, accidental assemblies of amassed molecules, regardless of their evident phylogenetic relationships.

However, Senapathy's claim does prove one important point: that evolution is *not* the only position of the atheist! Nor is it exclusively held by atheists. Consequently, evolution is not a doctrine of atheism, and creationists know it's not, but they say it is anyway.

These extremists often say that secular humanism is recognized as a religion by law, and since they wrongly think atheism and what they call "evolutionism" are the same thing, then by extension they think even rationalism should be considered a religion—that even antireligion is religious. But of course they're wrong again on all counts.

In the 1961 case of *Torcaso v. Watkins*, Roy Torcaso was denied his commission as notary public when he refused to declare a belief in God. "The point at issue," said Torcaso, "is not whether I believe in a Supreme Being, but whether the state has a right to inquire into my beliefs." At that time, the state of Maryland's Declaration of Rights required "a declaration of belief in the existence of God" as a qualification for any office of profit or trust in that state.

> **Art. 36.** That as *it is the duty of every man to worship God* in such manner as he thinks most acceptable to Him, all persons are equally entitled to protection in their religious liberty; wherefore, no person ought by any law to be molested in his person or estate, on account of his religious persuasion, or profession, or for his religious practice, unless, under the color of religion, he shall disturb the good order, peace or safety of the State, or shall infringe the laws of morality, or injure others in their natural, civil or religious rights; nor ought any person to be compelled to frequent, or maintain, or contribute, unless on contract, to maintain, any place of worship, or any ministry; nor shall any person, otherwise competent, be deemed incompetent as a witness, or juror, on account of his religious belief; *provided, he believes in the existence of God*, and that under His dispensation such person will be held morally accountable for his acts, and be rewarded or punished therefor either in this world or in the world to come.
>
> Nothing shall prohibit or require the making reference to belief in, reliance upon, or invoking the aid of God or a Supreme Being in any governmental or public document, proceeding, activity, ceremony, school, institution, or place.
>
> Nothing in this article shall constitute an establishment of religion (amended by Chapter 558, Acts of 1970, ratified Nov. 3, 1970).
>
> **Art. 37.** That no religious test ought ever to be required as a qualification for any office of profit or trust in this State, *other than a declaration of belief in the existence of God*; nor shall the Legislature prescribe any other oath of office than the oath prescribed by this Constitution. [Emphasis added]

The Supreme Court ruled that such requirements violated Article 6 of the U.S. Constitution, as well the 1st and 14th Amendments. Delivering the opinion of the court, Supreme Court Justice Hugo Black wrote,

> It was largely to escape religious test oaths and declarations that a great many of the early colonists left Europe and came here hoping to worship

> in their own way. It soon developed, however that many of those who had fled to escape religious test oaths turned out to be perfectly willing, when they had the power to do so, to force dissenters from their faith to take oaths in conformity with the faith.
>
> . . . There were, however, wise and farseeing men in the Colonies—too many to mention—who spoke out against test oaths and all the philosophy of intolerance behind them.

The official ruling also included a series of footnotes, called *obiter dictum*, or "said in passing." These notes represent only the personal opinions of the justice, with no official or legal significance. In a dictum footnote attached to his opinion, Justice Black listed secular humanism along with "ethical culture" and Taoism as religions that do not teach a belief in God. The footnote is not legally binding, which is fortunate since neither of them really count as religion. Yes, ethical culture societies exist, but imagine attending the church of ethical culture!

So secular humanism is not a religion in any sense, legal or otherwise, and neither is atheism. Religion must include a professed conviction, and simply being unconvinced as to the real-life existence of what they see as mythical characters hardly counts as that. So atheism alone is no more a religion than health is a disease. One may as well argue over which brand of car pedestrians drive.

Evolution is even less religious! It is the branch of biology that explains biodiversity. It doesn't permit supernatural explanations and has no doctrines, nor dogma, nor fables with morals; it has no rituals, traditions, or holidays, nor leaders or defenders of the faith, because it doesn't allow faith. It holds nothing sacred, there's no place of worship, no enhancements, no clergy, no fashion of garb, and it neither promotes nor discourages belief in gods or souls, and says nothing about how we should live, or what happens after we die. Evolution is therefore *not* a religion, and creationists *know* it's not—but they say it is anyway.

THE 6TH FOUNDATIONAL FALSEHOOD OF CREATIONISM

"Evolutionism" Part II: "Evolution Must Explain the Origin of Life, the Universe, and Everything"

In the late 1980s, the Comedy Channel aired *Mystery Science Theater 3000.* It showcased a lot of really bad B-movies from the 1950s, '60s, and early '70s, but with witty commenters making fun of each film all the way through. They were often surprisingly good gags, and I laughed a lot. Much of the subjected cinemas were old sci-fi films from the days when the future was looming and gloomy and filmed in black and white. Back then, I was already old enough that I had seen most of those gems in their original form on Saturday afternoon television, before they were ridiculed on MST3K.

When I was a little kid (circa 1973), I remember being really irritated at how bad the science always was on every one of those movies. They used to play old Flash Gordon reels at the local pizza place. I enjoyed watching them, but I couldn't help but complain: "His ship is obviously a rocket, so why does the engine sound like a piston-driven airplane? And if the ship is shown in outer space, flying from the left of the screen to the right, then why is the smoke from the exhaust going up? There wouldn't be any 'up' in outer space!" My parents never seemed to understand what the problem was.

I think that the best science fiction is where the *story* is fiction but the *science* is real, or at least as real as possible. If you're going to write good sci-fi, and you want me to believe the one wholly implausible idea that your story is about, then every other aspect of the film should be as seemingly reasonable as it possibly can be. That's what *Jurassic Park* tried to do. If I am to believe that an impossibly huge reptilian monster is destroying the city, then the back story of its origin ought to sound realistic enough to counterbalance that. That's what the first Godzilla movie tried to do, apart from that whole atomic breath thing. After that, filmmakers adopted the opposite strategy, so that everything else in the subsequent sequels had so many outrageous absurdities, each so insanely stupid, that the monster in the middle was the most reasonable element by comparison. As I wrote on my Reason Advocates blog,

> Of course you've got a 30 story-tall sapient dino-lizard with a phaser-generator in its mouth. Why not? How else is it going to fight the equally enormous bipedal beetle with drill-bit arms? Especially now that it's being mind-controlled by those aliens from another Japan in outer space. Someone needs to get those 3-inch tall identical twins to sing their song that wakes up their giant bug-god. I don't know what a 100-meter moth can do, but we need all the help we can get now that our toxic waste has come alive too!

As the newest incarnation of this monster demonstrates again, the situation has not improved in modern times. The monster in *Godzilla* (2014) is neither a dinosaur nor a lizard; now it's a god. It's millions of years old, and comes from a time that never was, back when the earth was so radioactive that giant monsters like him were common, despite their absence from the fossil record. They also apparently plucked nuclear devices from the trees and ate them like fruit, metallic peel and all. The director said that "today's audience is more sophisticated." No, they're obviously not.

Remember the movie *Avatar*? Not only did they blend *Pocahontas* and *Dancing with Wolves* and package it as an extraterrestrial *Ferngully*, but the movie also centered around the quest of a precious metal called "unobtanium." Un-obtain-ium?! Was this creation named with sophisticated audiences in mind? This was no more sophisticated than Rocky & Bullwinkle's antigravity element "upsidasium." Apparently James Cameron was unfamiliar with the periodic table and didn't know that you can't just add elements to it—not if you expect them to be stable enough to hold safely in a naked human hand as one of the characters did.

While I'm on this tirade, let me bring up one more cinematic sore spot: there were a lot of factual errors in the movie *Lucy*, but I'm only going to talk about the biggest one, being the central premise of the film.

First, the movie centers on a mystery drug called CPH_4 which has outrageous properties. Once again, movie producers are unaware of the periodic table; combining one carbon and one phosphorus atom with four hydrogens isn't that complex or mysterious, but it is hard to pronounce. In real life, it's a metabolic enzyme called 6-carboxytetrahydropterin synthase, and it doesn't have the properties that the movie gives it. The director insists he made up the name of the drug to hide the real name of a real chemical that is naturally produced "in tiny quantities" in the sixth week of pregnancy. He wants to keep the name of the real drug secret, because it might really do what he says it could. However, biologists discussing this online say that chemical isn't what the director says it is. No drug ever could do what he says that one could.

Anyway, because the director thinks the properties of this mystery molecule can't be known, then maybe they could really cause the outrageous development of the lead character's hyperactivated intellect. We can't prove that *wouldn't* happen, right? So we take advantage of the ignorance of the audience so that anything sounds plausible.

The back story isn't the real issue though. The problem is the premise, the old (and erroneous) adage that "the average human only uses 10% of their brain capacity, so imagine what we could do if we tap 100%?"

When questioned about this, the director said: "It's *totally* not true. Do they think that I don't know this? I work on this thing for nine years and they think that I don't know it's not true? Of course I know it's not true!" Thus science fiction becomes science fantasy.

In the movie, Lucy (Scarlett Johansson's character) accidentally gets a gargantuan overdose of this mystery drug, and one of the first side effects causes her to temporarily defy gravity. Then she somehow begins to access more and more of her brain's total efficiency. As she does, she acquires new powers—not just of perception or cognizance, but psionics. She can visualize the fabric of the universe as if reality were an illusion constructed by the computers from the Matrix. And like that other movie, Lucy develops all the seemingly miraculous powers of Neo. She can even turn back time manually. Consequently she becomes a sort of deity, complete with all the powers of Spock, the X-Men, and ET combined. None of that is even on speaking terms with reality. But that's still not my primary complaint. As I said, my issue is the premise of this film.

I always knew there was something wrong with that old saying. We only use 10% of our brains? That can't be right. It didn't make any sense, neither from a material nor a theistic perspective, unless brains are just really inefficient. Everyone seemed to believe that when I was in school, and no one could tell where that claim first came from.

So I'm sitting in my second-level college course of biology for science majors, and the lecture is on the brain. The basal portions regulate glands and bodily functions, process sensory input, enable motor control, and so on, but all our wisdom, intelligence, and personalities emerge from the "thinking" part of our brain, the cerebral cortex (or "grey matter"). So we use our entire brain. Every part of it has a known and necessary function.

Then the lecture mentioned that intelligence can be correlated with the number of neurons, and that our neurons represent only 10% of all the cells in the human brain, as if we only use 10% of our brain cells for thinking. Once I heard that, I knew *exactly* where that notion had come from. Not that anyone cares, because of course people would rather believe that if you unlock the mysteries of your own imagination, then you can wield telekinetic powers. Faith promises much the same thing, as in Matthew 21:21:

> Jesus answered them, "Truly I tell you, if you have faith and do not doubt, not only will you do what has been done to the fig tree, but even if you say to this mountain, 'Be lifted up and thrown into the sea,' it will be done.

Moviemakers understand the fantasies people want to believe in, but they don't understand biology and they don't understand chemistry. So they certainly don't understand biochemistry or the physical limits of natural processes in the real world. Now imagine a populace where there are few scholars; where most people only get whatever they "know" of science from fiction, and where that fiction only promotes fantasy. That's our society. No wonder we have so many creationists in this country!

It is fair to say that none of the people who deny evolution exhibit any understanding of what it is. But it's not always their fault because many of the people promoting evolution don't know what it is either. Hollywood producers certainly don't!

The TV show *Heroes* and Marvel Comics' *X-Men* are very similar ideas. In both, random youth in different walks of life in various parts of the world abruptly develop incredible super powers. Each of these abilities is biologically, physiologically, and well ... logically impossible, with no precedent in the animal kingdom nor technology either—often not even in physics. Yet these paranormal attributes are treated as scientifically plausible, being attributed to "mutations."

According to either story, a "normal" person has no mutations at all, because (in science fiction) a single mutation can cause a suite of sweeping changes throughout the subject's entire body, reconfiguring every cell at once. It's often a complete fundamental restructuring, but one that more often than not somehow leaves the subject looking and thinking exactly like the normal beautiful teenagers they were before this happened, so that regardless of how the fundamental structure is redesigned, the surface somehow remains unchanged. There usually is no disfiguring condition, nor any chronic disease or disability associated with these mutant genes. For most characters, there is no apparent downside, nor are the attributes ever realistic, either. They're never moderate on average or within reason. Rather, it is typically godlike awesomeness that looks great in spandex.

These mutations aren't described as something you're born with, but rather as something that happens to you. The sorts of changes shown (when they could be explained at all) would require many different mutations occurring simultaneously, and all of those that occur in any one character are concordant with each other—like the sudden development of a third set of limbs, which just happen to be fully avian flying feathered wings. There would be *lots* of different mutations required for that, and these should be accumulated over myriad generations with many variants due to trial and error. This particular result would be an obscene improbability in any case, mathematically impossible and with neither the necessary skeletal structure nor musculature nor respiration or anything else to make any of this possible. Yet in the story, all these mutations occur all at once, as a single unified event, and they're orchestrated perfectly, seemingly according to some predetermined plan. Of course they are; that plan came from the writers.

According to the script, these mutations usually do not occur or are not expressed until adolescence, and are not inherited either. Only in rare instances is there any familial relationship between any of these "mutants," and that is usually a lateral one like between siblings. So they couldn't have inherited these powers from a common ancestor like their parents, who typically do not possess any superhuman enhancements; the parents of teenagers never do. The powers the siblings have are usually different from each other too, so there is no common bond between them. In short, the way mutations are depicted in movies, TV, and comics couldn't be much more wrong.

When 20th Century Fox released their first film adaptation of *X-Men*, we heard Professor Xavier introduce it thusly: "Are they the next link in the evolutionary chain? Or simply a new species of humanity, fighting for their share of the world?" Then we heard Dr. Jane Grey say, "We are now seeing the beginnings of another stage of human evolution."

No. No. No.

It is difficult to imagine a worse understanding of evolution than what we saw in *Heroes* or *X-Men*. Nobody knows what a mutation is, so that explanation works for everything. The same goes for the *Fantastic Four*, *Hulk*, or *Spider-Man* for that matter. In each of those cases, mutations were brought about in adults by accidental exposure to "weird" radiation. The radiation didn't have to be weird; the audience doesn't even understand what normal radiation does. So that explanation still works just as well, especially since the one thing everyone knows about radiation is that radiation causes mutations!

So if some primitive-looking reptile survives the fallout of a nuclear detonation, then it might "mutate" into an unrecognizable monster the size of a battleship, and it might even hiss with radioactive breath. Why not? Maybe radioactivity *can* cause all that; we don't know for sure that it *can't*, right? That makes radiation a better explanation than magic, which is what it used to take to turn one thing into something completely different. That's what evolution is, isn't it? Sure it is! I saw it on television. The scriptwriters for the movie *Underworld Evolution* had a man "evolve" into a werewolf in a momentary transformation, right before our eyes, and that's okay because no one knows what evolution is either! Cartoons about Pokémon evolution made sure of that. Whenever a Pokémon reached maturity, it suddenly "evolved" into something else. No kids—that's metamorphosis, not evolution. If they saw that puerile DreamWorks movie with Julianne Moore and David Duchovny, they *still* wouldn't know what evolution is even though that was the one-word title! None of our popular media ever explains science correctly.

The word "evolution" simply means "change over time." We can trace its etymology to the Latin *volvere*, which means to roll or unfold or open out. It was a word used for the opening of scrolls or books, gradually expanding them so that new pages could be read and new information revealed over time.

The use of "evolve" stems from 1832 in Sir Charles Lyell's book on geology, in terms which contrasted catastrophism against a gradual process of evolving (unrolling or unfolding) change. Darwin later used it in the same sense to apply to biological organisms, having read Lyell's book. Like geological features, they can be demonstrated to change gradually in response to forces in their environment, to unroll slowly a new story from the scroll of living things.

Today, we use the term to describe all the ways that heritable changes occur in organisms, but evolution stands in contrast to sudden changes like those in science fiction that appear in a single generation without precedent. Evolution is the accumulation of those heritable changes over time. Sometimes this increases the diversity of extant species, but it could also describe a sudden reduction in diversity, as happened in the Great Extinctions at the ends of the Ediacaran, Cambrian, and Jurassic periods.

Unless otherwise specified, the word "evolution" only refers to an aspect of biology when used in a scientific context. Specifically, it is a process of varying allele frequencies among reproductive populations leading to (usually subtle) changes in their morphological or physiological composition, which—when compiled over successive generations—can increase biodiversity when continuing variation between genetically isolated groups eventually lead to one or more descendant branches increasingly distinct from their ancestors or cousins. Or more simply, it is how life forms diversify via descent with inherent modification.

To put it another way, it is the method by which cats branched into so many different breeds within several distinct species in a half-dozen genera, out of only two or three major divisions: *Panthera*, *Felis*, and the extinct *Machairodontinae*. It's a continuous process with a few parent groups diversifying into an increasing number of descendant groups. That's what "biodiversity" means, and it seems that no one who criticizes evolution understands that.

Mutations do happen for several reasons, but other than cancerous growths, they're rarely detectable or significant in adults. Our growth requires that cells divide and replicate. During that process, the DNA of the daughter cell is not always a perfect copy. Because whenever cells replicate or duplicate themselves, which most do regularly, any one of them can copy some snippet of their DNA incorrectly. That's what a mutation is, and they happen with some frequency despite controls that normally prevent this. In this case, we're talking about a point mutation, where a single nucleotide is inserted or deleted in the DNA sequence. Not every cell in the body is affected; it's just that one, and it's usually only the tiniest portion of that one. More cells may be derived from it in subsequent division/duplication (mitosis), but they're always going to be outnumbered.

We all have mutations. We're born with them, and they continue to occur occasionally throughout our lives, but they're usually neutral or trivial, not the overtly dramatic system-wide changes we see in the comics. Most mutations that occur as we grow don't get the opportunity to do anything to us that we would ever pay attention to because we're multicellular and we have a lot of other cells. So mutations are usually significant only when they're inherited.

Your children won't inherit every mutation that occurred in every cell you have. It's only your one rare mutant germline cell that happens to *become* your child. Gametes (whether sperm or egg) are germline cells that already have mutations within them. Further mutation can result from genetic crossover when gametes combine to form a zygote. That's the template cell. Whatever combined or crossed mutations exist within that first cell will be replicated with every new cell division as the embryo develops. Essentially every full cell in its body would likely copy and carry it, so those mutations are more likely to be expressed in the developed organism.

Already we can see that this is nothing like our favorite comic-book mutants—a bunch of adults or adolescents, with no discernible relationship to each other, all suddenly developing completely different features can't be called a "new species" of humanity. Rather, a species would be one related group where everyone shared the same trait(s), and those features would be definitive of that group.

Any particular trait derived from mutation will emerge within a single family group, first expressed in one child and then later in his or her children, and these mutations will usually be slight. Since only a small portion of our DNA contains genes, and since not every change in those genes causes a measurable difference in form or function, we would expect the vast majority of mutations to be largely silent. Of those mutations that are not silent, some will of course count as defects, but a few might also be slight improvements of some sort, depending on the circumstances. If it's a trait that matters in their situation, then it might become advantageous or desirable, at which point it's an issue of natural or sexual selection and has an improved probability of continued reproduction. Otherwise, even if it isn't "selected," it's still a matter of population mechanics whether that trait eventually spreads through every descendant line generations down the road.

Natural or sexual selection acts on these nonsilent mutations, slightly increasing or decreasing the offspring with that trait in each successive generation against a background of random genetic change. This random genetic change and the accumulation of nonselected mutations by chance is called genetic drift. Drift and selection are two opposite types of forces acting on population: one is the random accumulation of heritable changes, while the other drives increases or decreases in traits based on the advantage or disadvantage they confer on the organism. Taken together, randomness and direction produce what biologist Stephen Jay Gould called the "drunkard's walk." A person stumbling from a bar and deeply inebriated might have some sense of which direction home lies, but their path would be anything but direct. They might, if the randomness predominated, even be headed in the wrong direction.

Evolution includes both drift and selection, but the relative importance of each is something still debated by biologists. What is clear is that those nearly silent mutations can't be selected, but they can be accumulated by random chance, which is why it is described as a "drift." Deleterious or beneficial mutations, however, depending on how strong their effect, will predominantly be acted on by selection. We've already discussed that the majority of mutations will be silent, so the process of evolution can be described best by long periods of random change with only occasional but important moments of selection acting on novel traits. Contrast this with the very simple "arrow of evolution" advanced by most creationists. They ask why we aren't getting "bigger, stronger, smarter, and faster" every day.

In fact, creationists usually accept the nature of selection, and they may conflate it with genetic drift, but they rarely combine the two into one set of principles for hereditary change. There are a lot of variables, but as different descendant groups intermingle in succession, pairing the daughters of one line with the nephews of their neighbors, some genetic lineages will thin out over time and disappear, while others will become more robust and common and intimately connected with each of the other lineages in that group. This is how it is that long-isolated communities can be identified by their uniquely shared characteristics. When particular traits eventually affect whole populations, then it becomes evolution—descent with inherent modification. Evolution is a change over time, but it is always only ever a change in the genetics at the population level over the course of several generations.

Creationists usually accept that this happens, and they sometimes even accept *how* this happens. But they don't call it evolution, because they've been conditioned to execrate that word and to utter it only with a distasteful sneer, so they use other, safer words that don't mean quite the same thing. Sometimes they'll call it adaptation or genetics. These they say are not evolution, nor even a part of it. Genetics is obviously a factor of evolution. One could even say that evolution is genetics, but it wouldn't be fair to say that genetics is evolution; they're not synonyms.

Adaptation is also misrepresented. If a community moves from the tropics to some far northern country, trading places with those northern colonists who then move to the tropics, then over even a dozen generations, they should be visibly lighter or darker than their ancestors were. Dark skin protects against the intense solar radiation of the tropics, but we also derive a necessary vitamin D3 from sunlight. We won't get that if our skin is too dark in the northern climes, because of the lower angle of the sun through the atmosphere. So tropical people will get darker, while those closer to the poles will get lighter, and this doesn't take very long in either case. This form of adaptation is not the same as evolution. These people have not "evolved" to suit their climate, because the change is not genetic; it's epigenetic. It's not a change in the genes, but rather the environment is directly affecting how the genes are expressed.

However, the transition of lizards or turtles developing flippers instead of feet is another form of adaptation, but one which definitely does count as evolution, because it requires selected mutations. Thus it cannot be minimized or dismissed as adaptation *rather than* evolution; in that case, it is both.

Those who wish to denigrate evolution take what it really is and present it as something else whenever they're no longer able to deny that aspect of it. They'll misdefine any of the terms as necessary, but they'll also misrepresent the concept as a whole, usually with a straw man fallacy. They'll take some very different (and often absurd) idea and call *that* evolution to make it easier to ridicule. They have to, because they're forbidden to accept it, and if they understood it, they'd probably believe it.

The problem creationists have with evolution is not that it challenges belief in God, because it doesn't. Their problem is that evolution, like every other field of science, eventually inevitably challenges the accuracy and authority of the storybooks that creationists equate *to* God. Creationism is a defense of *myth*ology, not *theo*logy. Consequently, creationists tend to reject science almost entirely, and will often take aspects of all the various fields of study they perceive as threatening and lump them all together under one heading, which they then refer to as "evolutionism." It's an attempt to minimize the sheer volume of sciences allied against them.

The creationist's umbrella definition of the science of evolution apparently includes significant portions of each of the following:

- Uniformitarianism
- Methodological naturalism
- Hypothesis, theory, and peer review
- Big Bang cosmology
- Geology, paleontology, and plate tectonics
- Embryology and developmental biology
- Zoology and taxonomy
- Deep time (including the speed of light)
- Cosmology and astronomy
- Theoretical physics and quantum mechanics
- Atomic theory of radioactive decay
- Anthropology and archaeology
- Abiogenesis and biochemistry
- Ecology and environmentalism
- Population genetics and genomics, and phylogeny

Religious extremists also have problems with sociology and psychiatry, but for different reasons. They don't (usually) say they're against all of science in general. For reasons that are obvious even to creationists, they can't always deny science as a whole, but that is essentially what they're doing when they deny the fundamental philosophy of science and all of its methodology, except in the rare instances that it suits them. They call it "evolutionism" in an attempt to separate what they see as objectionable science from acceptable science. Presenting evolution and all the other sciences associated with it as though they collectively only amount to a single worldview is part of their intentionally erected illusion of equality; the idea that science and religion are both subjective beliefs allows them the impression that both might have comparable merit. Given a choice of what they imagine to be two equivalent options, all they have to do is ridicule the one they don't want to understand so that what they would rather believe seems to prevail, even though they don't understand *that* either.

Religious extremists are typically not very good at thinking hypothetically, nor will they consider that there even might be a third option. So it's an extreme position that if their legendary folklore isn't the absolute authority (being both literally and completely true), then God couldn't create or even exist any other way. They'll say morality only comes from God and ignore all the arguments and evidence to the contrary. The only god they know or can imagine is the one in their sacred fables, so if the fable is false, then God is false. They can't understand or realize how morality might be naturally advantageous in a societal species; instead, they think that the absence of God means everything figuratively goes to hell. So they'll say that if things aren't the way they believe it is, or if they couldn't believe that it was, then they'd all go mad and do terrible things to each other just for the fun of it, as if causing other people to suffer would be fun. It's a desperate and destitute delusion of dichotomy that if their legends aren't right, then *nothing* is right.

Creationists don't accept that this is simply a matter of fact versus fantasy; they think it is a matter of good versus evil, and that anyone who "believes in" evolution is a radical feminist, Communist, Nazi, racist, drug-abusing, pornographic pervert. They've actually published propaganda saying each of these things, and they have a poster declaring every slur in this paragraph all at once. They frequently posit that evolutionists (those who accept natural science) also embrace humanism and antitheism at the same time as paganism and New Age religions, despite the contradictions there. They say that "evolutionism" also leads to other moral challenges like drug culture, abortion, suicide, and euthanasia—despite the fact that so many god-believers are admittedly habitual alcoholic drug addicts. Euthanasia on the other hand is when life is ended out of mercy, often with consent or even at the request of the person in question. Religion traditionally prefers that suffering be prolonged. That's why euthanasia and suicide are the only forms of killing that the Bible doesn't endorse.

There's a negative statistical correlation between religiosity and what we typically think of as moral behavior. Convicted felons are commonly estimated to be overwhelmingly religious, while the estimated number of atheists in prisons is always much lower than among free-roaming citizens. The factions of dominant religion statistically have the highest crime rate, with special emphasis on hate crimes. Religious people are more likely to condone the killing or torture of prisoners, where nonreligious people are more likely to consider that morally wrong.

But it gets worse; the most religious countries also have the highest murder rate, and the same is true of the most religious areas of the United States. The higher the religiosity of a given populace, the higher the murder rate. Nations that are more secular show the opposite tendency, as the less religious they are, the more peaceful they tend to be. Here in the United States, evangelical Christians have the highest divorce rate. They also have the highest rates of teen pregnancy, which isn't surprising in areas that teach "abstinence only" instead of offering sex education. I live in Texas, where we have the highest rates of *repeat* teen pregnancy.

But it gets even worse. Students in private religious schools (where evolution is not taught) are statistically *more* likely to get an abortion than their peers in public schools where evolution *is* taught. This is a testament to their hypocrisy and shows what a colossal failure the policies of the Religious Right have always been.

But it gets even worse than that. Child Protective Services and other agencies report that a significant majority of child abusers and molesters identify as very religious, and that the more religious they are, the worse offenders they are. Yet religious people argue that the less religious *we* are, the less *moral* we are. Because without God, they say, there can be no objective moral standards. As (atheist) Scott Clifton said in *A Treatise on Morality*,

> A particular action or choice is moral or right if it somehow promotes happiness, well-being, or health, or if it somehow minimizes unnecessary harm or suffering or both. A particular action or choice is immoral or

> wrong if it somehow diminishes happiness, well-being, or health, or if it somehow causes unnecessary harm or suffering or both.

Finally, creationists also forget that many Christians accept evolution. That's a third option they refuse to consider. In the creationist mind-set, for evolution to be true at all, it must first disprove God, then utterly replace God and account for everything they used to attribute *to* God. So whenever they meet someone trying to explain or endorse evolution, the first thing creationists may ask is where "everything" came from—not just living things, but all matter and energy in the universe, as if evolution should account for the origin of life, the universe, and everything. In his ruling against the teaching of intelligent design in *Kitzmiller v. Dover*, U.S. District Judge John E. Jones III got it exactly right when he said:

> Both defendants and many of the leading proponents of Intelligent Design make a bedrock assumption which is utterly false. Their presupposition is that evolutionary theory is antithetical to belief in the existence of a supreme being, and to religion in general. To be sure, Darwin's theory of evolution is imperfect. However the fact that a scientific theory cannot yet render an explanation on every point should not be used as a pretext to thrust an untestable alternative hypothesis grounded in religion into the science classroom, or to misrepresent well-established scientific propositions.

By definition, religious belief is irrational, so it is defended irrationally. Creationists do deliberately misrepresent evolution many different ways in all their arguments. Even when they know better, they still say that evolution necessarily requires the godless origin of life from inorganic matter. But it doesn't mean that and it never did. Evolution doesn't necessarily exclude God, and it doesn't include the origin of life either, but neither would the origin of life have been from inorganic matter.

I mentioned before how it seems that religious fundamentalists are using a different dictionary than scientists do. In this case, when they say "organic," I think they mean "something with organs," or perhaps "something that is alive." I've seen both of these contexts applied in conversations with creationists. However, organic chemistry is really the study of carbon-based compounds (excluding carbonates, carbides, carbon dioxide, and a few other exceptions). Methane, for example, is an organic chemical. All the building blocks of life were already organic long before the first organism could be considered alive.

Scientists have even detected vast amounts of organic matter in deep space. They've reportedly discovered some suspiciously lifelike inorganic matter in space, too! As *Astrobiology Magazine* reported on August 21, 2007,

> Researchers have demonstrated the formation of microscopic strands of helical structures in plasma clouds. The researchers say that these structures undergo changes that are normally associated with biological molecules like DNA and proteins, such as dividing and forming copies of the original structure.

But creationists say that evolutionists believe that space evolved too, and that the Big Bang is part of the same evolutionary process as that which leads to the creation of higher chemicals, the origin of life, and the emergence of new species on earth. So they often say that evolution requires "something coming from nothing," which is ironic since creationists believe that themselves while evolutionary biologists typically do not.

I used to say that strict scientists never say that anything came from nothing, but then I met Lawrence Krauss, a theoretical physicist and professor of cosmology. In his book *A Universe from Nothing*, he posits a different understanding of "nothing" than I previously considered. He knows this subject very well, and I doubt if I understand his argument, but that's not surprising as he said it would confuse some people. So giving him the benefit of the doubt and all due respect, I can confidently and honestly say that I am confused by nothing, and that Krauss understands nothing. He certainly understands nothing better than I do.

It is a curious thing to me how believers point to whatever we don't yet know (or what they think we don't know) in order to deny what we really do know and clearly understand. Evolution is one of those things we definitely know. So our critics quickly change the subject to how life began, or how the universe began. These are things we don't know nearly as much about. If you push back far enough, you'll eventually get to something *no* one knows, and those are the few places left where their creator can still hide. But our not knowing what made life come alive does not change the verifiable fact that life does evolve and certainly has evolved.

Trying to deny evolution by questioning cosmology or the origin of life is rather like this: imagine one of those huge artistic arrangements of thousands of dominoes, where each are poised to fall in succession, along with the other lines around it, as part of some pattern demonstration. Now imagine that you're seeing this in really slow motion, so slow that you can't detect any movement. You can see the dominoes yet standing. You see the ones that have fallen, and you see the ones seemingly frozen in the act of being toppled. Now imagine that someone, for whatever reason, is determined to dispute whether any of this is even going on. They'll deny that any of these have fallen, are falling, or will fall, on the excuse that no one knows what knocked over the first one—as if knowing that would change what we can already prove to be true, regardless of what the explanation is.

Obviously, no aspect of biology has anything to do with how the earth formed or where anything else in the universe came from. It really doesn't matter how the cosmos came to be. It could be a steady-state universe (if that were possible), or a cyclic series of Big Bang and Big Crunch contractions, or a one-time eruption from a string theorist's dimensional rift. It could even be magically conjured by the gods of creation, or it could be some other method used by a more reasonable version of God. However the universe originated, it does not relate at all to how life evolved. Another important point is that life definitely *has* evolved, regardless whether any god exists or not.

Creationists habitually misdefine their terms and commonly insist that evolution means "life from nonlife." But of course, that's not right either. Evolution explains how life diversifies, not how it began. Since evolution at every level is by definition limited to the variation of allele frequencies inherited over generations of living organisms, then it obviously can't operate where no genomes yet exist. Descent with inherent modification requires that descendants have an ancestor to inherit from—the evolutionary process starts with genetics and can't start before it. So how the first genes came about may seem similar to evolution, and may even involve a form of natural selection in some way, but it is in fact a very different chemical process called abiogenesis.

We all agree there was a time when there was not yet life on this planet, and then later there was. How did it get here? Superstition says it was magical. Science says it was natural. But science only works with natural explanations, because nothing else *can* work. As Eric Rothschild, lead counsel for the plaintiffs in *Kitzmiller v. Dover*, stated,

> Methodological naturalism is the term used to describe science's self-imposed limitation, that it will only consider natural causes for natural phenomena. Science does not consider supernatural explanations because it has no way of observing, measuring, or testing supernatural events. It doesn't mean that supernatural events, including divine miracles, have not happened, just that science cannot properly make any statements about them.

Creationists also misrepresent hypotheses on the origin of life. They will use any parody they can to link it to evolution, and try to make both sound ridiculous. That isn't surprising considering how ridiculous their own position is. The most common fib they use here is to point to the complexity of any single eukaryotic cell and question how that could have poofed into being, in its current state as if by magic from a rock or from mud. The irony here of course is that they're the ones who believe in magic, and the book of Genesis has people being created by a golem spell from (essentially) mud. Consider the words of Kent Hovind, the convicted fraud and charlatan who deliberately misrepresents cosmology, abiogenesis, evolution, and science in general:

> Once upon a time, billions of years ago, there was nothing. Suddenly, magically, the nothing exploded into something. . . . It was magic rock. Inert and lifeless, but still magical. . . . Then one day some of these minerals magically formed into a kind of goo in the pool of water. . . . The goo magically became ALIVE. So anyway, this bit of magic goo magically found something to eat. Then, magically, it found another bit of magic goo to marry, and they had a whole bunch of magical little goos. . . . some of this goo magically evolved upwards and upwards, growing ever more advanced, bigger, stronger, smarter, until it became a magical hairless ape with thumbs.
>
> And do you know who those apes are? That's right! They're YOU and ME! We are the magic rock apes! And you know what else? Someday we'll evolve enough that we'll become the god we all know doesn't exist.

One lie that I've seen repeated a few times lately is the allegation that evolution is a fairy tale for grown-ups. This falsehood is ironic too, because if you look up the definition of a fairy tale, you'll see that it is a folklorish story, usually with a moral, but which includes dragons, witches, giants, magic spells, and/or animals that speak and act like human beings. This does not match the study of evolution in any sense, but it does exactly describe the biblical story of Genesis perfectly. Even if the Bible were true, it would *still* be a fairy tale by definition! Although my wife says that involvement of deities properly defines the Bible as myth, and some isolated portions of it might even qualify as legend, but that's about the best you could say about it.

Creationists often say that the evolutionist's explanation for the origin of life is spontaneous generation, where already-complex organisms somehow pop out of rotten filth fully formed. It's the idea that anything that was once alive still has "life" in it—that as it decays, the spiritual life force trapped within it decays as well, causing it to conjure up vile organisms like bacteria, maggots, and vermin.

This was a purely supernatural belief from the creative imaginations who told us about hexes, foul spirits, omens, and "bad blood." But this belief differed from a lot of other religious nonsense, because this one was testable. Eliminating the spontaneous generation of mice and maggots was relatively easy—they simply used sealed or covered containers. In 1859, Louis Pasteur performed the most famous refutation regarding the then-inexplicable emergence of microorganisms like mold and bacteria, which had only recently been discovered with the invention of magnification lenses. Guided by earlier experiments that were considered inconclusive, Pasteur provided a flask of water and chicken broth as a food source. These were left open to the air. However, the air was first drawn out of the flask and the water boiled to eliminate any microbes that might already be there. The neck of the flask was also bent into an S-shape so as to create a "trap" like the one under the drain in your sink, except for the additional curve requiring air to flow upward some distance to enter the tubing. This allowed air to slowly flow back in, but not to bring airborne microbes with it. Those were caught in the gravity trap.

Creationists delight in talking about how spontaneous generation was disproved, which it was—by experiments reliant on methodological naturalism, because it disproved the involvement of supernatural spirits. Yet somehow they think that evolutionary scientists still believe in the superstitious mysticism of the horse-and-buggy days!

Of course that's not the case. How could it be? How could the world's leading academic scholars really believe anything so preposterously silly? Even if that were possible, how could any serious researcher (much less all of them) have overlooked the fact that spontaneous generation had been disproved over 150 years ago? I have to consider that the only way creationists could think that we're that dumb is if *they're* that dumb!

Early scientists had no idea how complex living cells actually are. We couldn't go straight from basic chemicals to even the simplest cells. The actual perspective of modern science is obviously not remotely like that; it's a surprisingly intricate sequence requiring multiple processes over a series of different stages, from simple chemicals to polymers to replicating polymers, to a hypercycle, then protobionts and eventually the simplest cells like bacteria.

> **Spontaneous generation:** Proposed by Anaximander in the sixth century BCE, and disproved in a series of experiments from 1668 to 1861: the proposition that fermentation and putrefaction activate a latent "vitalism" (life force) in once-living matter, thus recycling organic refuse such as old meat, rotting vegetables, and feces into new forms of already complex, albeit vile, viruses and living organisms from bacteria all the way to animals such as flies and even mice.

> **Abiogenesis:** The current hypothesis replacing spontaneous generation as an explanation for the origin of life, proposed by Rudolph Virchow in 1855 and coined by Thomas Huxley in 1870: the proposition that the formation of life requires a prior matrix, thus genetic and metabolic cells must have developed through an intricate sequence of increasingly complex chemical constructs, each having been naturally enhanced by particular constituent and environmental conditions.

If you look up either of these words in any general usage dictionary today, you will not get these definitions. *Wikipedia* offers a good explanation for both, but otherwise, common dictionaries still list abiognesis and spontaneous generation as if they were one and the same. I personally wrote to two online dictionaries, *Merriam-Webster* and *Dictionary.com*, advising them of this error. But it doesn't matter that they know this is wrong; they're not going to correct it, and that is a big part of the problem with this whole alleged controversy wherever nonscientists are concerned.

Creationists obviously have no appreciation for how solid all our combined sciences are, whether it is genetic orthologues confirming ancestral phylogenies that were once only indicated morphologically, by determining derived synapomorphies, or whether it is the several different kinds of radioactive decay rates and other dating methods that cross-confirm each other to produce the same consistent results once any variables are accounted for, including

- Uranium-Lead
- Uranium-Thorium
- Potassium-Argon
- Argon40-Argon39
- Rhenium-Osmium
- Lutetium-Harnium
- Samarium-Neogymium
- Rubidium-Strontium
- Fission track
- Chlorine-36
- Luminescence
- Dendrochronology
- Varves
- Ice Cores
- Radiohalos

Creationists often cite the laws of thermodynamics as if they could somehow apply to the diversification of life on earth. They don't. Lord Kelvin, the nineteenth-century scientist who determined those laws, was a creationist himself. He was also a contemporary of Darwin's who was definitely opposed to evolution. He famously said, "Overwhelming strong proofs of intelligent and benevolent design lie around us." Of course, he also said that X-rays were a hoax, that radio had no future, and that heavier-than-air flying machines were impossible. But he was a mathematical physicist; in his area of expertise, he said that thermodynamics demands that the earth has to be on the order of at least twenty million years old even if the bowels of the world didn't continue to heat themselves radioactively, which of course they do. Once Kelvin realized that, it pushed the age of the earth back much further. In his 1871 presidential address to the British Association for the Advancement of Science, he discussed his views on the origin of life:

> How, then, did life originate on the Earth? Tracing the physical history of the Earth backwards, on strict dynamical principles, we are brought to a red-hot melted globe on which no life could exist. Hence when the Earth was first fit for life, there was no living thing on it. There were rocks solid and disintegrated, water, air all round, warmed and illuminated by a brilliant Sun, ready to become a garden. Did grass and trees and

> flowers spring into existence, in all the fullness of ripe beauty, by a fiat of Creative Power? Or did vegetation, growing up from seed sown, spread and multiply over the whole Earth? *Science is bound by the everlasting law of honour, to face fearlessly every problem which can fairly be presented to it. If a probable solution, consistent with the ordinary course of nature, can be found, we must not invoke an abnormal act of Creative Power.* [Emphasis added]

Importantly, Kelvin also said, "The hypothesis that life originated on this earth through moss-grown fragments from the ruins of another world may seem wild and visionary; all I maintain is that it is not unscientific." So if the author of the laws of thermodynamics concedes that he has no scientific argument prohibiting evolution, then why do those who neither understand his laws nor evolution suppose that either is opposed to the other?

Amusingly, a physicist named Jeremy England is now citing thermodynamics as one explanation for the origin of life. In his paper, he explains how the 2nd law of thermodynamics holds that abiogenesis is not only probable, but more or less inevitable.

Whether creationists accept any amount of proof against them or not, the fact is that everything we've learned about physics and thermodynamics since Kelvin demands that the earth be billions of years old. And according to every ounce of paleontological evidence anyone has ever dug up, there is every indication that the further back in time you look, the simpler and more similar living things appear to be until there are only single cells—and prior to that, there is no evident life of any kind at all. And for those who ask, why is there no evidence of abiogenesis occuring today, Darwin himself provides a compelling answer in his February 1, 1871 letter to J. D. Hooker:

> It is often said that all the conditions for the first production of a living organism are now present, which could ever have been present.— But if (& oh what a big if) we could conceive in some warm little pond with all sorts of ammonia & phosphoric salts,—light, heat, electricity etc present, that a protein compound was chemically formed, ready to undergo still more complex changes, at the present day such matter would be instantly devoured, or absorbed, which would not have been the case before living creatures were formed.

There were no primates 100 million years ago, no mammals 200 million years ago, no dinosaurs 300 million years ago, and no land animals whatsoever 400 million years ago. Five hundred million years ago, there weren't any insects or vertebrates with actual bones, and 600 million years ago, there weren't even the most primitive fish yet. We've never found any trace fossils for macroscopic life forms prior to 700 million years ago, but we do have bacterial microfossils covering another 2.8 billion years prior to the first multicellular anythings we've ever found a trace of. The only possible conclusion we can draw from all that is that the most advanced organisms were still only microscopic and microbial for the first 80% of the history of life on this planet.

It's very rare that anything is ever fossilized, and when that happens, it is usually only the hard parts that leave any impression. We can still find tracks and tunnels of worms and that sort of thing, but not everything leaves trace fossils. The smaller it is, the harder it is to leave a recognizable imprint. So if we're talking about single-celled microbes without any bones or shells or teeth, then circumstances have to be perfect, and we'd have to know exactly where to look or we wouldn't find any trace of them at all.

Stromatolites are a notable exception. Some microbes like cyanobacteria live in colonies and leave secretions that can catch grains of dust and concrete them, eventually building up a sort of city full of rocklike buildings in shallow waters. These account for some of the oldest fossils in the world, dating back 3.5 billion years.

Some stromatolite-building microbes use photosynthesis to produce their food, and they can live without oxygen, but they expel oxygen as waste product just like plants do today. The abundance of these types of fossils all over the earth for billions of years before the appearance of multicellular organisms implies that they produced a lot of oxygen, enough to envelop the whole world!

This oxygen was initially absorbed into the oceans, and then into exposed iron on land, turning it into iron ore, but rock layers from 600 million years ago show that the iron had finally absorbed all the oxygen it could. The rest filled in our atmosphere and our ozone layer relatively quickly. Prior to that, it seems the atmosphere was made of greenhouse gases that would have been poisonous to us and wouldn't have protected against solar radiation. However, oxygen is poisonous to most anaerobic organisms, and the new atmosphere would have killed off many primordial life forms. So if the air we breathe today was made by living organisms, then the atmosphere had to have been very different back before anything ever lived.

Even Christian biologists admit that at its most basic, life is simply chemistry, and living tissues conform completely to those guidelines. The ingredients that form the most basic cell structures create a phospholipid bilayer automatically upon contact with water, due to their combined polarity. Likewise the function of enzymes and transport vesicles and other minuscule but critical elements within a cell all conform to simple physical properties.

But we have to start with the right basic chemicals. In order to show whether life could emerge naturally, scientists first had to show that the environment could have derived complex organic compounds out of relatively simple inorganic ingredients, and that this could happen incidentally within that environment.

The original Urey-Miller experiment successfully produced amino acids in a contained environment of water, ammonia, methane, and hydrogen, which was consistent with early hypotheses of the original atmosphere of the prebiotic Earth. But that hypothesis was not well supported, and is no longer considered accurate. Even though the experiment was a success under those conditions, it was later determined that those weren't the *right* conditions. So Stanley Miller conducted another experiment based on a "volcano in a bottle." This one mimicked what scientists now perceive as the most likely environment more than three billion years ago when life apparently began, with the hypothesis being that amino acids could have formed when lightning struck pools of volcanic gas.

This is not remotely like the idea Ben Stein presented in his movie promoting intelligent design. Stein said that the evolutionary position was that lightning strikes a mud puddle and the mud becomes alive, or something along those lines. No, there is no mud puddle here; there are many steps and stages to this process, and some appear to be separate events. Remember also that abiogenesis requires a prior matrix, and first we have to have prebiotic chemistry to work with.

Stanley's second experiment didn't seem to produce the multiplicity of amino acids that he had expected, so that second experiment was not considered successful. However, after Miller's death in 2007, his original samples were rediscovered by one of his former students, a biochemist named Jeffrey Bada. Bada analyzed those samples using modern high-performance liquid chromatography, technology far more advanced and precise than anything available in 1954. Bada's team could detect what would have been invisible half a century ago. They found that Miller's original experiment had actually produced twenty-two amino acids, several of which Miller himself had never seen in his whole career. It turns out that experiment was successful after all.

They repeated the experiment with different gases, according to different hypotheses for the atmosphere of the prebiotic Earth, and all of them produced amino acids. Shortly thereafter, Sidney Fox studied the spontaneous formation of peptides, short chains of amino acid monomers, and he demonstrated that they developed into long strings called polypeptides as they dried, becoming increasingly complex.

Then it was discovered that these could even form in outer space. Fragments of the Murchison meteorite (a 4.5 billion-year-old chondrite that fell to Australia in 1969) were found to contain a great many different types of as-yet unidentified molecules of extreme complexity, in addition to simple organic molecules such as sugars, lipids, and nitrogenous bases. It was also found to contain more than eighty amino acids, some in large amounts. These couldn't be earthly contaminants because they were an equal mix of D and L isomers, whereas living organisms almost always produce only L isomers.

With that established, scientists then had to show how the most important ingredient came about, the extremely complex macromolecule known as DNA. There is another type of genetic molecule called RNA, and activated RNA actually builds DNA. Some viruses don't have DNA, but they do have RNA, and this led many scientists to hypothesize life beginning in an RNA world.

RNA and DNA are both made of many-times repeated components called nucleotides. It took a while, but researchers have shown how RNA nucleotides could have formed naturally in conditions now expected of the prebiotic Earth. In a Wired.com article, biochemist Adam Johnson, a coauthor of Bada's study, notes,

> The amino acid precursors formed in a plume and concentrated along tidal shores. They settled in the water, underwent further reactions there, and as they washed along the shore, became concentrated and underwent further polymerization events.

Following on the above quote from the previous study, we move to the next stage of the sequence. In 2009, organic chemist John Sutherland at the University of Manchester showed how a specific cocktail of relatively simple chemicals became increasingly complex after several cycles of repeated inundation, dehydration, and irradiation—again, according to conditions consistent with current estimates for the prebiotic Earth. Soon these produced what were described as "half-sugar, half-nucleobase molecules." Once the mix became sufficiently complex, after enough repetition, a phosphate was introduced, and the mix "spontaneously" transformed into ribonucleotides—the precursor of nucleic acids and the linking blocks of RNA and DNA. Sutherland remarked that his laboratory conditions were like the "warm little pond" that Darwin speculated might be how life first emerged, provided that the pond "evaporated, got heated, and then it rained and the sun shone."

Interestingly, the pathway from simple chemistry to complex chemistry to basic biology involves multiple mediums, repeatedly heating and cooling, dowsing and drying. Alternately this process could also occur, optimally in deep water underwater geothermal vents, or in volcanic pools near a shoreline.

Other scientists had already shown that dripping solutions of amino acids or RNA nucleotides onto montmorillonite clay produced polymers—nucleotide precursors spontaneously assembled into RNA strands, even without the help of enzymes or ribosomes. Another team of researchers at Harvard showed how RNA could even duplicate itself without the normally necessary enzyme. As the eighth edition of the textbook *Biology*, by Neil A. Campbell and Jane B. Reece, states,

> In 2009, a study demonstrated that one key step, the abiotic synthesis (non-living) of RNA monomers, can occur spontaneously from simple precursor molecules. In addition, by dripping solutions of amino acids or RNA nucleotides onto hot sand, clay or rock, researchers have produced polymers of these molecules. The polymers form spontaneously without the help of enzymes or ribosomes.

An enzyme is a catalytic molecule that reacts to other molecules to perform certain functions. They can metabolize food in digestion or help rapidly replicate macromolecules, or make new copies of RNA. In fact, the enzyme itself is actually made of RNA. When double-stranded RNA is heated, it divides, and one strand contorts into a ribozyme and the other becomes the template to be copied.

The discovery that RNA can actually construct ribozymes capable of catalyzing their own chemical reactions lent support to another of the key hypotheses among the associated stages and processes collectively known as abiogenesis, that of the hypercycle—a sort of autocatalytic information feedback loop that may be contained within a single phospholipid cell wall. Although these wouldn't yet qualify as living, they could reproduce in a very similar manner and would even generate quasi-species, being subject to some of the terms of natural selection. Scientists have even composed proto-cells with some degree of metabolic processes, something similar to this. So it seems like the whole hypothesis has already been substantiated on some levels, but not quite fully yet.

This is just a super simple summary of something that is really crazy complex. There is still the question of how such protocells would grow and divide without all the components we know of in modern cells, which the most primitive cells might not even need. Further, some people want to see the whole collection of processes as a single experiment, where basic chemicals are poured in, and something like bacteria comes out. We're not quite there.

Quite the opposite of spontaneous generation, there are literally dozens of concordant hypotheses within the field of abiogenesis research, almost all of which could be true at the same time. Some have even been partially substantiated, and even those that may not be necessarily relevant to the origin of life on Earth are still supported by compelling evidence. As Albrecht Moritz wrote in his article "The Origin of Life," posted on October 31, 2006 to Talk.Origins,

> The experimental study of the origin of life kick-started with Miller's 'prebiotic soup' experiment (Miller 1953, Miller-Urey experiment) which produced amino acids essential to life. In the following decades, much impressive chemistry on the building blocks of life has been performed, but for a long time many crucial questions had been without experimental answers that might give hope for firm future directions. Among these were the synthesis of nucleotides, polymerization of nucleotides to oligonucleotides, incorporation of a self-copying gene into simple 'cells'

> upon which natural selection could act, and the origin of homochirality of amino acids and sugars. Conceptually, the genesis of the protein translation system posed a fundamental problem. Yet in the last decade significant progress has been made in all those areas, even though details are still sketchy and problems persist on many issues. Proposed reactions in the 'metabolism-first' model, which assumes metabolism, not genes, at the origin of life, have also been rendered more promising by recent findings. Overall it can be said that puzzle pieces are starting to come together in such a way that the scientific assumption of a spontaneous origin of life from non-living [sic] matter finally has achieved plausibility on the level of experimental evidence.

Teams of biochemists around the world are still working out the long, complicated string of chemical combinations that began with simple and already self-replicating polymers and eventually lead to the first metabolic cells capable of maintaining some level of homeostasis, a balanced internal environment. That is the definition of life.

Viruses are not considered to be alive even though they can be killed, because they lack metabolism, which is an independent internally maintained chemical balance. Protobionts (which biochemists propose) would be quasi-biotic cells, a step up from hypercycles and superficially similar to viruses. But whether we're talking about fully living cells or not-quite-yet-living cells, they are both driven by the natural functions of enzymes, chemical reactions, and molecular polarity. If there is any other aspect to life, science has yet to detect it—and if there is a supernatural component to life, science will *never* be able to detect it.

Abiogenesis has a decent amount of evidence behind it, but nowhere near as much as evolution does. So far we still don't know which or how many of the current competing hypotheses posed for the origin of life are the most accurate, but if there's one thing the wisdom of the ages has taught us, it is that simply not yet knowing the real explanation is no reason to go and blame anything on magic. Besides, even if a god did appear and summon the first life into being billions of years ago, there is no question but that life has certainly evolved since then, and is still evolving now.

THE 7TH FOUNDATIONAL FALSEHOOD OF CREATIONISM

"Evolution Is Random"

Ever get the feeling you're being watched? We humans are a very self-conscious sort, so it is not at all unusual that we think others are paying close attention to us. An awareness of that possibility is a necessity when you're living in a wilderness area, where there may be large predators stealthily closing in, so it is easy to see how that feeling could have been secured in our psyche over the ages.

Of course, our psychology is much more complicated than that because we're also a social species. Even when no one is watching, it still matters what anyone who looks at us might see. At any moment, secret admirers may be gazing at us under cover. Rivals may be there too, sizing us up to find our weaknesses, or there may be concealed critics, ready to ridicule anything they notice. Our conduct and appearance are always subject to judgment, even when we think we're alone. Any mistake or embarrassing moment might have been seen by someone and maybe reported to someone else. Whatever we did today that we think is secret might come back to visit us tomorrow, and could change the course of our lives. We need to be aware that there may be eyes spying on us at any time. Just 'cuz you don't see them doesn't mean they can't see you, so it is safer to assume that you're never really alone—ever. Better to conduct yourself as if someone somewhere may be watching, and that it matters to them who you are and what you're doing.

If that sounds paranoid, it shouldn't. I think that's just how normal people normally behave. But there are also those who take it to extreme, who are convinced they're under constant surveillance at all times, and that all their actions—and perhaps even their thoughts—are being noted by someone somehow.

Each of us tends to place ourselves in the center of our world. In fact, we're individually so self-centered that even if we're not the center of attention, we complain that we *should* be, that we deserve more attention than we're getting. We naturally assume that someone potentially could be examining us at any time, or maybe even *all* the time. It is natural for us to feel that way, even when it is irrational.

We're also very good at recognizing patterns and making connections. We're still analytical even when we're irrational; it's just that the analysis may be focused on our primary suspicions and biased against any more logical options. Consequently, some patterns emerge in our minds, appearing as though they were orchestrated, as if what has happened before will happen again—not according to diminishing probabilities in each situation, but because that's what we think is *supposed* to happen.

If one person's string of successes begins to look like luck, we might imagine that person is blessed, that they have advantages being manipulated by outside entities. Similarly, one who endures several successive setbacks might appear cursed. In our minds, it might seem as if one of them is being punished and another rewarded. Why? Well, they must have done something to deserve whatever happened to them. It's the simplest concept of karma: "What goes around comes around," as they say. For people who think this way, there is no such thing as coincidence; everything happens for a reason.

One of my favorite shows as a kid was *The Twilight Zone* with Rod Serling. The pattern I noticed in that show was that everyone tended to get their just desserts; good-natured pleasant and charitable characters always seemed to do all right, because the author of that reality didn't want to torment kindly people and he knew his audience didn't want that either. However, those characters who were greedy, selfish, prejudiced, or cruel were going to face ironic consequences of a sort of poetic justice, as if the universe was concerned with balancing good and evil. Not only does life sometimes seem to be this way, but it actually *should* be this way! Wouldn't it be great if it *was* that way? That's why it feels so good to pretend that it is.

I'm sorry to say I'm as guilty as anyone else for holding this sort of irrational feeling. All my life, it seemed to me that I had established some sort of deal with the fates and that I would never have great wealth or fame because karma charges a price for that. By avoiding extreme benefits, I am spared having to endure tragedies of equal measure, things I couldn't bear or deal with. It didn't bother me that I wasn't rich or famous because I wasn't in prison either, and I wasn't suffering some chronic crippling condition. It didn't matter that I've always been poor, hounded by debt collectors, and that I never had nice things because throughout my youth I was good-looking, popular, and always in excellent health. That seemed to me to be a fair trade. It really did seem as if I had struck a deal with *something*. Things would never be great but they would always be good, because even in the worst of times I would be thankful that it wasn't as bad as I knew it could have been and that happier days were never far away.

These days, I realize that simply having a positive attitude can account for a lot of that. Otherwise I logically understand that things can't really be governed this way, but it still seems that way within the confines of my own imagination. It is an impression I find difficult to shake.

Consequently, the last religious belief I ever gave up was karma. I held onto that one as long as I could because I really wanted to believe there was some innate sense of justice in the world, especially when terrible people get everything they want despite exploiting or victimizing others. Making the situation worse, often the villain somehow feels both victimized and vindicated—because the ends justify the means according to what they see as righteous indignation, and they won't experience guilt or shame and never even realize they've done anything wrong. Instead, they feel they deserve whatever they've stolen, as if it were rightfully theirs to begin with. That's not justice! It's not enough to imagine them getting their punishment posthumously either. Leviticus 26 says we will be punished for our transgressions in *this* life, not the next. That is how it should be, especially once you realize there can't really be an afterlife anyway.

Alas, I have seen too many cold, cruel, and thoughtless people excel at the expense of others while the nice guys finish last. No one wants to mess with the sociopaths, so those without compassion or compunction too often prevail in oppressing honorable innocents who never did anything to deserve what was wrought upon them. Fortunately, I'm not talking about my own experience, but rather what I keep seeing happen to others, so I feel the observation is objective. Being kind and charitable, honest, fair, and noble doesn't mean that you'll be treated in accordance with your standards. It *should* mean that, and if karma were real—if there really were some righteous god or fair judge behind the workings of the world—it *would* mean that. If only there were some invisible Rod Serling writing the screenplays of our lives.

Morpheus:
Do you believe in fate, Neo?

Neo:
No.

Morpheus:
Why not?

Neo:
Because I don't like the idea that I'm not in control of my life.

This scene from *The Matrix* where Morpheus meets Neo is illustrative of my point. Whenever I hear someone say that "everything happens for a reason," I think about all the millions of variables inherent in every effect they imagine to be caused and how they think these were accounted for or orchestrated. By what means were they manipulated? I really don't think believers ever actually think about these things; it's hard to keep believing if you think about it. But *I* think about it, and it seems to me that the only way God could control everything the way they say he does is if he wrote the script. If life is but a stage and we are merely players, then God wrote the play. Believers will probably like the sound of that until they realize that means that we didn't come up with any of our own lines. We can't change the course of our lives, either. We can only do what we were programmed to do—what we are destined to do—by design.

Whenever I hear someone say that God exists "outside of our reality" and "beyond time," I think of that scriptwriting deity, that invisible Rod Serling, who can play back any of his favorite episodes. He's watching everything you do, and he controls everything that ever happens to you.

When you watch an old movie, you exist outside of their reality and their frame of time. You were there before it began, even if the film is older than you are. You remain after it's all over, and you can watch it all over again. You can predict the future because you *know* the future. For you, the future has already happened, and you know that there's nothing the characters can do to change it. Perhaps no scene in any film illustrates that concept better than this one from *The Rocky Horror Picture Show*:

Riff Raff:
Frankenfurter, it's all over!
Your mission is a failure!
Your lifestyle's too extreme!
I'm your new commander.
You now are my prisoner!
We return to Transylvania.
Prepare the transit beam!

Frankie:
Wait, I can explain!

Audience:
IT better be GOOD;
It got you KILLED last week!

Some say it is a contradiction to believe simultaneously in both prophecy and free will, for they can't really coexist. If anything a time traveler does in the past might dramatically alter the present, then anything done in the present could also change the future in ways no prophet or fortune-teller could foresee. We can examine events of the past because the past is set and cannot be changed. The only way we could ever see into the future would be if it too were set and couldn't be changed either.

"Everything is fixed, and you can't change it."

—Jesus, *Jesus Christ Superstar*

This would still be true even if you're a god with a brain like a supercomputer accounting for infinite variables. You can only gauge the future when everyone does what they're expected to do according to their particular physiology and psychology and conditioned reactions to the dynamics of their situation. Once anyone does something unexpected like exercising free will, then everything has to be calculated all over again. It has to be recalculated to account for every other person or additional factor each time that one lone mind uses that power. So if everyone has free will and we all use it repeatedly at random, then the entire network of predictable interactions falls apart, and calculations would be so chaotic that it wouldn't be possible even for a god to keep up with it. Not even God could predict the future anymore, not if we have a say in what our future is.

The apparent incompatibility of divine omniscience and prophecy versus the free will of humanity has lead some religious denominations such as Calvinists to believe in predeterminism or predestination, where everyone's future has already been determined by God according to an unbroken chain of interrelated events, and we humans can do nothing to change that. If there really are no coincidences (as these people believe), then that has to be the case. Other religions have a similar belief—that everything that ever happens and everything we think we've decided to do is controlled by the fates, as if we were merely characters in an old movie where someone already knows how it ends.

> *"I've never seen anything to make me believe*
> *there's one all-powerful force controlling everything.*
> *There's no mystical energy field controls my destiny."*
>
> —Han Solo, *Star Wars, Episode IV*

The irony here of course is that Han Solo didn't realize he was only a character in a movie—in which case, yeah, his destiny is being controlled by the writers, and by the director. Everything that ever happened to him was predetermined by intelligent designers who existed outside of his reality entirely. Even his mistakes, chance encounters, lucky breaks, and what he thought were random accidents were all done deliberately, though not by him. They were each prearranged by forces beyond his control.

Imagine you've misplaced your keys and you're running late for work trying to find them. The house phone rings at a time when you would normally have been gone already. It happens to be a radio station running a promotional contest where you'll win a prize if you can answer one obscure question—but oddly enough, the subject of that question happens to be something you learned all about from one of your friends just yesterday. So you win a dream vacation where you just happen to meet a wonderful person who just happens to fall in love with you and changes the course of the rest of your life. Maybe that was just a lucky break, something that could only happen once in a lifetime to one person in a million. But most theists would readily say that God wanted that to happen, and so he arranged for all these lesser events to lead up to that.

"Call it fate, call it luck, call it karma,
I believe that everything happens for a reason."

—Dr. Peter Venkman, *Ghostbusters*

M. Night Shyamalan's 2002 movie *Signs* has this mantra as its central theme. Shyamalan's character Ray Reddy hits the wife of Mel Gibson's character with his car, which pins her to a tree. She takes a long time to die, and as she does, her mind drifts into the twilight zone wherein she seems to be delirious. Gibson's character was a Catholic priest who lost his faith because such tragedies demonstrate that life has no meaning.

His surviving family all have quirks that prove to be important later. His brother would have been a baseball player, but he couldn't control his swing and always hit too hard. His son has a pretty serious case of asthma. His daughter has this weird habit of only taking one sip of water and leaving the rest of the glass on the table—and on the shelf, and the TV, everywhere to the point that there are always full glasses of water all over the house.

The film reveals that extraterrestrial invaders spray toxic gas from their wrists, but that doesn't affect a little boy who can't breathe during an asthma attack brought on by fear. Liquid water is somehow poisonous to the aliens, so they not only invaded the wrong planet, but the wrong house, because these inhabitants have dozens of full glasses to splash onto the intruders. (Strange that they have interstellar space ships but no other technologies like weapons or protective clothing.) Most absurd of all to me was that the guy who always hits too hard with a bat doesn't know what to do with it until Gibson's character remembers his dying wife's delirium and interprets her last words as a prophecy: "swing away." Well, duh.

The point of the movie is that Gibson's character realizes that everything happens for a reason. In the end, we see that his faith in God is renewed and he is a Catholic priest once again. Of course. Makes perfect sense if you've just gotta rationalize and restore your precious faith somehow.

I've seen a lot of really stupid movies in my life, and this was one of them. Amusingly, *Signs* follows the idea that crop circles are evidence of extraterrestrial visitors, even though many of the actual perpetrators of such patterns in the real world had already come forward to show exactly how they pulled them off. Well before production on the movie began, it was already common knowledge that crop circles are man-made hoaxes. But I guess if this movie made sense to you, then you would still believe in the alien origins of crop circles too.

So they'll tell you that if your brother struck out in baseball, and your daughter has an obsessive compulsive disorder, and your son suffers from a chronic and potentially deadly affliction, then that's all part of God's unified plan to protect you from hydrophobic aliens—a courtesy he curiously did not extend to any of your neighbors. If young children lose their mother in a tragic accident that leaves others traumatized for life, then God wanted that to happen too. No matter what happens or how it happens, God did it, and only God knows why.

Another film to use this mantra, and which I found equally unenjoyable, was *I Am Legend*. Will Smith plays a soldier who believes himself to be the last surviving human after a proposed cure for cancer had the accidental side effect of turning everyone into mindless cannibalistic vampire-werewolf-zombie-mutants. Eventually, Smith's character meets a woman and her child—the only people he's seen in months who are still human—only they think they're on a mission from God. Smith plays an atheist in this movie, and he tries to make the logical argument that this could not be God's plan if "*Everybody died! EVERYBODY died!*" He abruptly changes his mind in the end without explanation, and for no reason that was ever apparent to me. I guess it doesn't matter how bewilderingly inane the situation is; you can always smile like a dead-eyed mannequin and say it was all part of God's plan.

"It was strange the way it happened.
Suddenly you get a break!
All of the pieces seem to fit into place.
What a sucker you've been, what a fool!

The answer was there all the time.
It took a small accident to make it happen, an ACCIDENT!"

—Frankenfurter, *The Rocky Horror Picture Show*

"There are no accidents."

—Tom Frost, *Naked Lunch*

"Because 'accident' implies there's nobody to blame."

—Nicholas Angel, *Hot Fuzz*

This is an important point: if everything without exception happens for a reason, including every incident we would otherwise have to attribute to random chance, then despite the absence of intent on the part of anyone involved, even accidents must always happen on purpose. That's why fortune-tellers can toss tea leaves or chicken bones and read the "meaning" of how they landed. Anyone else should agree that was totally random. The same goes for Tarot readers laying out a deck of cards that were shuffled by the patsy who paid for the reading.

Look at the state lottery for example, where you're supposed to pick which six numbers will be selected by a blind machine. Everything about that process and the mechanisms devised for it are designed to be as random as anything possibly could be, so that we can be sure that anyone who wins does so purely by chance. Yet many of the people who have won believe that this random event of pure chance was directly guided by God just for that purpose. However, some of the same people who believe that would not accept that God could also have guided evolution.

"What matters is that whatever happened,
happened for a reason."

—Agent Smith, *The Matrix Reloaded*

"And it would be a leap of faith
far beyond belief in Jesus or Buddha or Allah
to think it just 'accidentally' got that way."

—Bill O'Reilly, *The O'Reilly Factor*

Yet such orchestration is impossible in our reality. There is no mechanism imaginable that could guide *everything*. We can't just assume that anything is possible either, because there is a lot that is not. In trigonometry, can $\cos(\Theta) = 3$? Not if you need to provide a *rational* answer, and there is no rational explanation possible in this case. That's one of the many reasons why God and science are wholly separate topics that do not overlap.

If they want to argue that god could still do that, then he could have directed evolution too! Yet creationists *will* say how God could *not* have done it, even though God can supposedly do whatever he wants. Creationists differ from mainstream Christians in that they impose limits on their God, insisting that the one element of nature that God cannot control is evolution. Their black-or-white perspective only permits one of two extremes: either God cannot use natural processes creatively, or naturally creative systems cannot be governed by God, and thus somehow count as evidence *against* God.

Conversely, those who don't understand how complex orderly systems can be derived by natural means tend to conjure creators to compensate. For example, twentieth-century astrophysicist Fred Hoyle wasn't a Christian, and he wasn't a creationist in the typical sense: he didn't believe in a particular god, and he didn't hold his positions in defense of doctrine. He didn't believe in the Bible either, but he did argue for an intelligent designer behind the universe, and he did propose a type of creation science opposed to mainstream concepts regarding cosmology and abiogenesis.

Hoyle was popularly credited with having invented the name "Big Bang" to explain the inflation of the cosmos from a quantum singularity, although that was actually a Catholic priest named Georges Lemaître. Hoyle didn't believe any of that himself; he used that term as a pejorative. Instead, he argued that the universe should be in a perpetual "steady-state" maintained by a continuous sequence of miniature creation events to produce new hydrogen atoms out of nothing. In addition, he figured that the chance of obtaining the required set of enzymes for even the simplest living cell might be one in $10^{40,000}$. His assumptions (based on erroneous notions, a basic misunderstanding, and a lack of more recent data) led him to argue that "[a] common sense interpretation of the facts suggests that a super intellect has monkeyed with physics, as well as with chemistry and biology, and that there are no blind forces worth speaking about in nature. The numbers one calculates from the facts seem to me so overwhelming as to put this conclusion almost beyond question." So this man so often painted as an atheist confessed his belief in an immortal intellect with inexplicable powers. Hoyle believed in a god.

More importantly, to demonstrate just how poorly he understood evolution and abiogenesis, he also wrote the following criticism of both processes, which he blended into one, the way creationists typically do.

> A junkyard contains all the bits and pieces of a Boeing 747, dismembered and in disarray. A whirlwind happens to blow through the yard. What is the chance that after its passage a fully assembled 747, ready to fly, will be found standing there? So small as to be negligible, even if a tornado were to blow through enough junkyards to fill a whole universe.

Creationists seized on this of course. Here they had a respected professional scientist, whom they describe as a "humanist" yet who also rejected the Big Bang—who confused evolution with abiogenesis and rejected both, and who instead pleaded for creation events and a supernatural intelligent designer. Consequently, the fallacious absurdity that Hoyle described above has been parroted many ways over the years. Sometimes they'll trade the tornado for some other destructive randomizer, or they'll swap the flight-ready 747 for a shiny new red Lamborghini. There are many different variants of this gross misunderstanding being mass-communicated all over our public media. Consider, for example, this one from Creationism.org,

> The evolution of the automobile has been due to intelligent design input, advanced engineering skills and management oversight. If one just throws a bunch of parts together or puts all the car parts into a giant tumbler along with four quarts of oil and 10 gallons of gas and begins agitating the lot for eight months, the end result will not be anything resembling a car.

Those opposed to evolution/abiogenesis misunderstand it and mischaracterize it, as if "evolutionists" believe that simple elements just accidentally fell together into what just happened to be extremely complex working configurations. Sometimes they'll admit that time is involved, but they never understand how or why because they don't comprehend how these configurations don't all have to be assembled at once, or that the emergence of complexity is achieved through incidental incremental steps. Nor do they permit that this is even possible.

> Random mutations and natural selection cannot in principle explain the origin of any system which requires many integrated parts to perform a function. That is because evolution is blind and has no foresight. It must work by slight successive modifications, where each step provides an advantage significant enough so that it can be selected for. The ATP motor is built with interdependent parts that are finely engineered and which are much more efficient than anything we humans could ever create. Evolution CAN'T CREATE THESE MACHINES because these machines can't be built in "slight successive modifications," or one mutation at a time.

The above contention was posted and directed toward me in an online discussion forum by a YouTuber named OnceForgivenNowFree. That same person also argued against the straw man that "a bomb in a junkyard can create a spaceship."

The challenge he issued was for me to show a biochemical pathway to building the first ATP motor. This is a highly conserved molecular motor found in various forms throughout the biosphere in cells from every taxonomic kingdom. In real life, they look like wibbly-wobbly molecular masses fluctuating according to thermal noise, but in diagrams, they're simplified to look like brand-name rotary motors on sale at the home improvement depot, so they're another favorite of intelligent design proponents.

My opponent actively avoided every topic that he knew I could explain very well, because he doesn't want to understand; he wants to believe, and he knows I can destroy his illusion, so he had to find something he was sure I wouldn't be able to explain. He issued this specific challenge to me only because he had read religious websites that assured him that science *couldn't* explain these structures—that they were indicative of divine design, because they could not be constructed "accidentally" nor with any succession of one mutation or slight modification at a time.

I replied first with a correction: each step doesn't need to provide an advantage significant enough to be selected for; it only needed *not* to be so detrimental as to be selected against. Then I cited two different studies from German and American laboratories, each independently documenting the evolution of ATP synthases and ATPases. One of them (A. Mulkidjanian, Osnabrück) included an illustrated step-by-step progression. The other (S. Cross, Upstate Med.U.) described in detail how there was a single point-mutation at one stage followed by a loss of function in this other gene, then the subsequent duplication/fusion of a different gene followed by another loss of function in this other gene. Thus I showed how science *could* explain the sequence of how the structure was built piece by piece, one mutation at a time, and how it changed functions at certain stages but remained functional at every stage.

My opponent didn't know how hypotheses are employed in science, so he dismissed these peer-reviewed explanations as mere stories. However, these explanations also included a phylogenetic orthologue and referenced analyses of stoichiometry to confirm how the described steps were evident both in the structure and in the genome. Of course, none of that matters when you're forbidden to change your mind. Then one must use any excuse necessary, even if that means distorting your opponent's position to the point that it doesn't make sense anymore.

For example, review this conversation from the documentary *The Root of All Evil* between Pastor Ted Haggard of New Life Church and Professor Richard Dawkins, the famous zoologist from Oxford University.

Haggard:
Sometimes it's hard for a human being to study the ear or study the eye and think that happened by accident.

Dawkins:
I beg your pardon—did you say "by accident"?

Haggard:
Yeah.

Dawkins:
What do you mean, "by accident"?

Haggard:
That the eye just formed itself somehow.

Dawkins:
Who says it did?

Haggard:
Well, some evolutionists say it did.

Dawkins:
Not a single one I've ever met!

Haggard:
Really?

Dawkins:
Really!

In the mid-1990s, a graphic artist with a master of science degree from MIT named Karl Sims inspired a line of computer programs designed to evolve virtual creatures. His and subsequent programs encoded various evolutionary rules into a population-based metaheuristic optimization algorithm for a simulated application of mutations against natural selection. This worked with an artificial neural network of virtual sensors and virtual muscles connecting the cuboid "limbs" of his virtual creatures. The whole system was designed to solve functional problems in a 3D world of simulated physics, gravity, and friction. There was even an objective to access and control a particular cube, generating competition. Different lines of descendant designs were optimized automatically, purely through preferential selection of the more functional variants according to the rules of population genetics and without any prior assumptions of "fitness." He also eliminated any influence of bias or desire and let the computers calculate this purely on their own, only according to the rules imposed on evolution. Even without an end goal, once simple rules are laid out, simulated life forms will effectively design themselves through an unconscious practice of trial and error over several generations. They'll even refine and improve those designs literally without trying, because intent was deliberately removed from the equation.

Even when we consciously try to make evolution into a random accident, it still isn't one; instead, it's an autonomous process, directing itself—however blindly—to continue and eventually improve however it will. Consequently, creationists may say evolution is random, but "evolutionists" do *not* say that regardless of whether they believe in a god.

Evolution does depend on mutations, and these do appear to be random despite an apparent regularity, but each cumulative mutation may become significant factors for that organism once pitted against the dynamics of the environment in which they are introduced. Remember that they're only slight variations; most will be trivial and inconsequential. Some can be problematic and will thus be eliminated quickly, but a few may be improvements depending on the situation, and those will usually be optimized, having statistically preferential success. Thus, natural selection isn't random; it's deterministic.

Many creationists will even admit this. And as some computer models have already shown, natural selection can actually even exceed the skills of human designers. In fact, natural selection can be so deterministic that it often leads to innovations which some perceive as evidence of intelligent design, and which even rationalists describe as though modified for intended benefit.

Whether it is deliberately guided or not, there is definitely a system of design. But there doesn't actually have to be any apparent intent or intelligence involved, nor any goal to be achieved either, because while our normal intuition might be to imagine one governing body issuing authority from the top down, a new field of science called emergent complexity uses computers to trace numerous patterns in nature that are all "emergent" from the bottom up. These are controlled or constructed by an intricate interrelated array of the lowest components working together in unison, each according to a few relatively simple rules.

"Their strength and their speed are still based in a world that is built on rules."

—Morpheus, *The Matrix*

"Infinite complexity can be described by simple rules."

—Mathematician Benoit Mandelbrot

Both of the simulations generate patterns of increasing complexity out of basic components acting in accordance to a few simple rules, just like in nature, and what they illustrate acts a lot like nature. It can be expressed either in the fluid patterns of motion like that of a flock of birds wheeling and turning in unison, or in random alterations over generations of algorithmic optimization of simple modifications being incidentally designed.

Another computer simulation deals with random mutations occurring in different sections of a population. These are neutral and trivial, inviting no selection either way. Over thousands of generations, the system shows these unselected mutations "drifting" throughout that population, such that each of the once-identical groups are now not only distinct, but remarkably different species—just like we see in nature.

In either case, the universe does not seem to be dependent on any authoritarian lawgiver, but rather on integral components acting on their own according to the rules imposed by chemistry and physics and population mechanics. If any god exists, he wouldn't need to design each individual organism if he could instead design a whole system that would automatically generate all these complex components for him. Which method implies greater wisdom?

Using computers to study emergent complexity reveals some ways in which order *can* come from disorder, and how chaos can also achieve balance; illustrating how even the origin of life is as much chemical as it is mathematic.

"Your life is the sum of the remainder of an unbalanced equation inherent to the programming of the matrix."

—The Architect, *The Matrix Reloaded*

Regardless of what field or subject we're talking about, anything that is regularly analyzed or revised naturally tends to become more complex as those processes wear on, and we know that environmental pressures on population genetics are no exception. This is especially true when we're talking about replicating populations where subtle changes may be introduced in any lineage at any time, and every lineage over enough time. Even before computers existed, we already knew that natural selection can and often will produce results that look like trial-and-error experiments, including elements of seemingly intentional fine tuning. That is what we see throughout nature, but we should never see that if everything was designed and coordinated by an infallible omniscient being who knows already what he's trying to make in the end. For all the implications of apparent design, there is never any indication of any intended goal or final product, nor any hint of infallibility on the part of the designer.

The problem creationists have is that once one accepts that evolution happens at all, then just comparing the phylogenetics network to the arms race of predator and prey could easily remove any idea of an intended goal of design from what we see happening instead. Theistic evolutionists—even the ones who hold "traditional" religious beliefs—obviously don't interpret it this way. When I still believed in God, I didn't see it that way either, but that was because my god couldn't see the future; he didn't predetermine anything, and he employed natural systems and processes working on their own. That's the only way to explain all the obvious errors, inexcusable waste, and purposeless features with no possible function.

I was told that God deliberately formed every facet of every animal's body in order to serve a particular purpose. Some vestigial traits counter this idea. Now, most vestigial features that I'm aware of actually do have a function; it may be a greatly diminished capacity of its original function, like our diminutive human canines or pathetic fingernails, or it might have a completely new function, like the way penguins use their now-flightless wings as flippers. But there are some vestiges that serve no possible purpose whatsoever—the dewclaws of a dog or the arms of an emu, for example.

Emus are large flightless ratite birds from Australia. They're very primitive-looking paleognaths, as tall as a man. Anyone trying to argue that birds did not evolve from dinosaurs has to ignore these and other ratites because they retain so many dinosaur features. Emus do not have wings—they have tiny arms like a tyrannosaurus. Their arms are so small that they're hidden by their weird hairlike feathers. Unlike "modern" neognath birds, ratites still have claws on their wings/arms. An emu's arm is a spindly scrawny stem of a thing, with only one finger that ends in a claw. The claw is a perfectly bladed crescent-shaped talon just like it should be, but it is attached to an arm that has no musculature at all; the only way an emu's arms move is by gently expanding and retracting with its ribs as it breathes because the arms are completely atrophied, so that claw is inarguably without any possible function. If God "created" the emu rather than allowing it to evolve, then he would not have given it that claw without any muscles to use it. That would have been a mistake; and it's certainly far from the only one, even just in birds. In fact, so many errors of so many types are known that even if there was an unnatural architect using miraculous means instead of natural ones, then it seems that entity must either be blind and barely competent or there are whole teams of designers working on separate lines competing against each other.

Natural selection even mimics the experiments of human designers when new technologies emerge. For example, when humans first achieved powered flight, there were myriad marvelously imaginative contraptions all at once collectively trying to set the standard for what airplanes should be. Eventually, planes followed a more standardized pattern as many of the fancier designs were discontinued and more functional tried-and-true contrivances remained. Significant improvements occasionally appear, but there are no more of the wild and weird designs like the pioneer planes built when aviation was new and less understood.

The same sort of thing occurred when life moved up to the multicellular level in the late Vendian Period some 600 million years ago, leading into the Cambrian explosion. The first multicellular animals had no skeletons or organs or sensory systems of any kind; once primitive drafts of these began to develop over the course of the next 160 million years, the oceans went from being basically a sea of sponges and plankton to a relatively sudden proliferation of new forms even more dramatic than when they began to move onto land some 70 million years or so later. Many of those early experiments were unsuccessful because there were more phyla to emerge in that era than we have left today. Some of the earliest forms were so bizarre that we can't even make sense of them because they're so alien and incomparable to anything still around. Then—as with early designs of airplanes and automobiles—once other possibilities were explored, the more functional lines continued to diversify while less practical derivations thinned into extinction, and a few of the basic designs that worked well enough are still out there now.

Another common argument against evolution refers to the Cambrian explosion. The Cambrian is the first period of the Paleozoic era stretching out about 50 million years, beginning roughly 541 million years ago. This is where the fossil record shows a sudden proliferation of a wide variety of multicellular organisms where there was previously no sign of life at all. At least that's how it seemed in Darwin's day. The pseudoscientists say that every taxonomic phyla we have today appeared all at once, without any precursors, and that those same phyla haven't evolved since then. I guess the argument is supposed to leave one with the impression that geologists don't understand the geologic column, and that it reveals to the faithful the moment in the fossil record when God created life on Earth.

However, we now have evidence of earlier organisms indicated in trace fossils and in the genome, as well as complete fossils of Precambrian precursors of later orders. There were at least a few multicellular animals in the previous period, including Aysheaia, grandfather of all Arthropoda; Halkieria, a 525 million-year-old proto-brachiopod; Wiwaxia, an apparent predecessor of mollusca; and *Spriggina flindersi*, an almost-trilobite from the Vendian era. Spriggina's resemblance to trilobites may be superficial. Several Spriggina fossils are known, but none are detailed enough to be certain. It has been classified as an arthropod ancestor of trilobites, but could instead be an annelid worm or even Proarticulata, a whole phylum now entirely extinct. How's that for intelligent design?

Yes every animal phyla we have today appeared in that initial proliferation, but that doesn't account for every "kind" of animal, even by creationist standards. Phylum is a very large taxonomic category; one single phyla, Chordata, includes all by itself everything most of these folks would recognize as animals—all fish, amphibians, reptiles, birds, and mammals belong in this set. Yet the only organisms representing this entire group at the dawn of the Cambrian were tiny sort-of proto-fish; they were little more than specialized swimming worms with a noto-cord, the very beginnings of a spinal cord. Yet they didn't have spines, because they didn't have bones yet. That's how primitive they were. Not only were there no mammals, birds, or other vertebrates at that time, there weren't even any proper fish yet, much less any insects or recognizable plants.

Most of the forms that existed at the dawn of the Paleozoic era were extinct by the end, but there are a few surviving shapes from that age, the most advanced of these being shrimp. They're not the same species as we have today, but they're still obviously shrimp. Why didn't they change completely? Well, the first reason is that they don't have to. The second reason is that they *did* change; they just diversified into different species of shrimp, as well as lobsters, crabs, and so on, and then new species of those too.

Remember, we're not talking about a process that follows a direction or seeks ascension. Changes can occur in every direction at once, and any time a new progressive form diverges from its kin, the other conservative relatives are still there. Some will change significantly, some hardly at all, and all of them will still have relatives from the old school which won't take that same path, nor do they have to.

In any environmental niche, there is a perfect shape, one especially efficient form which, once obtained, need not be substantially modified until the environment changes or the animal moves on to new circumstances. For example, the shape and job of the crocodile has been used by several extinct predecessors including a couple different lines of giant Permian amphibians, at least one type of lunged fish, Triassic phytosaurs, one cretaceous dinosaur (Spinosaurus), and even the precursors of whales when they still had legs. Each of these actually looked a lot like crocodiles, and apparently acted the part. The iconic shape of a shark has also been employed by bony fish as well as ichthyosaurs, dolphins, and even some mosasaurs. Each time that convergent shape was determined by the conditions streamlining for efficiency at speed—there was even a Mesozoic crocodile adapting that shape after having moved on to the open ocean. The role of "lion" has been played by everything from fossil marsupials and Miocene creodonts to the lion-sized gorgonopsids, mammallike reptiles who lived before the dinosaurs. So some styles can be preferred by conditions that allow them to become classic motifs. These even encourage convergence to the point that each of these lines following similar pursuits begin to look and act like the others in that niche. Then it's just a matter of which ones can avoid getting killed off in the next catastrophic extinction event.

"Life finds a way."

—Dr. Ian Malcolm, *Jurassic Park*

"I suddenly had this feeling that everything was connected."

—Detective Finch, *V for Vendetta*

Anything or everything could seem like the result of an undirected and incidental string of random chance accidents, right? Or are they? Theistic evolutionists understand that many drugs and medicines are being discovered or derived according to our growing understanding of the effects of evolutionary principles in biology. That means that throughout the entirety of human history, until these medicines were just discovered, our ancestors suffered and died without them, because God didn't provide them, and he didn't show us how to make them. Yet in any life-saving operation—even when theistic believers know that the natural explanation is the right one—they still credit God over the skilled surgeons using state-of-the-art equipment and the advantages of cutting-edge medical breakthroughs. It doesn't matter how seemingly disconnected or coincidental any series of occurrences may be; those who believe in destiny will still suggest otherwise.

"We have not come here by chance.
I do not believe in chance."

—Morpheus, *The Matrix*

In this chapter, I've quoted a few movies that I remember referring to this general theme. Foremost among them are *The Matrix* and *Signs* because I think these two illustrate what the real problem is. Granted, both movies are actually on the same side of that argument, trying to promote faith-based beliefs. When have you ever seen a movie or heard a song promoting rational skepticism instead? To borrow from the classic Journey song, it's always "Don't Stop Believin'." But there is one scene in *The Matrix* that I have to interpret differently.

When our two heroes first meet, Morpheus offers Neo "the truth." He offers it as a choice between two pills: Neo could take the blue pill and believe whatever he wants to believe, which in this world was a comfortable illusion, but the red pill means awakening to a disorienting reality. Neo takes the red pill, choosing to see the world for what it is, no matter how harsh. Ironically, however, even this situation ends up promoting faith, because the movie paints reality as the illusion. It promotes the idea of mind over matter—that if you believe "hard enough," you can alter reality with the power of pure will. That is the core of faith-based assertions, because some people prefer fantasy over reality. Reality is such a bitter pill for some people that we later hear one of the villains in the movie say, "Why, oh why didn't I take the blue pill?" That is the problem here! Some people would rather believe what they do than know what is really true.

That is essentially what Mel Gibson's character is like in *Signs*. At one point, he explains that people break down into two groups, believers and unbelievers. When unbelievers get a lucky break, they think that's all it is—mere coincidence. But believers see it as more than that; they take that lucky break as a "sign" that someone is watching over them, because believers do not believe in coincidence.

"I do not see coincidence; I see providence."

—Morpheus, *The Matrix Reloaded*

"There are no coincidences, Delia. only the illusion of coincidence."

—V, *V for Vendetta*

In *Signs*, Mel Gibson's character goes on to explain that when unbelievers are presented with a potentially threatening situation, they get scared when they realize that they're on their own, whereas believers facing the same situation imagine it to be a miracle and take it as sufficient reason to assume that some god will be there to protect them.

To me, one group is brave enough to take reality as it is, while the other group isn't—and they don't; they're pretending. They need that imaginary security blanket around their minds because they're the ones who are scared. They're simply projecting their own fear onto those who aren't really all that afraid.

"Denial is the most predictable of all human responses."

—The Architect, *The Matrix Reloaded*

Many Christians would say that evolution is one of God's creative methods, but creationists reject that possibility outright, because the issue for them is not whether their God is true, but whether their dogma is true. It can't be in any case; even if current concepts of evolution were proven wrong tomorrow, biblical creationism still couldn't be true either because it has already been disproved many times in many ways, and the various fables in it collapse on their own lack of merit. Those stories once disproved cannot be resurrected, revived, or restored. But of course, literalist believers can never admit that.

Many people—not just creationists, but most people in fact—feel that if reality is not manipulated in mysterious ways for our benefit, then life is without purpose. Since creationists think evolution most directly disproves the purpose they choose to believe in, then they say that evolution must be without design by design simply because it is natural. Of course, if there is such a thing as supernatural providence, then all these seemingly undirected evolutionary advances were obviously destined to happen.

"I believe it is our fate to be here. It is our destiny."

—Morpheus, *The Matrix*

THE 8TH FOUNDATIONAL FALSEHOOD OF CREATIONISM

"There Are No Beneficial Mutations"

Religion is big business, especially in the United States. A history of rural revivals has made forms of religious extremism like fundamentalism and creationism alarmingly popular. Every English-speaking country has at least one organization dedicated to producing and publishing creationist propaganda, whether as online articles, books, videos, or magazines. But of all of them, the highest grossing marketer of *myth*information is Answers in Genesis (AiG), based in Kentucky. Each of these other organizations is either connected with AiG or uses it as a resource.

AiG reportedly brings in more than $20 million in revenue annually, and continues to expand. Since 2007 the organization has boasted a $27 million animatronic Creation Museum dedicated to the idea that dinosaurs and people lived together, just like in *The Flintstones*. In July 2016, it finally launched a long-planned Noah's Ark theme park estimated to have cost over $100 million, with the vast majority of that coming from private investors. I think it terribly funny that even with hundreds of skilled construction workers using modern industrial hydraulic equipment, they couldn't do even in a few years what mythology says one 600-year-old man did with three untrained laborers, with no trucks or transportation systems, using only primitive tools from the Stone Age.

AiG has hundreds of regular employees, including teams of professional scientists with legitimate degrees. These people are not hired to do science, but to undo science in favor of promoting belief in religious dogma. Those interested in using their degrees to push this propaganda must first prove their compliance with a list of Bible-believing references, including at least one current pastor. They must sign the organization's statement of faith, indicating adherence to both the religious and political beliefs of its six-figure-salary CEO, Ken Ham. Otherwise, all that matters is that they undermine science and secularism to promote the extreme views of young-earth science denialists. What they know about science is irrelevant, which is why most of the "famous" creationists with actual doctorate degrees are dentists.

One thing mass-marketing organizations like this have achieved is that they have effectively fooled the undereducated masses to a measurable extent. In 2006, there was an international poll of thirty-four countries showing various levels of the public's acceptance of evolution. Apart from the United States and Japan, all of the countries polled were in Europe, and the majority of these countries also held a majority of respondents who reported that evolution was "definitely true." The United States was one of only a half-dozen countries where this was not the case; only 14% of Americans confidently accepted evolution, while a third rejected evolution with equal conviction. In such cases, acceptance is typically correlated with education. Other polls show that one is more likely to accept evolution the more educated one is, while rejection is typically correlated with the importance of religious belief. There are polls showing that older Americans are more religious and more likely to reject science. Otherwise the notable change in statistics was that the number of Americans who were "uncertain" about evolution rose from 7% to 21% in the past couple decades—this despite all our arrogance as an educated nation in the age of enlightenment. This rise is almost certainly due to creationist propaganda mills like Answers in Genesis, the Institute for Creation Research, the Discovery Institute, and several similar organizations.

The United States would have been at the very bottom of this list except for one other country, Turkey. Why was Turkey the only country in the poll to reject evolution even more than the United States? Reasons are varied, but just as we have Answers in Genesis, the most prolific purveyor of propaganda in all of Christendom, Turkey hosts the headquarters of the Islamic equivalent. Creationist author and video producer Adnan Oktar, also known as Harun Yahya, offers almost identical methods and merchandise marketed the same way. The only thing he doesn't have is an amusement park. He also doesn't have any original ideas; all the arguments posed by his organization are exactly like those of American creationists, just without the bit about being saved by Christ. Instead, they attribute everything to Allah.

One example of Harun Yahya's marketed diatribes is a video titled "The Collapse of Evolution 14—Imaginary Mechanisms of Evolution 2—Mutations." It features a young woman wearing a lab coat to disguise her incompetence of the subject. She begins:

> Mutations are genetic disorders that occur in the DNA in a random and unconscious manner. Like all accidents, they cause harm and destruction. . . . The changes affected by mutations can only be like those experienced by people in Hiroshima, Nagasaki, Chernobyl; that is death, disabilities, and the freaks of nature. Evolutionists claim these distortions cause organisms to evolve. However, scientific findings reject this claim because all observable occasion of mutations cause only harm to living things.

Notice you have all the keywords focusing on the godlessness of any evolutionary idea. It has to be random, unconscious accidents deteriorating what they imagine to be an originally perfect creation. You're supposed to be shocked and repulsed, so the images flashed onscreen include people with too many digits, or missing digits, or a baby with a hair lip, or a cow with a partially absorbed twin (which isn't even a result of mutations to begin with). You're expected to associate these birth defects with evolution and be sick at the thought of it.

In addition to this emotional manipulation, we also have a deliberate misrepresentation of science by saying that "scientific findings reject this claim." A quick search of peer-reviewed literature will immediately prove otherwise.

Creationists insist that mutations are very rare and are usually if not always harmful, but the fact is that the vast majority of mutations are completely neutral. They'd have to be, because according to the National Center for Biotechnology Information, there is an overall average of 128 mutations per human zygote! So apparently, in creationist terms "very rare" means "more than a hundred per person right from the point of conception," because those are just the mutations we start out with. Our cells will mutate again perhaps thirty more times over the course of our lives, and some of these subsequent mutations can be passed on to our children too, usually with no more effect than those we recognize as family traits.

However, for multiple reasons, one cannot study the evolution of humans in real time—certainly not in a laboratory environment. Gauging mutation rates or evaluating the percentage of advantageous, detrimental, or neutral mutations usually come from studying fruit flies or other rapidly reproducing organisms with short life cycles. This is a matter of necessity when any scientist tries to study evolution over many generations. For reasons that should be obvious, this is easier when we restrict our analysis to single-celled organisms, so our most profound research in this area usually comes from studies of bacteria.

A favorite example of beneficial mutations is the "nylon-eating bug." Nylon is an artificial polymer invented by Dupont chemical company in 1935. Nothing like it had ever existed before; not even the linkage of its subunits is found in nature. It was completely alien to this biosphere, so there was at first no organism capable of using nylon as a food source. But after forty years of chemical factories dumping nylon into the ecosystem, one or two strains of bacteria were discovered to have developed the ability to hydrolyze the new substance. This completely new function was identified as the result of a frame-shift mutation and serves as an excellent example of positive selective pressure.

Of course, spin doctors were hired by Answers in Genesis to try and find some excuse for how it wasn't really a mutation, or how it wasn't natural selection, or that it wasn't new "information," or that it wasn't "random chance," because they're not allowed to admit to any of it. They *had* to find some way to discount and dismiss this example because that's what apologists do: make up excuses to justify an *a priori* position or rationalize away all arguments and evidence against it. But in this case, there was no excuse that was good enough. All they could manage were weak criticisms that would only sound convincing to the laity.

AiG complained that the mutation doesn't count because it was found only on plasmids (nonchromosomal DNA), as if that meant that it couldn't really do what they were already proved to be doing. Then they complained that the nylon genes lacked stop-codons, but that's not really a problem. So they tried to minimize it by saying that all the transposable elements were identical, but they weren't. Not only did that not matter either, but it's blatantly untrue; some were duplicated and others inverted, so they tried to mischaracterize the scientists, saying they expressed consternation when they actually expressed excitement. The best AiG could do was to suggest that maybe the information was already there and always had been? Dave Thomas, president of the New Mexico Academy of Science, answered such claims thusly:

> Scientists have studied both the original (pre-mutation) plasmid and the novel (post-mutation) plasmid, in great detail. It turns out that the novel plasmid's mutated DNA for production of nylonase is *almost* identical to a non-coding repetitive DNA sequence on the original plasmid; the difference is the single nucleotide that triggered the Frame Shift. This mutation did not exist 60 years ago. If this gene was always there, whether in a plasmid or not, we can reasonably wonder why a bacteria would have a gene for hydrolyzing an artificial polymer that did not exist until just a few decades ago; and why, in the absence of such a substrate, was the gene not mutated to uselessness over the millennia?

Ken Ham said that an eyewitness is always better than evidence and math. Good thing he's not a judge! Doubtless he says this is because the evidence is never on his side and his claims never add up, which is why one of the anti-evolution arguments he apparently invented is the imaginary distinction between "historical" science and "observational" science. Usually, his staff of professional apologists would complain that even if evidence proves that a mutation caused this, it doesn't *really* prove it because no one was there to see it happen, because you can't duplicate evolution in the lab. But on this occasion, they did! Under controlled conditions, this exact adaptation was directly observed to occur again! But AiG's creation "scientists" aren't allowed to admit any such error. They can't admit any confirmation that blind and unguided processes could ever produce new and beneficial functions. So they tried to make it sound suspicious by complaining that it happened too fast, that it took only nine days to duplicate in the lab. (The whole process actually took three months, and under the specific conditions applied, that could be considered slow.) Regardless, Answers in Genesis failed to dispute the nylon-bug, and of course they're not going to admit that either. As we've already seen, they're forbidden to.

It's hard to find one rigid set of numbers from any laboratory for a constant rate of how many mutations are beneficial versus those that are detrimental, because they're situational and determined by variable environmental conditions. Still, there is a general consensus that they're nearly equal with deleterious mutations being slightly more common and more often profound. But there are plenty of cases where a definite advantage has been identified and positively linked to a specific mutation. An August 10, 2007 report in *Science* titled "Adaptive Mutations in Bacteria: High Rate and Small Effects" writes,

"Our estimate implies that 1 in 150 newly arising mutations is beneficial."

Some textbook examples of such adaptive mutations can be found simply by searching "Examples of Beneficial Mutations and Natural Selection" on Google. They include:

- Adaptation to high and low temperatures by E. coli
- Adaptation to growth in the dark by Chlamydomonas
- Selection for large size in Chlamydomomas
- Adaptation to a low-phosphate chemostat environment by a clonal line of yeast
- Evidence of genetic divergence and beneficial mutations in bacteria after 10,000 generations
- Adaptation of yeast to a glucose limited environment via gene duplications and natural selection
- Molecular evidence for an ancient duplication of the entire yeast genome
- Evolution of a new enzymatic function by recombination within a gene
- Changes in the substrate specificities of an enzyme during directed evolution of new functions
- 12% (3 out of 26) random mutations in a strain of bacteria improved fitness in a particular environment.

These are significant if you're sincerely interested in understanding the subject. But if you're afraid to understand it, and are more concerned with dismissing even the possibility of such mutations, then these examples aren't going to matter to you.

This list continued, showing among other things the evolution of new metabolic pathways. But since they're still only talking about bacteria, who cares? There was even one citation of a single-celled organism becoming multicellular under controlled laboratory conditions! It happened spontaneously too, just like creationists typically think that it should, but they still wouldn't be satisfied until we show a single bacteria turning into a human.

As we'll see later in this book, whenever creationists demand evidence that we should reasonably expect to see, they get it. So the only way to avoid being repeatedly disproved is by demanding evidence we should not reasonably expect to see. They demand absurdities that evolution could never provide. They'll only make realistic demands that they feel are safe, expecting that there is little chance we could ever show them what they asked for. They've learned to tailor the question such that it eliminates most of the appropriate answers.

They know we can't experiment on humans. It's not just a matter of ethics—it's impractical to do lab studies of the evolution of any multicellular organism so long-lived that a single generation can last decades. But humans are all they care about, so when they ask for evidence of beneficial mutations, they have to include a caveat that they're not going to accept any studies of lower life-forms; they'll only demand to see what (they think) we can't show. They want to see it in people.

Harun Yahya's lab-coat lady continues:

> All mutations that take place in humans result in mental or in physical deformities such as albinism, mongolism, dwarfism, or diseases such as cancer.

When I saw this video for the first time, back around 2008, I knew I had to reply with something they didn't expect. I could give them exactly what they asked for, but I wasn't going to take my examples from the textbooks they already knew about and had already made up excuses for. I figured I could find evidence of positive mutations in humans, and I wanted each of my citations to be easy to look up and easy to understand yet unexpected at the same time. So I went to an unlikely source to find them: mainstream news media.

For example, I found that kinfolk in the village of Limone Sul Garda in northern Italy have a mutation which gives them a better tolerance of HDL serum cholesterol. Consequently this family has no history of heart attacks despite their high-risk dietary habits. This mutation was traced to a single common ancestor living in the 1700s, but has now spread to dozens of descendants. Genetic samples from this family are now being tested for potential treatment of patients of heart disease.

Another example of new variance was a mutation of two genes identified in some Tibetans that allows them to endure prolonged periods at high altitudes without succumbing to apoplexia (also known as altitude sickness). As Peter Tyson states in *Inside NOVA*,

> Scientists have found evidence that, over the thousands of years that Tibetans have lived on the Tibetan Plateau, natural selection has been working on their genes, causing evolutionary changes that enable Tibetan peoples not only to survive but to thrive at altitudes of 13,000 feet or more.

A different but similar mutation was identified in high altitude natives in the Andes.

Another example of a beneficial mutation is the CCR5-delta 32 mutation. About 10% of white people of European origin now carry it, but the incidence is only 2% in Central Asia, and is completely absent among East Asians, Africans, and tribal Americans. It appears to have suddenly become relatively common among white Europeans about seven hundred years ago—apparently as a result of the Black Plague, indicating another example of natural selection allowing one gene dominance in a changing environment. It is harmless or neutral in every respect other than its one clearly beneficial feature. According to Science-Frontiers.com, if one inherits this gene from both parents, they will be especially resistant (if not immune) to AIDS.

Similarly, population genetics is being credited as one reason why incidence of sickle-cell genes in African-Americans is apparently decreasing over time.

Then there's the account of a family in Germany who were already unusually strong, but one child in the family happened to be born with a double copy of an antimyostatin mutation carried by both parents. The result is a Herculean kiddo, who was examined at only a few days old for his unusually well-developed muscles. By four years old, he had twice the muscle mass of normal children and half the fat. Pharmaceutical synthesis of this mutation is being examined for potential use against muscular dystrophy and sarcopenia. An MSNBC report from June 24, 2004 states,

> Somewhere in Germany is a baby Superman, born in Berlin with bulging arm and leg muscles. Not yet 5, he can hold seven pound weights with arms extended—something many adults cannot do.

And then there's a family in Connecticut who've been identified as having hyperdense, virtually unbreakable bones. A team of doctors at Yale traced the mutation to a gene that was the subject of an earlier study in which researchers showed that low bone density could be caused by a mutation that disrupts the function of a gene called LRP5. This clued them that a different mutation increased LRP5 function, leading to an opposite phenotype—that is, high bone density. According to their investigators, members of this family have bones so strong they rival those of Bruce Willis' character in the movie *Unbreakable*. According to Professor Richard P. Lifton, M.D.,

> If there are living counterparts to the [hero] in *Unbreakable*, who is in a terrible train wreck and walks away without a single broken bone, they're members of this family. They have extraordinarily dense bones and there is no history of fractures. These people have about the strongest bones on the entire planet.

All of these are examples of specifically identified mutations that are definitely beneficial, and that have spread through the subsequent gene pool according to natural selection. These are just some of many indisputable proofs of evolution in humans.

When I posted my reply to Harun Yahya's video, it got the attention of a Canadian creationist named Ian Juby. Juby tends to dress like he's on safari, as if he were a field scientist. He is reportedly a member of MENSA and allows himself to be introduced as a "genius," a "professor," and a "paleontologist"—all despite having no advanced degrees and not having taught at any university. Without the costume, I doubt even other creationists could take him seriously.

For the most part, Juby's attempt to dispute my indisputable proofs of human evolution could be summarized as saying, "Maybe; maybe not." He argued that (1) Italian villagers may have high cholesterol tolerance even without the Milano mutation; (2) white Europeans may be resistant to AIDS, but maybe not immune; (3) having more muscle mass might come at a loss somewhere else (though no such detriment was indicated in any of the studies I read); and (4) the side effects of having hyperdense bones were particularly severe.

Juby showed images of horrible conditions often associated with most types of increased bone density, implying that any change we might call an improvement would come at a terrible cost, so that we still couldn't admit such a thing as a beneficial mutation. However, the *American Journal of Human Genetics* study I cited regarding the family with unbreakable bones was one in which there were no associated symptoms at all—just a solid benefit with no downside.

As L. Van Wesenbeeck et al. found in their study titled "Six Novel Missense Mutations in the LDL Receptor-Related Protein 5 (LRP5) Gene in Different Conditions with an Increased Bone Density," published in the journal in February 2003,

> [A] gain-of-function mutation (G171V) in the LRP5 gene was described in two kindreds with an enhanced bone density. . . . the family described by Little et al has no features other than very dense bones, whereas patients from the other kindred also suffer from a wide, deep mandible and a torus palatinus.

Similarly, a study by Michael Whyte et al. in the *New England Journal of Medicine* titled "High Bone-Mass Disease and LRP5," published in May 2004, concluded the following:

> This autosomal dominant condition seemed benign, enhancing skeletal mass without causing complications of osteoporosis. Genealogic studies showed no relationship with the healthy kindred with high bone mass described in 1997 by Johnson et al. and documented as having the identical LRP5 V171 mutation by Little et al. These findings increase to eight the number of different mutations in LRP5 that cause high bone density.

Regarding my citation of two online news sources, I don't know whether there is any validity to Juby's criticism, but there is no reason to think there is any—and considering his track record, I rather doubt it. Even so, I have since learned not to cite news media for anything scientific. For any extraordinary claim promoted in the media, I recommend that you use the media story as a guide to look up the actual peer-reviewed articles and refer only to those. I have three examples as to why I say this.

First, in 2009, the news announced the discovery of *Darwinius masillae*, a 47 million-year-old monkey-lemur affectionately named Ida. She was hailed as a human ancestor and the missing link between humans and apes. I was shocked at the headlines because I knew what she really was: Ida bridged a gap between prosimians and true simians, not humans and apes. She was a missing link because she was the first (and still only) fossil showing traits from both prosimians and true simians. Her fossil is exquisitely preserved and she is a significant find, but she is not what the media played her up to be. Tabloid reporters called her the 8th Wonder of the World. One scientist was quoted saying that news of her would be as profound as an asteroid hitting the Earth. Sky News broke the story with highly inappropriate sensationalism, calling her "the anthropological equivalent of the Holy Grail." They said, "She could confirm Darwinian theory and debunk creationism. She could even challenge religion itself." Bullshit! What they're calling Darwinian theory has already been confirmed, and religion is virtually impervious to evidence or reason. Ida was the last link necessary in an otherwise already sufficiently complete chain joining all members of the primate family tree. But she was not a human ancestor, and we didn't need all that autofellating fanfare.

A second example of sensationalism souring science came in 2010 with the reported discovery of an "arsenic bug," a microbial organism that uses a normally deadly poison as part of its DNA in place of phosphorus. It was discovered in Mono Lake, California, a surreal setting rich in arsenic and poor in phosphorus, and a sufficiently alien environment ideal for the seemingly alien life-form said to be living there. It was even discovered by an astrobiologist working for NASA. Who even knew that there was such a thing as an astrobiologist? That seemed to be where the hype began. The microbe, known as GFAJ-1, was originally presented as unrelated to anything else on earth—the result of a completely different (and apparently much later) abiogenesis event. Now *that* would be huge!

Once upon a time, people thought that all life on earth was either eukarya or bacteria. I remember when they discovered Archaea, a second sort of prokaryote. It looked like bacteria but had vastly different genetics, enough to qualify as a third "domain" of life. That was huge news in the world of biology in the 1970s, but it was nothing compared to this claim about the organism found in Mono Lake! Imagine finding an organism that is utterly unique and not related to anything else on Earth.

Alas, it was not true. Genomic sequencing revealed that it was a form of bacteria, not unrelated, and not even as exotic as Archaea. Later tests revealed that it didn't even use arsenic in its DNA. The whole thing was apparently a mistake. This is the benefit of the peer-review process. When you make an outrageous claim challenging the status quo, don't think the world won't put that claim to the test.

The third and final reason for me to distrust news articles is that Ian Juby noted something I actually did get wrong. In my YouTube series on the foundational falsehoods of creationism, I mentioned an emerging population of tetrachromatic women who could reportedly see into the ultraviolet spectrum. When I read that story, it seemed completely plausible. I mean, we have bees that see ultraviolet light, so if a woman has a four-cone array in her eyes rather than the usual three, that doesn't seem far-fetched at all. However, I was not reading peer-reviewed science; I was reading popular press, and I fell victim to the same sort of sensationalism mentioned above. RadioLab on NPR found several women who had a fourth "yellow" cone in their eyes and at least one could see differences in shades yellow that no other human could. Turns out, tetrachromats can see *up* to the ultraviolet range, not *into* it.

As if that weren't embarrassing enough, Ian Juby pointed out a second error in my research. When I posted about Tibetans' ability to survive at high altitudes, I cited the wrong paper and credited a Glycophorin A somatic cell mutation. I should have paid closer attention. Juby thought that I had misread the paper I cited, and he argued that there was no mutation allowing any group of Tibetans or Peruvians any unusual resistance to apoplexia. However I immediately posted an errata video wherein I showed three peer-reviewed articles defending my initial claim. They were

"Hemoglobin Concentration of Pastoral Nomads Permanently Resident at 4,850–5,450 Meters in Tibet" by C. M. Beall and M.

C. Goldstein, published in the *American Journal of Physical Anthropology* in 1987; and

"Ventilation and Hypoxic Ventilatory Response of Tibetan and Aymara High Altitude Natives" by C. M. Beall et al., published in the *American Journal of Physical Anthropology* in 1997;

"Analysis of the Myoglobin Gene in Tibetans Living at High Altitude" by L. G. Moore et al., published in *High Altitude Medicine and Biology* in 2002.

These studies refer to different genes being naturally selected for a novel advantage for increased blood oxygen saturation, promoting the Tibetans' adaptation to high altitude as compared to another population at lower elevations.

I never heard from Ian Juby again, and that is not surprising. In my errata video, I showed where another creationist organization was aware of this discovery and was desperate to discredit or dismiss it. That's what pseudoscience apologists must do, even when they know full well that they're wrong.

Case in point: my son was taking high school biology at that time. His teacher told the class that evolution was "just a theory" and that there had never been a beneficial mutation. But my son knew better already, and he had his textbook open to show him. "But teacher, beneficial mutations are listed right here."

For that, he was sent to the principal's office. I was outraged of course, but my son's mother asked that I not make that teacher famous while my son was still in school.

This is one of many ways to know that creationists are deliberately lying; they know what the truth is, but they don't want to believe it, so they don't want you to know what the truth is either. We'll see many more examples of that later in this book, and they will be of the sort that can be easily looked up and confirmed.

It's important to note that we've identified beneficial mutations in many other species besides humans. That myostatin mutation doubling muscle mass has occurred in domestic dogs and cattle. In each case, including humans, the mutant is simply stronger than normal, and there still is no indication that I'm aware of to imply any disadvantage to that. But this is something a lot of people don't understand and it requires some explanation.

As I said in a previous chapter, mutations in real life aren't like they are in *X-Men*. There is a sort of economics in biology. You don't get a super power without having to pay the price. Of course with severe mutations, there's going to be a trade-off. It's the same sort of give-and-take we expect of any modification or specialization of pretty much anything. But the vast majority of mutations are subtle and slight, usually causing no more than an increase in variety of physical or chemical proportions in descendant groups, which is how evolution usually works. If any lineage survives severe mutations for many generations, then the fact that we average well over a hundred mutations per zygote means that future mutations have an increasing probability to enhance or improve the new condition. These are also cumulative in the descendant genome, such that it is possible to construct or confirm an evolutionary phylogeny by reverse-sequencing compiled mutations. This is another way to prove that humans have evolved, that we are much older than the story books say, and that our genome is not being "degraded" the way creationists like Ian Juby want us to believe.

Juby even said that all mutations, even those that lengthen the genome and add receptors to see in more colors, should be detrimental, or "a *loss* of information," for which he had no adequate or applicable definition, despite his claims. According to his own explanations, two existing chromosomes being fused into one counts as a *loss* of information, such that it actually takes less "information" to create a human than it does any other species of ape.

Humans only have 23 pairs of chromosomes, whereas all other apes have 24. It is thought that the complete loss of any pair would be lethal and highly improbable, even when we're already talking about things that are extremely rare. An abrupt addition of a new chromosome shouldn't be possible either. So how could humans be descended from apes when we don't even share the same chromosomes? Turns out, we do. Once both human and chimpanzee genomes were sequenced, it was shown that human chromosome 2 is actually two ancestral ape chromosomes fused together. As PBS Evolution Library states,

> In 2005, a peer-reviewed scientific journal published results of the tests. It turns out that chromosome 2, which is unique to the human lineage of

> evolution, emerged as a result of the head-to-head fusion of two ancestral chromosomes that remain separate in other primates. Three genetic indicators provide strong, if not conclusive, evidence of fusion. First, the banding (or dye pattern) of human chromosome 2 closely matches that of two separate chromosomes found in apes (chimp chromosome 2 and an extra chromosome that does not match any other human chromosome). Second, a chromosome normally has one centromere, or central point at which a chromosome's two identical strands are joined. Yet remnants of a second, presumably inactive centromere can be found on human chromosome 2. And third, whereas a normal chromosome has readily identifiable, repeating DNA sequences called telomeres at both ends, chromosome 2 also has telomere sequences not only at both ends but also in the middle.

Creationist apologists try to come up with whatever excuses they can to dismiss or discredit any examples of beneficial mutations discovered by science.

Yet, creationists don't do their own experiments. Even when they publish their own journals, so they can pretend to be peer-reviewed, they still don't take part in the scientific process, because there's no way to falsify their findings and call their primary assumptions into question. They certainly can't refute what has already been peer-reviewed, because even if a given peer-reviewed finding is later refuted, it would take science to do so. All that creationists can do when we cite peer-reviewed journals is to try to misrepresent what has been said. So I like to cite examples that refute their arguments very clearly. For instance, a February 10, 2010 research article, "Human and Non-Human Primate Genomes Share Hotspots of Positive Selection" by David Enard et al., published in the peer-reviewed *Public Library Of Science Genetics*, precisely defines and positively identifies beneficial mutations:

> An advantageous mutation spreads from generation to generation in a population until individuals that carry it, because of their higher reproductive success, completely replace those that do not. This process, commonly known as positive Darwinian selection, requires the selected mutation to induce a new non-neutral heritable phenotypic trait, and this has been shown to occur unexpectedly frequently during recent human evolution.... different approaches now conclude that a substantial

> proportion of non-deleterious mutations are indeed weekly to strongly advantageous.

Indeed, as demonstrated in a July 13, 2004 *PLOS Biology* research article, "Lineage-Specific Gene Duplication and Loss in Human and Great Ape Evolution" by Andrew Fortna et al., some of these have been shown to increase complexity.

> The evolution of genomes has been primarily driven by single basepair mutation, chromosomal rearrangement, and gene duplication (Ohno 1970, Samonte and Eicher 2002), with the latter being the key mechanism for generating new genes and biological processes that facilitated the evolution of complex organisms from primitive ones (Li 1997).

Some of these have even been confirmed to add new genetic information, as discussed in a March 26, 2010 *PLOS Computational Biology* research article by Chuan-Yn Li et al., "A Human-Specific De Novo Protein-Coding Gene Associated with Human Brain Functions":

> Many mechanisms for the origin of new genes are known, such as tandem gene duplication, retrotransposition, exon shuffling and gene fusion. By these mechanisms, the origination of new protein coding genes involved 'mother' genes that served as blueprints for the new genes. However, recent comparative genomic analysis identified a few 'motherless' or de novo genes in fly and yeast, which originates from non-coding DNA sequences.

But beneficial mutations were the result of deletions too. Referring to the March 10, 2011 *Nature* letter by Cory Y. McLean titled "Human-Specific Loss of Regulatory DNA and the Evolution of Human-Specific Traits," we see that scientists were able to isolate 510 deletions from the ancestral primate genome. These also caused significant regulatory changes including among other things improvements in the sexual characteristics of monogamous mates. Even the expansion of specific regions of the human brain can be correlated to a deletion of genes from the primate genome. Ian Juby's response to such findings would likely be little more than the same one he gave me: "First of all, this is a loss of a gene, a loss of information. Gee, notice a trend here?"

Ken Ham, the aforementioned CEO of Answers in Genesis, was the keynote speaker at a convention of homeschool teachers in Houston, Texas. My wife, a former old-earth creationist Christian school teacher now turned atheist activist science teacher, was outraged at this. She made a video criticizing Ham over his position and mentioned one bit of evidence that Ham thought he could refute, at least to the satisfaction of the sycophants in his massive social network. In reply, Ham made a video ridiculing my wife wherein he said he couldn't believe that she would "regurgitate" what he calls a "ridiculous, out-dated, and false idea" that wisdom teeth are evidence of evolution.

As a result of Ham's attack, my wife and I were invited to give multiple presentations, including one to a maximum-capacity crowd of hundreds at the Houston Museum of Natural Science, where we proved Ham was wrong about everything he had just said. Our citations included a April 11, 2006 *Proceedings of the National Academy of Sciences* article, "Natural Selection and Molecular Evolution in Primate PAX9 Gene, a Major Determinant in Tooth Development" by T. V. Pereira et al., which reads as follows:

> A reduced number of molars may be advantageous from a human evolutionary perspective. Because of the dramatic lifestyle and diet shift experienced since the discovery of fire and the development of cooking utensils, third molars, which could have been essential for the survival of earlier hominids, became not only functionless but also an important cause of morbidity for modern humans. *Dental arches have been reduced over hominid evolution.* As a result, third molars became frequently impacted or malpositioned, preventing the teeth from attaining a functional position. Furthermore, because of the difficulty of cleaning them and keeping them free of disease, impacted or malpositioned third molars lead to a higher susceptibility to periodontal disease, such as infections, carious lesions, cysts, tumors, and destruction of adjacent teeth and bone. These problems led some authors to raise the possibility of a complete elimination of these teeth in humans, and studies for intentionally stopping third molar growth for possible clinical application are being carried out. [Emphasis added]

Third molars are traditionally known as wisdom teeth. They have a high rate of impaction, 85%, as too many teeth are crowded into too small a jaw. What this study shows is that the formation of the third molar, or wisdom tooth, is controlled by a HOX gene called PAX9, which is widely conserved among primates and some other mammals. In fact, ours is almost exactly identical to that of gorillas.

The study notes that a particular heterozygous transversion mutation of the PAX9 homeobox gene, substituting alanine for proline, will result in the agenesis of the third human molar, meaning that wisdom teeth will simply never develop. This is why not everyone has them. Note that this means that even a deactivated or dysfunction mutation can be definitely beneficial! We humans would like to prompt this mutation across humanity so that no one has to risk deleterious morbidity if their jaw isn't quite big enough for all those teeth. That's the way any omnipotent benevolent intelligent designer would do it.

But evolution doesn't have a goal in mind. It uses blind mechanisms of random mutations emerging amid naturally selective patterns imposed by population mechanics. What you get from that are variable probabilities of this mutation according to however favorable genetic propensities are across a wide range of given demes, and this is what we see.

To determine whether this gene is polymorphic in humans, they sequenced 86 individuals sampled from Asian, European, and Native American populations, and this was the range that they found: third molar anagenesis (no wisdom teeth) was 0.2% in Bantu-speaking Angolans to 100% in Mexican "Indians." This indicates an evolutionary pattern with no indication of intelligent design.

Then there's the case of a gene called MYH16, discussed in a January 20, 2004 *Nature* letter by Hansell H. Stedman et al. titled "Myosin Gene Mutation Correlates with Anatomical Changes in the Human Lineage":

> Powerful masticatory muscles are found in most primates, including chimpanzees and gorillas, and were part of a prominent adaptation of *Australopithecus* and *Paranthropus*, extinct genera of the family Hominidae. In contrast, masticatory muscles are considerably smaller in both modern and fossil members of *Homo*. The evolving hominid masticatory apparatus—traceable to a Late Miocene, chimpanzee-like morphology—shifted towards a pattern of gracilization nearly simultaneously with accelerated encephalization in early Homo. Here, we show that the gene encoding the predominant myosin heavy chain *(MYH)* expressed in these muscles was inactivated by a frameshifting mutation after the lineages leading to humans and chimpanzees diverged. Loss of this protein isoform

> is associated with marked size reductions in individual muscle fibres and entire masticatory muscles. Using the coding sequence for the myosin rod domains as a molecular clock, we estimate that this mutation appeared approximately 2.4 million years ago, predating the appearance of modern human body size and emigration of Homo from Africa. This represents the first proteomic distinction between humans and chimpanzees that can be correlated with a traceable anatomic imprint in the fossil record.

Now what does all that mean? Suffice to say that the MYH16 gene is common to all primates (monkeys, gorillas, lemurs, and so on), meaning we have it too. Except there's no reason why we should have it if we were created in the image of God. Like everything else in biology, this only makes sense in light of evolution.

We have in our genome a primate pseudogene; it's broken. It's a dysfunctional molecular vestige of our evolutionary ancestry. It makes a protein, but the protein doesn't work. When this primate gene is disabled, as it is in humans, what happens is that much of the massive muscles that would have given gorillas tremendous bite pressure only allow us the barest fraction of that. An associated symptom of that is that if we no longer have these huge jaw muscles, then we wouldn't develop a sagittal crest either. (That's the ridge along the top of a gorilla's skull that is the anchor point for those massive jaw muscles.) The cranial pressure there apparently prevented the brains of other simians from enlarging the way our brains have, so one of the reasons that we are so smart is because we have so many useless monkey genes that don't work. How's that for being created in the image of God?

Time to check in again with Harun Yahya's lab-coat lady:

> Another reason why it is impossible for mutations to cause living things to evolve is that mutations do not add any new genetic information to an organism. Mutations cause existing genetic information to be randomly reshuffled similar to playing cards. In other words, no new genetic information is introduced by mutations.

Remember that at least some of the earlier citations in this chapter prove her wrong, but that there is no way to get creationist apologists to admit that. Faith means never having to be accountable and never admitting any error. My intention is to reason with those who still care to understand what is true, and there is another level of misunderstanding that I can explain with an analogy.

The evolution of life is analogous to the evolution of language. For example, there are several languages based on the Roman alphabet, which in English has only twenty-six letters. Yet by arranging these in different orders, we've added several hundred thousand words to English since the fifth century, and many of them were completely new. Experts estimate there were originally 50,000 to 60,000 words in Olde English and nearly a million words now! (The count from the Global Language Monitor is 995,112 words.)

The principle is the same in genetics. There are millions of named and classified species of life, all of them based on a variable arrangement of only four chemical components. These are adenine, guanine, cytosine, and thymine. For ease of documentation, these are abbreviated to their first initial, so that when geneticists "read" DNA, it appears as codons of three letters: AGC TTC GAG CTA, and so on. These aren't actually letters of course, they're chemicals—but calling them letters actually helps my analogy.

We know that Spanish, Italian, French, and Portuguese all evolved from Latin, a vernacular that is now extinct. Each of these newer tongues emerged via a slow accumulation of their own unique lingo, occasionally acquiring new slang, abbreviations, or terms adopted from other sources. These are analogous to mutations. Thus, the original Latin diverged into new dialects and eventually distinct forms of gibberish such that the new Romans could no longer effectively communicate with either Parisians or Spaniards.

In sexually reproductive species, speciation is determined when two members of the same genus no longer interbreed or can no longer produce viable young. The longer two groups are isolated, the more mutations build up between them and the less chance they have of producing fertile offspring. There will come a time when they can only produce infertile hybrids, and then when they can't produce anything at all. This would be analogous to two cultures having increasingly distinct dialects until they reach a point that they can't understand each other anymore.

Similarly, if we took an original Latin-speaking population and divided them, with one group sequestered in complete isolation over several centuries, they might still be able to understand each other, though their independently derived local jargon will most likely be unintelligible to all foreigners, including other Latin-language speakers. They each may develop a new Latin-based language, but they won't start speaking Italian or Romanian because identical vocabularies aren't going to occur twice.

The same situation is true in biological evolution. For example, look at old sci-fi movies where the aliens look just like us, but are a different color or have weird ears. How could unrelated organisms with unique origins on different worlds each evolve into humanoid people?

The absurdity of this is illustrated nicely in an old episode of *The Simpsons*. The family is abducted by a flying saucer, and the squidlike aliens speak with a familiar accent.

Marge:
You . . . you speak English?

Kang:
I am actually speaking Rigellian.
By an astonishing coincidence,
both of our languages are exactly the same.

With an eye toward the history of world literature, this type of analogy can even help explain what is meant by transitional species. For example, *Don Quixote* is old enough that in its original form, it isn't yet Spanish, but it isn't quite Latin anymore either; it is an intermediate form.

Most evolution is defined by *cladogenesis*, where one ancestral type branches into multiple others. In this case, Latin would not have to be extinct before French, Spanish, and Romanian could have developed around it. This happens when groups of people are isolated long enough that unique differences filter into one or both cultures over time and are not still sharing the same cultural influences/genetic material.

However, some evolution is defined by *anagenesis*, where an ancestor is completely replaced by a descendant. In this analogy, that would be Latin slowly becoming Italian. In this case, of course, Latin does go extinct, and the next generations weren't even aware of it.

Importantly, notice that there was never a first person to speak Italian. There was never a time when there were Latin speakers and one guy speaking Italian, nor were there ever old Latin people living among young Italians. The old fogies may not understand the kids today, but they can still communicate when they have to. Using the analogy of language is the best way I know to get people to understand how evolution is a gradual change—and one that happens on the population level, not on the level of individuals. The individuals wouldn't even notice what is going on.

Creationists don't like my analogy, because it makes sense, and they don't want to hear anything that makes sense. As an antiscience protestor proclaims in an episode of *Futurama*, "I don't understand evolution, and I have to protect my kids from understanding it!"

That's why they turn to shysters like Ray Comfort of Living Waters Ministries. Ray makes millions of dollars every year misleading wanna-believers who don't want to understand evolution. So he gives them the worst imaginable distortions of it with explanations no one could understand. As he said to Pat Robertson on *The 700 Club*,

> Let's pretend I'm a believer in evolution for a moment. There's a big bang, life form begins, and over millions of years, a dog evolves. It's the first dog. He's got a tail, legs, teeth, eyes. And it's good that he's got eyes, because he needs to look for a female. He's been blind for millions of years; now he can see. He's gotta find a female. She's gotta be evolved at the right place at the right time with the right reproductive organs and a desire to mate. Because without a female, he's a dead dog. There's no species; you've gotta have a female.

Notice that in Comfort's deliberately distorted delirium, there's no ancestral phylogeny, no descent with modification, and not even the possibility of transitional forms. There is no ancient world either. The Big Bang seems to have caused life to begin, as if both happened on the same day. Of course in reality, the dog in this example wouldn't have been an amorphous mass of meat, blind for millions of years before it suddenly turned into a dog. It had ancestors who were not yet dogs, but who already had tails, legs, teeth, and eyes that could see long before we would recognize them as dogs. It didn't just have direct ancestors either, but distant cousins too, and plenty of available females to breed with in every generation.

Comfort of course knows all of this. He's a street preacher. People explain evolution to him every day just so he can play dumb, usually with word games to try to frustrate honest people. There is no way anyone could be this stupid. Comfort is a professional liar who doesn't care what the truth is. I once had a debate with him, and I told him that his only strength was in pretending to misunderstand simple concepts.

If we apply the analogy of language to Comfort's fallacy, we would have a guy who just appeared out of nowhere and needs to find a woman who also appeared out of nowhere and who also happens to speak French. Of course, the reality is that there was never a first guy who spoke French, meaning there was never one person who time travelers could point to and agree that he was the first Frenchman. By the same token, there was never a first dog.

Mutations are degrees of variation that are usually quite subtle but cumulative, normally harmless, and occasionally advantageous. Whether we're talking about genetics or language, any change in information in either case is different information not already present, and therefore can only be considered "new." But of the many types of mutations known to occur, there are additions and duplications as well as deletions and the rest. So yes, genetic material can be added or taken away—but as to whether "information" has been added as opposed to lost, we can't really tell because creationists won't tell us what they think "information" is or how to measure it. They'll readily state (as if it had somehow been confirmed) that it takes more "information" to make a bird than it does a dinosaur, but if you ask them *how much* more, they'll shut right up, and if you demand to see the data that justifies how they could even make that claim in the first place, they'll try to change the subject.

There is an excellent example of how natural selection can create a sort of arms race with combined mutations in multiple species. Some American species of rough-skinned newts are harmless to hold, but after handling them just be sure to wash your hands before you eat, rub your eyes, or even touch your lips! Their skin can sweat enough tetrodotoxin to kill twenty men! One cute newt, *Taricha granulosa*, is the most poisonous amphibian on Earth.

Lots of amphibians have toxic skin, but why would a small American salamander be toxic enough to kill an elephant? Because its natural enemy is a garter snake (*Thamnophis sirtalis*), which keeps building up its tolerance of TTX poison.

Now let's hear from Harun Yahya's lab-coat lady one more time:

> While a very simple structure of bacterium comprises 2,000 different types of proteins, a human organism has 100 thousand proteins. Exactly 98,000 new proteins have to be discovered for a bacterium to evolve into a human being.

How should I respond to that? First, I would like to point out that humans didn't evolve from bacteria, and I'll explain why not in another chapter. While researching her claim, I happened across a video called, "How Big Is Your Genome? Strange DNA." It was made by a graduate from Texas A&M University identified only as Crystal who offered the perfect counterargument to Harun Yahya's spokesmodel.

> Actually, we have about 250,000 proteins being coded by 25,000 different genes. But you know what? A grain of rice actually has forty to fifty thousand genes! Can you imagine? A little grain of rice has fifty thousand genes, and we have twenty-five? What's up with that? . . .
>
> Now when scientists first started to study genetics and the human genome, we actually expected to find three times more of the number of genes than we actually found. . . . the twenty to twenty-five thousand genes that we did find is only about twice the number that a roundworm has. And in fact, there's a variety of an amoeba that has 200 times the number of genes that a human being has.
>
> A puffer fish has approximately 21,000 genes. The puffer fish (isn't it adorable) actually has no junk DNA, which is interesting because humans are actually made up of 95% junk DNA.
>
> Well, one thing that researchers are particularly interested in is finding out, what is junk DNA? And does it actually do anything? One recent study suggests that, it really doesn't. Researchers took 2.3 million letters of DNA out of a mouse, and compared it to a mouse with a full-length version of DNA, and found out that there were no differences whatsoever.

As appealing as this video presentation was, I have learned not to trust unverified sources and to cite actual scientists instead. So rather than rely on this college student, I now prefer to refer to an acquaintance with more experience and expertise in molecular biology. There are no peer-reviewed journal entries with the specific intent of disproving creationism, so I'll cite another venue where that was the intended purpose. In a lecture I attended at Skepticon IV in Columbia, Missouri, P. Z. Myers, associate professor of biology at the University of Minnesota–Morris, set out to disprove the following creationist claim put forth by Stephen Myer of the Discovery Institute,

> Thus, far from being dispersed sparsely, haphazardly, and inefficiently within a sea of nonfunctional sequences (one that supposedly accumulated by mutation) *functional genetic information is densely concentrated in the DNA molecule*.... Far from containing a preponderance of 'junk'—nonprotein-coding regions that supposedly perform no function—*the genome is dominated by sequences rich in functional information*. [Emphasis added]

P. Z. confirmed a lot of what Crystal said about the puffer fish and simple plants with way more genes than humans have, and that the human genome has only 20,000 genes among 3 billion base pairs in the total genome. Of that, he said that only 1.5% of our genome is coding DNA while another 3% is regulatory. If we include all other categories of functional DNA, including ribosomal, transfer, and microRNA genes, it still amounts to only 5% of the total genome. So Crystal was right about that, too.

P. Z. then explained how another 10% of the genome is essentially random gibberish created by a buggy enzymatic process; 21% is parasitic viral copies; and 13% is copies of copies, while 8% of the genome was made by endogenous retroviruses. This is when a virus inserts its own DNA into a gamete cell, which is inherited by all the descendants of that organism. So the summary is that only 5% of the human genome is functional DNA, 10% is structural DNA, and 45% is known to be useless parasitic DNA. The remaining 40% he described as "job security for molecular biologists," because it isn't all yet understood. But he added that some of that remaining percentage is already known to be pseudogenes. Finally, he referred to two dueling predictions made by another creationist, William Dembski of the Discovery Institute:

> Thus on an evolutionary view we expect a lot of useless DNA. If, on the other hand, organisms are designed, we expect DNA, as much as possible, to exhibit function.

P. Z. had clearly shown that only the first prediction was correct, thus disproving creationism. With no intelligent designer evidently involved, natural selection weeds out detrimental mutations and selects for beneficial mutations, but the neutral ones, having neither cost nor ill effect may freely accumulate as "junk."

These mutations, though seemingly random, can be interpreted as occurring at a more regular rate when examined over a span of many generations. The relatively fast mutation rate of mitochondrial DNA provides further confirmation of the extensive periods of time in an organism's matrilineal history. DNA can also help us establish genealogy precisely because it contains a sort of record of inherited mutations that can be compared and matched to more distant relatives.

In 1998, a Y-chromosome analysis effectively ended a controversy that had flamed for two hundred years regarding Thomas Jefferson and one of his slaves, Sally Hemings, long believed by many to have been his mistress and to have given birth to six children fathered by Jefferson. A definite genetic link was established between the Jefferson male line and a descendent of Hemings' son Eston, who was born in 1808. Nowadays it is possible to confirm or discover your genetic ancestry with a DNA test costing less than $100. That test analyzes acquired mutations unique to certain ethnic demes to determine what percentage of which ethnicities you're composed of and which regions those ancestors came from.

Just like a court-ordered paternity test can positively identify your immediate lineage, a more in-depth genomic sequence analysis can also determine more distant ancestry. The more in-depth it is, the further it can trace, even where it pairs different species to one genus or different genera to one taxonomic family, collective families of one order, and so on.

Judgments of law are based on a preponderance of evidence, and genetic evidence is so reliable it can get a life or death sentence even without need of other types of evidence to corroborate it. So the fingerprints of mutation in molecular phylogeny not only provide profound evidence of evolution, they also amount to legal proof of it.

THE 9TH FOUNDATIONAL FALSEHOOD OF CREATIONISM

"No Transitional Species Have Ever Been Found"

In 1764, the fossilized remains of a huge animal were found buried ninety feet deep in a quarry in the Netherlands. It was an unusual creature and thus difficult to identify. Its clearest feature was a jawbone lined with uniform conical teeth. It was first said to be a "breathing fish," otherwise known as a whale. How did a whale end up buried so deep in the ground? What little they knew of stratigraphy back then led some experts to believe that the world had endured a series of successive catastrophic floods—not just the flood of Noah, but a repeating sequence of inundations of colossal scale.

A few years later, fragments of another skull of the same species was uncovered nearby, and this time people mistook it for a crocodile. Now *this* was strange! So far as anyone knew, there had never been crocodiles living in the Netherlands, and this was a really big one! How could this be? At the time, people had been finding fossils all over the world which were usually thought to represent species that were still alive somewhere else. No one yet knew that continents moved nor how one environment could change over time. Up to this point, most people thought that the world was exactly as it had been ever since the beginning. There was as yet no concept of an exotic lost world existing before human history when there were unfamiliar species different than we have today.

By 1798, a Dutch paleontologist named Adriaan Gilles Camper read detailed descriptions of this animal and recognized it as a gigantic monitor lizard. This is a group of long-necked, forked-tongued mildly venomous lizards of the genus Varanus that are found only in the tropics. There aren't any varanids in Europe either, and this one was much bigger than they ever get anywhere! Big monitors can be two meters long. We now know that giant exotic ones can be as much as three meters long, but Komodo dragons hadn't been discovered yet. The largest varanid we know of so far is the Australian fossil of Megalania, a massive monitor seven meters long. That too wouldn't be discovered for a few more decades, and there was still no comparison. The lizard in the Netherlands would have been thirteen meters long, or roughly forty feet, a lizard the size of a whale!

No one had ever found anything as profound as this. The specimen was significant enough that it was brought to the world's leading expert on anatomy, Jean Léopold Nicolas Frédéric [Georges] Cuvier, professor of natural history in the Collège de France. Upon inspection, Cuvier agreed that it was a lizard and that it was very similar to a monitor, but that it was not a monitor. This lizard was fully adapted for swimming, with flippers instead of feet. Mosasaurus, as he called it, was a prehistoric monster unknown in the world of men.

Despite the old tales of sea monsters in the open ocean, Cuvier doubted whether such a thing as this had ever been seen before, and that was hard to explain. If everyone commonly knew about whales and sharks and sea turtles, then how is it that no one knew about something as big as this? An air-breathing lizard would have to stay near the surface, where it could be seen. With sailors and whalers on every sea, surely many people would have seen several of these tremendous and terrible man-eating lizards if there were any left to find.

His contemporaries believed that any species created by God was saved on the Ark and thus still enjoyed divine protection, so the very idea that any species could be entirely wiped out was an abomination. When I was a boy, I actually met people who still believed such things even in the twentieth century! These were the sort of people who think that whales are fish, and who either deny that dinosaurs ever existed or insist they're still out there somewhere. Such was the mind-set in the late 1700s.

Cuvier would one day be known as the father of paleontology. He knew about Nicolas Steno's work a century earlier, establishing stratigraphy. Prior to that, people believed that fossils were never alive, but that they just grew out of the ground somehow and only coincidentally happened to look like living things. In 1659, Steno explained how fossils were slowly lithified casts of once-living organisms from a long time ago, and that different layers of strata could be identified by the types of fossils found in each layer. People had been overlooking fossils forever, rarely recognizing them for what they are. After Steno, they paid more attention, and this started a flood of fossil finds.

A century after Steno, Cuvier knew that the fossil record shows different species existing at different times, but that no modern mammals were ever found below a certain level, and importantly, many ancient species known only from fossils couldn't be found anywhere else anymore.

A dramatic example of that was his analysis of elephants. He examined bones from African and Asian elephants and determined they are in fact different species, and that they differ from each other more than "the horse from the ass or the goat from the sheep." Then he compared their bones to similar fossils brought in from North America and Siberia. Finding fossil elephants in the frozen wastes of Siberia was just as strange as finding giant tropical reptiles buried deep under Holland. The Siberian specimen was identified as a wooly mammoth. Cuvier showed that it wasn't just a hairy elephant; it wasn't an elephant at all. Likewise, the animal from Ohio was named "mastodon"; again very similar to an elephant, but still distinctly different and not the same species.

These were both huge herd animals, and the biggest land animals yet known by that time. Their remains were found in areas where they should have been reported repeatedly, yet they were unheard of, having never been seen alive. Cuvier had seen sketches and depictions of many other fossilized megafauna such as a giant sloth from South America, and he concluded, shockingly, that these animals no longer existed anywhere—that they had vanished from the face of the earth.

Regarding the mammoth and the mastodon, he said in 1796:

> What has become of these two enormous animals of which one no longer finds any [living] traces, and so many others of which the remains are found everywhere on earth and of which perhaps none still exist? The fossil rhinoceros of Siberia are very different from all known rhinoceros. It is the same with the alleged fossil bears of Ansbach; the fossil crocodile

> of Maastricht; the species of deer from the same locality; the twelve-foot-long animal, with no incisor teeth and with clawed digits, of which the skeleton has just been found in Paraguay: none has any living analogue. Why, lastly, does one find no petrified human bone? All these facts, consistent among themselves, and not opposed by any report, seem to me to prove the existence of a world previous to ours, destroyed by some kind of catastrophe.

Thus Cuvier introduced two revolutionary ideas: first, that the world we know conceals the remains of lost worlds of the past, of which we were not a part; and second, he also introduced the concept of extinction—and not just one such event, but in several successive waves, after each of which the world was repopulated by new species that hadn't existed before and which don't exist anymore. How could they explain this?

Although Darwin hadn't even been born yet, there were other scientists already proposing some form of evolution, or at least something like it. For example, Georges-Louis Leclerc, Comte de Buffon and his protégé, Jean-Baptiste Pierre Antoine de Monet, Chevalier de Lamarck, as well as Lamarck's colleague, Étienne Geoffroy Saint-Hilaire, all of whom were Parisian contemporaries who would have been familiar to Cuvier, but they were not like-minded!

Today Buffon is credited for his work on the forty-four volume encyclopedia that he called *Histoire Naturelle*, as well as having some novel ideas for his time, but he was often criticized and not much appreciated by the leading minds of his day, whether theologians or naturalists. Buffon was one of the first to defy the idea of a young earth of less than 10,000 years old. He argued for a world formed out of molten rock, the result of a catastrophic cosmic collision, and cooling over a period that he estimated to be roughly 70,000 to 75,000 years.

Like many at that time, he believed that life-forms appeared spontaneously, but he didn't think tiny simple organisms advanced to become larger or more complex. Instead, he thought that even large animals emerged as spontaneous and perhaps supernatural creations.

Where his ideas parallel evolution is that he suggested that each species could diverge into different variants as a consequence of the conditions of migration. Specifically, he said that as the earth cooled and the poles froze, animals that had been living all over the globe migrated to the tropics to escape the cold. For example, he said that African and Asian elephants were both derived descendants of wooly mammoths. He considered all of these to be the same species and no more different than comparing breeds of dogs. Buffon did not believe that one species could evolve into a distinctly different species however, and neither did Cuvier.

Just as Cuvier conceived a sequence of cataclysmic floods, where only the last of these was mentioned in the Bible, he also imagined a series of repeated creation events, as if God kept wiping his slate clean and starting fresh with new creatures. This idea later influenced the anthropologist Louis Agassiz and also Sir Richard Owen of the British Museum of Natural History.

Owen would become Cuvier's successor as the world's foremost expert on anatomy and paleontology. After examining fragments of Megalosaurus, Iguanodon, and Hylacosaurus, Owen realized they shared common characteristics indicative of a single taxonomic grouping. In 1842, he classified them all together as "Dinosaurs," meaning "fearfully great or terrible lizards." The addition of dinosaurs more significantly revealed the shadows of their prehistoric world, but that was just the beginning.

When Charles Darwin published his landmark observations in 1859, he lamented that the fossil record was still quite poor at that time. It was only in the last century or so earlier that anyone had even proposed the possibility that a single species could completely die out, and the first dinosaur wasn't discovered until Darwin was a boy. Fossils were known by previous generations of course, but extinct and therefore unfamiliar varieties were often mistaken for the fanciful monsters of mythology, if they were recognized at all, which usually requires a well-trained perception of both geology and animal morphology. That's especially rare when you're talking about an organism no one has ever seen alive.

For example, ancient Greeks usually wouldn't have known about elephants, and may have imagined that mammoth fossils belonged to cyclops because the olfactory cavity looks like an eye socket while the actual eye sockets look more like cheekbones. Likewise, the way narwhal canines twist around each other were sometimes said to be the spiraled horns of mythical unicorns when discovered disconnected from the rest of the skull. If ideas like these came from looking at the bones of modern animals, imagine how confusing the fossils of unfamiliar forms must have been.

When something dies, it is usually disassembled, digested, and decomposed. Only rarely is anything ever fossilized, and even fewer things are very well preserved. Because the conditions required for that process are so particular, the fossil record can only represent a tiny fraction of everything that has ever lived.

For example, *Tyrannosaurus rex* is the largest of several tyrannosaur species. There were only thirty T-rex skeletons ever found, and only three with skulls still included. The only complete one (known as "Sue") was discovered in 1989. Creationists accuse us of pure speculation about fossils we *don't* find. But we know there had to be more than thirty of these alive at one time, and we know their population must be multiplied over several generations—and despite creationist criticisms about our assumptions regarding missing fossil fragments, like the absence of most of their skulls, we're still pretty sure that all of them must have had heads. I don't think that counts as speculation.

In *On the Origin of Species* Darwin provided many environmental dynamics explaining why no single quarry could ever provide a continuous record of biological events:

> All geological facts tell us plainly that each area has undergone numerous slow oscillations of level, and apparently these oscillations have affected wide spaces. Consequently formations rich in fossils and sufficiently thick and extensive to resist subsequent degradation, may have been formed over wide spaces during periods of subsidence, but only where the supply of sediment was sufficient to keep the sea shallow and to embed and preserve the remains before they had time to decay. On the other hand, as long as the bed of the sea remained stationary, thick deposits could not have been accumulated in the shallow parts, which are the most favourable to life. Still less could this have happened during the alternate periods of elevation; or, to speak more accurately, the beds which were then accumulated will have been destroyed by being upraised and brought within the limits of the coast-action. Thus the geological record will almost necessarily be rendered intermittent.

He also explained why it would be unreasonable to expect, and impossible to find all the fossilized ancestors of every lineage.

> Only a small portion of the surface of the earth has been geologically explored, and no part with sufficient care, as the important discoveries made every year in Europe prove.

> I am now convinced that all our ancient formations, which are rich in fossils, have thus been formed during subsidence.
>
> In fact, the nearly exact balancing between the supply of sediment and the amount of subsidence is probably a rare contingency.
>
> These intervals will have given time for the multiplication of species from one or some few parent-forms; and in the succeeding formation such species will appear as if suddenly created.
>
> If such gradations were fully preserved, transitional varieties would merely appear as so many distinct species.
>
> [T]hese links, let them be ever so close, if found in different stages of the same formation, would, by most paleontologists, be ranked as distinct species.
>
> [W]e have no right to expect to find in our geological formations, an infinite number of those fine transitional forms.

But despite this, he predicted that future generations, having the benefit of better understanding, would discover a substantial number of fossil species that he called "intermediate" or "transitional" between what we see alive today and their taxonomic ancestors at successive levels in paleontological history.

In fact, in the century and a half since then, we've found millions of evolutionary intermediaries in the fossil record, much more than Darwin said he could reasonably hope for. Some of these illustrate a smooth and continuous transition from eukaryote cells through every taxonomic level of descent into modern humans. We will cover this in detail in the following chapter.

There are three different types of transitional forms, as explained in the following quote from the Transitional Vertebrate Fossils FAQ featured on TalkOrigins.org:

> "General lineage":
> This is a *sequence of similar genera or families*, linking an older group to a very different younger group. Each step in the sequence consists of some

fossils that represent a certain genus or family, and the whole sequence often covers a span of tens of millions of years. A lineage like this shows obvious morphological intermediates for every major structural change, and the fossils occur roughly (but often not exactly) in the expected order. Usually there are still gaps between each of the groups—few or none of the speciation events are preserved. Sometimes the individual specimens are not thought to be *directly* ancestral to the next-youngest fossils (i.e., they may be "cousins" or "uncles" rather than "parents"). However, they are assumed to be closely related to the actual ancestor, since they have intermediate morphology compared to the next-oldest and next-youngest "links". The major point of these general lineages is that animals with intermediate morphology existed at the appropriate times, and thus that the transitions from the proposed ancestors are fully plausible. General lineages are known for almost all modern groups of vertebrates, and make up the bulk of this FAQ.

"Species-to-species transition":
This is a set of *numerous individual fossils that show a change between one species and another*. It's a very fine-grained sequence documenting the actual speciation event, usually covering less than a million years. These species-to-species transitions are unmistakable when they are found. Throughout successive strata you see the population averages of teeth, feet, vertebrae, etc., changing from what is typical of the first species to what is typical of the next species. Sometimes, these sequences occur only in a limited geographic area (the place where the speciation actually occurred), with analyses from any other area showing an apparently "sudden" change. Other times, though, the transition can be seen over a very wide geological area. Many "species-to-species transitions" are known, mostly for marine invertebrates and recent mammals (both those groups tend to have good fossil records), though they are not as abundant as the general lineages (see below for why this is so). Part 2 lists numerous species-to-species transitions from the mammals.

Transitions to New Higher Taxa:
As you'll see throughout this FAQ, both types of transitions often result in a new "higher taxon" (a new genus, family, order, etc.) from a species belonging to a different, older taxon. There is nothing magical about this. The first members of the new group are not bizarre, chimeric animals; they are simply a new, slightly different species, barely different from

> the parent species. Eventually they give rise to a more different species, which in turn gives rise to a still more different species, and so on, until the descendents are radically different from the original parent stock. For example, the Order Perissodactyla (horses, etc.) and the Order Cetacea (whales) can both be traced back to early Eocene animals that looked only marginally different from each other, and didn't look *at all* like horses or whales. (They looked rather like small, dumb foxes with raccoon-like feet and simple teeth.) But over the following tens of millions of years, the descendents of those animals became more and more different, and now we call them two different orders.

We have ample examples of each of these categories, but creationists still insist that we've never found a single one because what they usually ask us to present are impossible parodies that evolution would neither produce nor permit. For example, child-actor turned creationist evangelist Kirk Cameron went on Bill O'Reilly's Fox News program, arguing against evolution and holding a painting of a crocodile's head attached to a duck's body. He said, "You've got to be able to prove transitional forms; one animal transitioning into another. And all through the fossil record and life, we don't find one of these: a crocoduck. There's just nothing like it!"

In fact, Darwin explained in detail why we should *not* find anything like this.

> I have found it difficult, when looking at any two species, to avoid picturing to myself, forms directly intermediate between them. *But this is a wholly false view*; we should always look for forms intermediate between each species and a common but unknown progenitor; and the progenitor will generally have differed in some respects from all its modified descendants. [Emphasis added]

Such blends are only permissible when both species are closely related (as within the same genus), and never when they're in distant families of separate taxonomic classes (as with crocodiles and ducks). Ducks are birds, living descendants of dinosaurs, which were themselves a sister group to crocodiles, but older than crocodiles, and thus not descended from them. So Cameron's crocoduck couldn't be much more wrong.

We're not looking for a blend of two species that both currently exist. Such a thing would actually go *against* evolution. Instead, Darwin said, that if his theory were true, then what we should find would be a basal form potentially ancestral to both current species. In this one case alone, we've found dozens of them in a nearly continuous lineage dating beyond the dawn of the Mesozoic era.

The most famous one was the first ever recognized as such. *Archaeopteryx lithographica* was discovered in 1860 and was the first of many lines of evidence revealing that birds had evolved from dinosaurs. So Darwin's theory was already vindicated while he was still alive. Of course, creationists will never accept that, and still complain that Archaeopteryx can't be intermediate because we can't prove it's the single crown species from which all other birds emerged. But it doesn't have to be, and that's not what transitional means.

"Transitional" does not mean a necessarily "ancestral" species. It means the subject has a mosaic of features intermediate between a line of organisms before and others after, or linking two potentially related groups, providing evidentiary support that interrelated lineages are fully plausible. That's what Archaeopteryx is: an intermediate morphological form "linking" Mesozoic dinosaurs to modern birds.

In biology, species can be precisely identified genetically. In paleontology, however, they're determined morphologically, according to their physical characteristics. Creationists argued that Archaeopteryx still doesn't qualify because it's 100% bird, but they're difficult to pin down as to why they say this. Because this animal (like all other quasi-birds of that age) lacks many definitive features of modern birds, and it retains so many distinctly saurian features, that when the last Archaeopteryx was found in the 1960s, the traces of its feathers weren't immediately evident and it was thus mistaken for a small Coelurosaurid dinosaur called Compsognathus. Coelurosaurs are small chicken- to turkey-sized therapods. So Archaeopteryx is apparently a coelurosaur with feathers.

Every "100% bird" has a beak, but Archaeopteryx didn't! No modern bird has teeth, but Archaeopteryx had reptilian teeth. Since there is so much resistance to accepting Archaeopteryx as a transitional species, I'll include a list of features making these distinctions more clear.

Exclusively avian traits found on Archaeopteryx but not on coelurosaurs:

- Opposable hallux or "big toe" (not diagnostic, not common to all birds)

Exclusively avian traits not shared with Archaeopteryx or other coelurosaurs:

- Trunk vertebrae fused
- Hand bones fused, wrist joint inflexible
- Some foot bones also fused

Transitional avian traits shared with Archaeopteryx as well as other coelurosaurs:

- Feathers
- Wishbone
- Claws on three unfused digits on "wing"
- First toe reversed
- Elongated pubis directed backward
- Four-chambered heart
- Gizzard
- Hollow bones with air sacs
- Parabronchial pulmonary (lung) system

Exclusively dinosaur traits present on Archaeopteryx but not shared with "100% birds":

- No beak
- Reptilian teeth
- Long tail with unfused vertebrae
- Neck attaches to skull from rear as in dinosaurs, not below as with birds
- Fibula length equal to tibia in the leg (fibula is shortened and reduced in birds)
- Pubic shafts with platelike and slightly angled transverse cross-section
- Center of cervical vertebrae have simple concave articular facets of archosaurs, rather than saddle-shaped surface in aves
- Sacrum occupies only six vertebrae (half the minimum for birds)

- Unreduced preorbital fenestra
- Ventral ribs

Anomaly:

- Uncinate processes have not been found on Archaeopteryx, but have been identified on theropods and are absent on some birds.

Conclusion:

1. Archaeopteryx is a transition between Mesozoic dinosaurs and modern birds.
2. There are no diagnostic traits to distinguish birds from dinosaurs.
3. Consequently, birds are dinosaurs.

Creationists continue to use every excuse they can think of to dismiss Archaeopteryx as an intermediate species. For example, they had long complained that dinosaurs were divided into the "bird-hipped" ornithischians and "lizard-hipped" saurischians, and that the line leading to birds was on the wrong side. However, when Harry Sealey named these groups in 1887, he didn't have a lot of examples to compare. He based his classification on the position of the pubic bone, and several examples of the theropod lineage, viewed in sequence, show the pubic bone apparently moving into the typically avian position. So the avian ancestors were on the right side after all.

Other creationists complained that Archaeopteryx's lungs weren't right to be transitional, because birds breathe through a system of hollow pneumatic bones and interconnected air sacs whereas crocodiles had a more standard hepatic-piston system; presumably dinosaurs did also. So Adnan Oktar (Harun Yahya) and others argued that "it is impossible for this structure to have evolved from reptile lungs, because any creature with an 'intermediate' form between the two types of lung would be unable to breathe."

That argument collapsed in 2005, when scientists compared specific bones from a sarus crane and a recently discovered theropod dinosaur called *Majunotholus atopus*. The comparison illustrated similarities in pneumatic features. It was already known that many dinosaurs had hollow bones like birds, but then someone noticed these bones also had all the minute indicators of an avian system of interconnected air sacs. They found the same indicators on Archaeopteryx. So they took another look at more familiar dinosaurs and noticed the same indicators on allosaurus and a host of others. As Jonathan Codd, a lecturer in the Faculty of Life Sciences at the University of Manchester, noted, "These dinosaurs possessed everything they needed to breathe using an avian-like air-sac respiratory system."

Creationists have even tried to imply that every such fossil found so far were fakes! These claims were common. Some people said the fossils were either buried all over the earth by "the Devil" or by God—but that in either case, their purpose was to test our faith.

Another guy thought he had exposed the whole paleontological "hoax" when he visited a natural history museum. He told me he reached over the ropes and scratched one of the "bones" and found that it was made of dental plaster. Thus he assumed that every fossil was a man-made forgery. He didn't care about my explanation about plaster casts being lighter and less dangerous to mount overhead in museums and how casts would enable a rare skeleton to be displayed in multiple museums at once, nor would he take my suggestion that he ask to see the real petrified bones downstairs in the lab.

One of the more famous allegations of fossil forgery came from the once-respected physicist named Fred Hoyle. As I explained in a previous chapter, Hoyle certainly wasn't your average creationist, but he did argue for a supernatural intelligent designer performing a series of recurrent creation events. He famously objected to Big Bang cosmology, and he was openly hostile to evolution. Hoyle said Archaeopteryx fossils were forgeries with faked feather impressions fraudulently glued onto a true dinosaur, and that Archaeopteryx was just Compsognathus dressed up like a bird.

Other more traditional creationists argued that the feathers were real, but that Archaeopteryx was a "true" bird and too different from dinosaurs to have evolved from them. How can the same fossil be 100 percent dinosaur to one creationist, and 100 percent bird to another? That doesn't matter as long as neither admits that it is truly *both!*

Their situation has become ever more desperate now that so many other dinosaurs have been confirmed to have feathers. In recent years, there has even been a cache of new fossils showing even finer gradations between Archaeopteryx and subsequent lines of increasingly more advanced birds, as well as earlier, less birdlike, but still definitely plumed or downy theropods.

There are so many feathered dinosaurs now that one wonders what excuses Fred Hoyle would have to make up to continue denying these facts if he were still alive.

They think any excuse will do, and they've done the same attempting to systematically dispute every additional intermediary ever seen since. No matter what, creationists will not admit that anything we ever find fulfills Darwin's prediction of transitional intermediates.

For example, I was in a radio debate with Ray Comfort of Living Waters Ministries when he denied there were ever any transitional species. I pressed him to say what traits would confirm whether any fossil was or wasn't transitional between humans and apes. What traits would he look for? He answered, "I wouldn't, because I'd know that I was made in the image of God."

I knew he couldn't answer the question honestly, because he knows that if he did, I would show multiple examples that met all those criteria—and he can't let himself be proven wrong on public radio.

Cameron and Comfort are partners, pushing the crocoduck idea together. Another amusing thing about that is that we have a long series of actual transitions between crocodiles and ducks, including Archaeopteryx and a series of other fossil birds leading to ducks. Among them are beaked birds like Ichthyornis, which still had teeth, and Confuciusornis, which still had unfused wing fingers.

On the other side, the fossil record also includes things like *Montealtosuchus arrudacamposi*, which Brazilian paleontologists described as a "missing link" leading to crocodilians. This is only one of several known intermediates between the first crocodiles and the earliest ancestral archosaurs such as Euparkaria, which were basal to both crocodiles and ducks. Again this is no short list of large gaps, but a fairly fluid gradation of many identified species in both directions. It's like a jigsaw puzzle that is so complete you have to examine it closely to notice where they might be any pieces still missing.

Creationists now know better than to ask for examples that evolution really requires, predicts, and might provide. Evolutionary biologists have often predicted what the fossil record should reveal and which fossils should exist only if their hypotheses are correct, and the geologic column keeps revealing exactly what they asked for, right where they expected it to be in geologic history.

Creationists make predictions too, but they're always wrong. For example, one I heard often as a boy was that scientists would never find a fossil fish with feet. This is one that evolution predicted, and actually required. If vertebrates moved out of the water and onto the land, then at some point in that process, there would have to be a fish with feet. This idea led to the Darwin fish that you sometimes see on bumper stickers. If you haven't seen it, it's the same as the Christian ΙΧΘΥΣ (Ichthus) symbol, sometimes called a "Jesus fish," except that the Darwin fish is facing the opposite direction and standing on L-shaped feet. Throughout my youth, whenever I was challenged to defend my "belief" in evolution, I had to admit that we didn't have that fish with feet—until they finally found it in 1987.

To be fair, they've actually found a whole sequence of them, from fish with lobe-fins or legs without feet to fish with partially developed fingers in their fins, and on into fish that could actually crawl on still-developing legs. I and others expected that the first fish to leave the water must have flopped across the land in an exaggerated crawling motion, the way some catfish and snakehead fish do today. I thought that legs and feet would have been a subsequent adaptation, but it turns out fish were already walking before they left the water.

According to other specimens already acquired, this momentous occasion must have occurred after fingerless leggy fish like Panderichthys and Elginerpeton, but before more advanced and adapted tetrapods like Ichthyostega or Seymouria. That put it in the late Devonian period. Our Darwin fish was found in the final stage of that epoch, right where it was supposed to be.

Acanthostega gunnari, discovered by Jennifer Clack, is the living embodiment of the Darwin fish. She found it in someone else's private collection. It had actually been found way back in 1933, but the guy who found it didn't examine it well enough to realize what he had. Imagine something that looks like a freshwater dewfish, jewfish, or eel-tailed catfish mixed with primitive salamander like an axolotl (sometimes known as a mudpuppy). It was essentially a fish with internal gills and a ray-finned tail, just like any other fish, but it looked more like a salamander because it also had four legs, complete with knees and elbows and ending in eight fingers on each "foot," though it didn't yet have any wrists or ankles. It had otherwise complete legs, but neither its legs nor its fishy skeleton were yet robust enough to support the animal out of water. That's one of the traits that makes it transitional. Later incarnations were better adapted. Tiktaalik, for example, even had a neck. Everybody made a big deal of Tiktaalik. I like Acanthostega more, because it had better feet and because my Darwin fish decal doesn't have a neck.

Of course creationists couldn't admit they were wrong, that they said we'd never find it and then we did. They're not allowed to admit things like that. So their apologists had to conjure whatever arguments they could to negate the significance of this Darwin fish. A pair of geologist apologists working for Answers in Genesis ministries both wrote evaluations intended to discredit Acanthostega. They each came to the required conclusion, but for opposite reasons.

The first, Andrew Snelling, wrote,

> The fossil record has also revealed many types of amphibians, including Ichthyostega and Acanthostega, in which the limb bones are firmly attached to the backbone and clearly *designed for* bearing the weight of the body while *walking*. [Emphasis added]

Then Paul Garner wrote,

> This conclusion is supported by the morphology of the fore and hind limbs which are difficult to interpret as load-bearing structures; rather they appear to be *designed for swimming*. [Emphasis added]

Both men argue that Acanthostega cannot be transitional because, according to Snelling, its legs support its body so well that it is "clearly designed" for walking, while Garner says it can't be transitional because its legs can't really support its body at all, and it is therefore apparently designed for swimming. Of course the reason he gave for why it could not be transitional is actually the very reason that it *is* transitional. I guess they wouldn't accept it as transitional unless it was inept at everything. I say that the best way to know whether a species qualifies as transitional is when creationists can't agree and each say that it is "100%" on either side of whichever division they want to imagine. This is our second example of that.

For another example of a failed creationist prediction, Josh McDowell's ministry was one of many to say that there "has never been a creature found with half-formed feet or a half-formed wing or feather." However, in the apparent lineage to early proto-tetrapods, there have been a series of sarcopterygiian proto-tetrapod fish fossils with as-yet incompletely evolved feet, and we can see their continued evolution just by looking at these fossils in a chronological sequence in a cladogram.

As for half-formed wings, we now know of several maniraptoran dinosaurs whose arms were plumed like half-wings—useless for flight, but very efficient for incubating a larger clutch of eggs. Anything that makes reproduction more successful also accrues a huge selective advantage for their evolution. Universal's *Jurassic Park* and Disney's *Dinosaur* were movies made before we discovered that velociraptors and oviraptors were both covered in feathers—including half-sized wings. We have others like Microraptor that have half-sized wings on their arms and also on their legs, such that they could glide when spread-eagled like a flying squirrel.

As for half-formed feathers, there again we have a parallel of feather development in the embryo being repeated in fossil transitions, represented by early dinosaurs like Sinornithosaurus, Sinosauropteryx, and so on. Feathers first appear as tiny spikes, which then erupt into hairlike follicles or divided fronds, which we call down feathers. Feathers develop further when the downy fibers fuse into a rachis. Some paleognath feathers consist of a rachis with downy fibers streaming freely off of it, but they aren't organized. Neognathe birds also have barbules zippering these fronds into a vane. Flight feathers are the most adapted by being asymmetrical. Each of these stages of feather development in embryo is seen duplicated in the appropriate geological sequence of fossils, so that some dinosaurs looked spiky or hairy, some were fluffy or downy, some had only a rachis, and so on. This is according to another one of the laws of evolution that we'll discuss in a later chapter.

For example, although significant mutations appear to occur with almost regular frequency, environmental dynamics rarely apply dramatic selective pressures such as the transition from water to land. How often does that happen? Whenever a new environmental niche opens up or a new survival challenge occurs, then new pressures are applied. Transitions are forced to result relatively quickly, according to Stephen Jay Gould's principle of punctuated equilibrium, which also holds that intermediate states would not continue to be favored and would consequently be comparatively rare in the fossil record and harder to find.

Evolution has to adhere to a number of natural laws like this, but creationism doesn't. It's just a belief in magic, but since it's also all about pretend, it requires that believers deny the evidence however they have to in order to maintain confidence of conviction. This is why creationists demand only monstrous absurdities or issue challenges they know still couldn't be satisfied no matter how true evolution may be; because they know already that whatever they insist on seeing today may be found and shown to them tomorrow, and if that happens, they'll have to make up new excuses for why it still doesn't count.

Some creationists say that to witness the origin of new species is not enough. They want to see the evolution of higher taxa, as if they're expecting the sudden appearance of some freak of science fiction. But every taxonomic level still begins with the subtle divergence of two very similar and closely related species. For example, palaeopriondon and haplogale might have looked like two similarly spotted civetlike kits of the same litter some 30 million years ago, but they were each basal forms of distinctly different families: Felidae (cats) on one side and on the other, Hyaenidae and Viverridae, being civets, genets, bearcats, meerkats, etc.

That's Feloidea, the cat side of the taxonomic order Carnivora—or some of it, anyway. Similarly, on the dog side, Hesperocyon was a postweasel precedent of undifferentiated bear-dogs from the Oligocene era 36 million to 39 million years ago. Dogs are digigrade, running on their toes, but bears and weasels are plantigrade, walking on the whole foot. Thus, their common ancestor was plantigrade like bears but still had a full tail like dogs, and consequently it looked rather like a wolverine. In fact, the whole of Canoidea is derived from a common ancestor that still had functional hands (like their shrewlike ancestors), and it looked a bit like a raccoon.

Transitional species were seldom the dramatic blend of unfamiliar extremes that evolution deniers demand, so they won't request to see anything the science actually requires. They usually won't define any criteria they would accept either, because they already know they won't accept anything even if we show them everything they ever ask for. Instead creationists might say, "If we evolved from fish, then how come they ain't found no fossils halfway between 'em to prove it?"

This of course ignores the staggeringly long list of fossil intermediates between those ranks of taxonomic clades, which we'll explain in the next chapter. Sometimes they'll say that if evolution were true, then we should see dogs giving birth to cats. This absurdity ignores the evolutionary law of monophyly, which we'll explain in the next chapter after that. More often they'll make only the most absurd demands imaginable, like trees giving birth to horses or bacteria turning into people, since everything and everyone is related anyway. Worst of all, we often get that sort of bewildering ignorance from legislators!

Often I hear complaints that evolution means that an individual fish decided to grow a pair of feet because it wanted to leave the water, so why don't animals today simply grow wings whenever they need to escape from predators? Yes, their denial *is* that desperate, and it really *is* that dumb. The willfully ignorant are capable of astonishing stupidity. It doesn't help that they won't look at what they don't want to see, either. Many people think there are no transitional species because the only fossil forms they're aware of at all are a handful of plastic pieces in a prehistoric play set. They've no idea how rich the fossil record is! We know of several hundred species just within dinosaurs, to say nothing of the thousands of examples of each of hundreds more taxa apart from that.

So far, we've found fossil fish with lungs, legs, feet, toes, hip bones, and even a neck. We've found turtles on the half-shell or with no shell at all, and in various stages of development between these, all the way from lizardlike basal reptiles to complete turtles. We've got half-reptilian semimammals too, lots of them! We've even got a robust family tree for elephants and elephant-like things tracing back to a basal form that is almost identical to the original ancestor of manatees.

Sirenians are fully aquatic mammals, like dolphins in slow motion. They're not as derived as dolphins, and they still bear many traits of their terrestrial past. For example, they don't just have fingers in their flippers, but elbows too, and their fingernails—the last remnants of their former hooves—still show from the outside of their flippers even though they can't serve a purpose anymore. No creator would design something like this! This only makes sense in light of evolution. The sirenian evolutionary lineage was also completed with the addition of *Pezosiren portelli*, the fossil of a walking manatee. According to paleontologist Daryl Domning,

> This is the most primitive fossil found so far. We've found others that couldn't support the animal's body weight. But this is the first whole skeleton with legs that could support the animal's body weight out of

> water, yet has clear adaptations for aquatic life. We essentially have every stage now, from a terrestrial animal to one that is fully aquatic.

Horses are another good example. While the equine family tree does have side branches, it also shows a direct descent from the 55 million-year-old 4-toed eohippus to the 40 million-year-old 3-toed miohippus. From there it continues to the 25 million-year-old merychippus, which had one primary toe and two vestiges (toes that are reduced and no longer touch the ground). Then we have the 10 million-year-old pliohippus, which had only one toe left, a trait all modern equines inherited.

There is now a fairly complete sequence for the marine transition and development of ichthyosaurs, mosasaurs, and whales, including whales that still have four legs, and earlier ones that could even walk. There is now also a complete record of the transition of lizards to snakes, with a few fossil snakes that still had itty bitty completely useless legs. The 120 million-year-old Tetrapodophis still had four fully developed albeit absurdly tiny legs, but the 95 million-year-old Haasiophis had only partially formed remnants of its no longer functional hind legs. Today all that remains of serpentine legs are a pair of tiny hind claws found only on male boa constrictors. It is important to note for any creationists who might be reading this that while we have confirmed that snakes once had legs, we're still absolutely certain they could never talk.

Experts estimate that all the collective genera still roaming around now only amount to about 1% of all the species that have ever lived. Practically everything there ever was ain't no more. Every animal species living today has definite relatives both extant and extinct, and evident in the fossil record. And in one sense, all of them, even the things still alive, count as transitional species.

Of course creationists don't accept that, and insist on a much more restrictive definition. That's fine. But in order to determine for certain whether anything does or doesn't meet the requirements, we have to know what those requirements are. Only one creationist website that I know of is brazen enough to post a definition of transitional species which is also correct according to evolutionary biologists, so at least we can verify there is a common set of criteria both groups can agree upon. The definition offered by the site, WasDarwinRight.com, is as follows:

> What is a transitional fossil? "A transitional fossil is one that looks like it's from an organism intermediate between two lineages, meaning that it has some characteristics of lineage A, some characteristics of lineage B, and probably some characteristics part way between the two. Transitional fossils can occur between groups of any taxonomic level, such as between species, between orders, etc. Ideally, the transitional fossil should be found stratigraphically between the first occurrence of the ancestral lineage and the first occurrence of the descendant lineage."

However, this site also says that no such evolutionary links have ever been found, but then it goes on to list several that have been found in an attempt to dismiss extant examples of single-cell to multicellular transitions, the successive phylogeny of insects, the emergence of vertebrates and of whales, amphibian fish, therapsid "mammallike" reptiles, and acquired adaptations for flight in dinosaurs, pterosaurs, insects, and bats.

I wrote to the webmaster of this site, Luke Randall. I pointed out all these items in their list of things never found actually *have* been found, and I explained how all of them meet every one of the criteria he himself laid out. He wrote back saying he knew that of course, but that he wouldn't make any corrections on the excuse that he could even ignore his own rules if he needed to.

I said:

I said that he [Darwin] predicted that many transitional species would be found, and as I have already shown you, they have been. But your site still says they haven't. Now is that honest?

He said:

I take your point about transitional fossils, but belief in them does rely on the assumption that one form changed into another.

No it doesn't.

Back around Y2K, Kathleen Hunt, then a zoologist with the University of Washington, produced a list of a few hundred of the more dramatic transitional species known so far, all of which definitely fit every criteria required of the most restrictive definition. Each of these also included detailed explanations. Among these were Enaliarctos, a soon-to-be seal, but one that could still walk on land, and a late-Oligocene transition between modern sea lions (themselves a transitional species) and otterlike canoids. Here's how she described it:

> Pinniped relationships have been the subject of extensive discussion and analysis. They now appear to be a monophyletic group, probably derived from early bears (or possibly early weasels?).
> Seals, sea lions & walruses:
>
> - *Pachycynodon* (early Oligocene)—A bearlike terrestrial carnivore with several sea-lion traits.
> - *Enaliarctos* (late Oligocene, California)—Still had many features of bearlike terrestrial carnivores: bear- like tympanic bulla, carnassials, etc. But, had flippers instead of toes (though could still walk and run on the flippers) and somewhat simplified dentition. Gave rise to several more advanced families, including:
> - Odobenidae: the walrus family. Started with *Neotherium* 14 my, then *Imagotaria*, which is probably ancestral to modern species.
> - Otariidae: the sea lion family. First was *Pithanotaria* (mid- Miocene, 11 Ma)—small and primitive in many respects, then *Thalassoleon* (late Miocene) and finally modern sea lions (Pleistocene, about 2 Ma).
> - Phocidae: the seal family. First known are the primitive and somewhat weasellike mid-Miocene seals *Leptophoca* and *Montherium*. Modern seals first appear in the Pliocene, about 4 Ma.

I'm not going to share her whole list, because she describes hundreds of examples representing all the major groups of terrestrial animals, extant or extinct, and that list was compiled a long time ago; they've found a whole lot more since then!

At this point, I can only think of a couple lineages that almost nobody cares about where we still have any obvious "missing links." For example, I suspect that scorpions aren't really arachnids; I think that they and arachnids emerged separately, with scorpions being closely related to Mixopterecean eurypterids. There's another line of eurypterids that look like they might have given rise to spiders too, and all eurypterids look to me like modified trilobites, although I understand that these appearances are likely just a superficial coincidence of convergence. I might well be wrong about the origin of scorpions because we're still missing a couple necessary intermediates for that lineage, but we've got them for pretty much everything else.

Suffice it to say that myriad transitional species have been and still are being discovered—so many in fact that many biologists and paleontologists now consider that list innumerable especially since the tally of definite transitions keeps growing so fast! Several lineages are now virtually complete, including our own.

Of course creationist websites and videos would never acknowledge that. They have to say that the missing link is *still* missing. No, it isn't, and it hasn't been for a long time now.

There was a missing link in 1859, back when there were only two species of humans yet known in the fossil record (*Homo sapiens* and Neanderthals) and no intermediate fossils to link them with any of the other apes we knew of at that time. Since then, we've found the fossils of thousands of individuals of dozens of hominin species, many of which provide a definite link to the other apes (see table 1). By 1976, the British Natural History Museum estimated its cache of fragments from still unidentified fossil hominids represented as many as 4,000 individuals.

Evolution deniers often try to minimize evidence by saying that all the fossils of ancient humans would fit in a single coffin. That was true several decades ago, but not anymore. According to a paleoanthropology textbook published in 2009, *The Human Lineage*, "All known fossils of ancient humans would fill a railroad boxcar."

Back in Darwin's day, there were only two particular pieces predicted to complete the puzzle. It was never supposed that we evolved from any ape species still alive today. Instead, the theory held that chimpanzees and humans were sibling species, daughters of the same mother. Remember that we're talking about populations, not individuals. So the first link we needed to find was an ancient ape apparently basal to both of us to prove there was a potential progenitor of both groups. We had already found that link in Europe five years before Darwin went public, so we already had an evident "chain" of transitional species from which only one more link was needed.

Why was there only one "missing link" in Darwin's time when we knew that the evolution of men would require more than that? Because modern apes are so similar to us already that if there was one found with features roughly halfway between humans and chimpanzees, then they'd be so like

Table 1. Fossils of ancient humans, by hominin species

Date range	The hominin tribe, a subset of fossil hominids, including early humans and their closest relatives	Minimum number of individuals
7.0–6.0 mya	*Sahelnthropus tchadensis*	6 individuals
6.3–5.6 mya	*Orrorin tugensis*	5 individuals
5.8–5.2 mya	*Ardipithecus ramidus kadabba*	5 individuals
4.4 mya	*Ardipithecus ramidus ramidus*	50 individuals
4.2–2.96 mya	*Australopithecus afarensis*	~120 individuals
4.17–3.9 mya	*Australopithecus anamensis*	10 individuals
3.5–3.0 mya	*Australopithecus bahrelghazali*	1 individual
3.3 mya	*Kenyathropus platyops*	3 individuals
2.9–2.4 mya	*Australopithecus africanus*	~130 individuals
2.7–1.9 mya	*Paranthropus aethiopicus*	8 individuals
2.5 mya	*Australopithecus garhi*	4 individuals
2.5–1.4 mya	*Paranthropus boisei*	43 individuals
2.3–1.6 mya	*Homo habilis*	25 individuals
2.0–1.5 mya	*Paranthropus robustus*	28 individuals
1.9 mya	*Homo rudolphensis*	5 individuals
1.9–0.4 mya	*Homo erectus/ergaster*	210 individuals
1.8 mya	*Homo georgicus*	4 individuals
250–25 kya	*Homo neanderthalensis*	77 individuals
900–800 kya	*Homo cepranensis*	1 individual
800 kya	*Homo antecessor*	5 individuals
700–100 kya	*Homo heidelbergensis*	60 individuals
600–125 kya	*Homo rhodesiensis*	7 individuals
160 kya	*Homo sapiens idaltu*	3 individuals
130 kya-present	*Archaic Homo sapiens*	154 individuals
94–13 kya	*Homo floresiensis*	8 individuals

Note: mya = million years ago; kya = thousand years ago

us that they'd almost *be* us. It would be impossible to classify us separately. If we found it in the Pliocene epoch, it might as well be us because there is no reason why such a blend of traits should ever exist except in the case of evolution—not unless God just liked to manufacture forgeries to try and fool us on purpose.

The theory required that another extinct hominid be found in strata chronologically between the Miocene *Dryopithecus fontani* and the earliest known human species, which from 1891 to 1961, was *Homo erectus*, ancestral to both sapiens and Neanderthals. So we already had that transition before the final "missing link" was predicted. These ancient "ape-men" were definitely human, but they were so apelike with large brow ridges and small brains that they sent creationists into fits of denial, including many accusations of fraud.

We've found lots of candidates among the ancient apes, as many as fifty species that are now all extinct. But more than that, the theory also demanded that we find one "halfway" between humans and other apes in terms of morphology. We found exactly that too, way back in 1974; *Australopithecus afarensis* (nicknamed "Lucy") lived roughly 3.2 million years ago. One of her curious features was the location of the foramin magnum, the hole into which the spinal cord goes. It's moved forward halfway between humans and other apes, meaning that the spinal cord enters the skull from below rather than from behind. This proves she was a habitual biped like us, and also that she and her kind had much more recently converted to upright walking. So Lucy proved to be a fully bipedal ape whose hands, feet, teeth, pelvis, skull, and other physical details were exactly what creationists challenged us to find, yet they're still pretending we never found it! Worse than that, we didn't just find that one. According to the Smithsonian National Museum of Natural History, scientists have so far found roughly three hundred individuals just from that one species alone, all dating between 3.85 and 2.95 million years ago.

In 1977, three years after we discovered the no-longer-missing link in the human evolutionary lineage, Harvard paleontologist Stephen Jay Gould mentioned an "extreme rarity" of other clear transitions persistent in the fossil record until that time, and his comment taken out of context remains a favorite of creationist quote-miners to this day. But in the nearly forty years since then, there has been a paleontological boon such that we now have way more transitional species in many more lineages than we ever needed or hoped for. Now the problem for evolution is that there are too many contenders, while a compounding problem for creationists is that not even one of them should exist if their story was true. And yet they do, by the bushel! Despite all their complaints to the contrary, the intermediate gradations in the human evolutionary line are now so fine that paleoanthropologists can't agree whether they're all different species or merely mildly modified varieties of the same ones, such that there are no more so-called missing links needed in any stage of human evolution anymore.

It is important to note that if one examines the skulls of various breeds of dogs, they will see far greater variety and much bigger differences between breeds of that one species than there are differences between humans and the proto-human hominids in the fossil record. Experts consider that many of them would be chemically interfertile with us too, if they were still alive. So there is no determinable taxonomic division between us anymore. Even so, creationists still say we've never found anything that was "half-ape and half-human." Adhering always to black-or-white absolutes and being thus unwilling to admit any degree of variance other than 100% or zero, they make sure to divide every find into one of two boxes, even when they can't make up their minds which side of that imaginary partition each one belongs to.

Amusingly, three professional creationists were separately shown the same unidentified fossil and asked for their analysis of it. Scientists figured that a skullcap labeled KNMR-ER 1470 either belonged to *Homo habilis*, *Homo rudolfensis*, or *Kenynanthropus rudolfensis*. In 1979, Duane Gish of the Institute for Creation Research identified it as 100% human. Six years later, he was shown the same fossil again, but this time contradicted himself, saying it was 100% ape. In 1992, another professional creationist, Marvin Lubenow said it was 100% human, while one of his associates A. W. Mehlert disagreed in 1996, dismissing it as 100% ape. This proves what I said twice before, that the best way to know whether a species is transitional is when creationists say that it's "100%" on both sides of any division, even if they can't agree which side that is. This is the third example of that in this chapter.

Demanding to see an ape-man is actually just as silly as asking to see a mammal-man, or a half-human, half-vertebrate. How about a half-dachshund, half-dog? It's the same thing. One may as well insist on seeing a town halfway between Los Angeles and California, because the problem with bridging the gap between humans and apes is that there is no gap because humans *are* apes, definitely and definitively. The word "ape" doesn't refer to a species, but to a parent category of collective species that includes us.

> **Ape**: n.
> Any member of the taxonomic family, Hominoidea, including the subset Hominidae "great apes" (humans, chimpanzees, orangutans, gorillas, and their extinct relatives) and the "lesser apes," or Hylobatidae (gibbons and siamangs).

This is no arbitrary classification like the creationists use. It was first determined via meticulous physical analysis by Christian scientists a century before Darwin, and has been confirmed in recent years with new regulations in genetics.

In 1747, Carl Linnaeus, who would come to be known as the father of taxonomy, wrote:

> I demand of you, and of the whole world, that you show me a generic character—one that is according to the generally accepted principles of classification—by which to distinguish between Man and Ape. I myself most assuredly know of none. I wish somebody would indicate one to me. But, if I had called man an ape, or vice versa, I would have fallen under the ban of all ecclesiastics. It may be that as a naturalist, I ought to have done so.

In 1766, in his *Histoire Naturelle*, Comte de Buffon said that the ape "is only an animal, but a very singular animal, which a man cannot view without returning to himself." In 1794, philosopher Delisle de Sales said that the apes "seem to form an intermediate line between animals and human beings," and in 1817, Georges Cuvier described primates as "the order closest to man."

Still, conservatives objected as they typically do and created an arbitrary taxonomic family Pongidae, with the sole purpose of keeping men separate from other apes. In recent decades, however, genomic sequencing has confirmed that this grouping is and always was inaccurate. It has now been restructured, such that Pongo is a genus including only orangutans and extinct Asian relatives while humans are now included in the family Homindae as one of the "great" apes. We now know and can show that man is definitely an ape, both by definition and derivation. As Jared Diamond wrote in *The Third Chimpanzee*, published in 1991,

> Humans differ from both common chimps and bonobos in about 1.6% of DNA and share 98.4%. Gorillas differ somewhat more, by about 2.3% from us and from both of the chimps. Humans differ from orangutans by 3.6% of DNA and from gibbons and siamangs by 5%.

Charles Darwin anticipated such confirmation in *The Descent of Man*, published in 1871,

> It has often and confidently been asserted, that man's origin can never be known: but ignorance more frequently begets confidence than does knowledge: it is those who know little, and not those who know much, who so positively assert that this or that problem will never be solved by science.

Furthermore, it is impossible to define all the characters exclusively indicative of every known member of the family of apes without describing our own genera as one among them. Consequently, we can and have proven that humans are apes in exactly the same way that lions are cats, and iguanas are lizards, and whales are mammals. So where is the proof that humans descend from apes? How about the fact that we're still apes right now?

THE 10TH FOUNDATIONAL FALSEHOOD OF CREATIONISM

"The Evolutionary 'Tree of Life' Is Not Implied in Biology or the Fossil Record"

As explained in the previous chapter, creationists typically criticize evolution on the assertion that no transitional fossils have ever been found, even though we now know of at least several hundred distinct examples that fit even the strictest definition of that term. However, those transitions were largely unknown in Darwin's day. He predicted them and he expected them to confirm his theory, but when he said that, they either hadn't yet been discovered or weren't yet recognized as such. So how did he know they would be? Without knowledge of any transitional species whatsoever in his time, how could he have even conceived of evolution in the first place? What reason did he have to think of that? It turns out that the real evidence of evolution isn't in the fossil record; the most compelling case for common ancestry is indicated by the classification of life forms, originally known as the science of taxonomy. It makes such a compelling case that our current understanding of it would still effectively prove evolution even if we had never found any fossils at all.

The first attempt to classify all living things was *Systema Naturæ*, first published in 1735 by Swedish doctor and scientist Carl Linne, better known as Carolus Linneaus. As a physician, as well as a botanist and zoologist, Linneaus probably had a better understanding of comparative anatomy of all known life forms than most anyone else up to that time. He studied myriad organisms and documented a suite of traits for each, attempting to categorize all of them according to their distinctive physical characteristics, otherwise known as morphology. He was also a devout Christian. As Henry Morris, founder of the Creation Research Society and the Institute for Creation Research, notes,

> He was a man of great piety and respect for the scriptures. One of his main goals in systematizing the tremendous varieties of living creatures was to attempt to delineate the original Genesis kinds.

Linnaeus lived a century before Darwin and had no concept of evolution. He believed that all life had been created by God, but his classification scheme failed to indicate any created "kinds." Instead, every species he classified also belonged to a series of parent categories, all of which included sister sets and each with their own descendant groups full of other things apart from the ones he was trying to classify. He initially organized organisms by genus and species, but found that he had to create a collection of higher ranks too (kingdom, phylum, class, order, family, etc.), because everything was in some way connected to everything else. It's as if he was trying to fit everything into separate boxes of unique creations, but each of his boxes looked more like a set of Russian matryoshka dolls because they all indicated a nested hierarchy for reasons he couldn't explain. This didn't make sense from a creationist's perspective.

When he classified plants, he discovered nested hierarchies again. Some sort of flux derivation from a common ancestral origin was the only explanation for it, but Linnaeus wouldn't have considered that because it was thought to be impossible. In his time, it was believed that one might produce variations within a species or cross two breeds to get a hybrid, but they knew of no way to derive any new taxa. Species were believed to be immutable. So why did his collective taxonomic categories already look like a tree of life? Linnaeus was left with a mystery.

When Charles Darwin sailed around the world on the HMS *Beagle*, he caught, preserved, and collected samples of many small animals from the various locations they visited. When he brought his collections home, he was surprised to learn that what he thought was a random assortment of birds from the Galapagos Islands were all actually different species of finch.

He could identify the common variety finch known from the mainland, so he compared that to all these other birds that were each unique to their respective islands and existed nowhere else on earth. Looking at all of them together, he could see at a glance how one original species evidently evolved into several new ones, apparently through a series of slowly accumulated subtle changes. He already knew this could happen between different breeds, like with dogs or pigeons, but here he was seeing the same thing happening between different species. He didn't yet know how this happened, but at least he now knew that it *could* happen. If it could happen with one genera of birds, why not every family of birds? Why not all living things? As he was already in the process of classifying his birds, Darwin turned to *Systema Naturæ* with this new information, and all those nested hierarchies of Linnaean taxonomy suddenly made sense! (Or, as Russian Orthodox Christian and pioneer geneticist Theodosius Dobzhansky directly stated in the very title of his 1973 essay,"Nothing in Biology Makes Sense Except in the Light of Evolution.")

In the late 1600s, before fossil hunters discovered evidence of a hidden prehistoric world, Dutch lens crafter Antony van Leeuwenhoek had already used his specially designed microscopes to discover another hidden world inside a drop of water. At first he thought these tiny swimming things were microscopic animals, just like familiar animals, only smaller. Some of them were, but it turned out that many more were a whole other category of previously unknown life that didn't fit into the expected options of animal, mineral, or vegetable. Once it was realized that molds, fungi, algae, and such were something different than plants, and that protists were something else altogether, then taxonomists added more "kingdoms" of life: Animalia, Plantae, Fungi, and Protista (which really ought to be several kingdoms all by itself; scientists are still working on that one).

Leeuwenhoek also discovered bacteria, and that was yet another hidden world because they're everywhere—even on dry surfaces and in the air. They're also much different than every other form of life yet known, fundamentally different such that merely adding another kingdom wasn't really enough. So in the 1970s, when scientists discovered Archaea, a second distinctly different form of bacteria, they took that opportunity to reclassify both prokaryote groups as two separate "domains." All the previous kingdoms of life were grouped together within a third domain called Eukarya.

Since prokaryotes are so much smaller and simpler than eukaryotes, it would make sense that they came first and that perhaps eukaryote cells had evolved from them, but as scientists studied the structure of the cell, they encountered some surprises. In 1883, German botanist Andreas Schimper was studying chloroplasts in plant cells and noticed that they closely resembled free living cells of cyanobacteria. He also noticed that chloroplasts divide and reproduce on their own, rather than being generated by cells like any other organelle. He wrote in the footnotes of a paper he published about it that perhaps chloroplasts were bacterial, and that maybe they had come to live inside plant cells through a process called symbiosis. In 1910, a Russian botanist named Konstantin Merezhkowski described the idea as "endosymbiotic theory."

Later scientists realized that mitochondria also looked like a form of Rickettsia bacteria, and that it too reproduced by mitosis independent of the surrounding cell. By 1978, it was confirmed that chloroplasts and mitochondria both had different DNA than the cells they lived in, and that their DNA matched the profile of bacteria. Virtually all eukaryote cells have mitochondria, including humans. We now know that the human body has roughly ten times as many bacterial cells living inside us than we have of our own cells, but endosymbiosis shows that even our own cells partially consist of bacteria too, and that it is appropriated bacteria that power our cells.

Evolutionary biology often refers to a tree of life, representing different branches or ancestral lineages that can be genetically traced, but the tree analogy only works consistently with eukaryotes. Prokaryotes, like bacteria and Archaea as well as viruses, can sometimes exchange genetic material with other unicellular organisms on contact in a process called horizontal gene transfer.

Because we're not talking about an ancestor-descendant relationship, then it's not really evolution. Evolution is descent with inherent modification; horizontal gene transfer is a different situation and an added complication. The result is that the tree of life for prokaryotes has branches growing back into each other in confusing ways. That's a lot harder to do with multicellular organisms, but it still happens on occasion.

Some viruses called retroviruses insert their own genome into the DNA of the cells they infect. If this happens in a sperm or egg, the developing organism will inherit the viral DNA as part of its own genetic makeup. These copies of viral DNA are called endogenous retroviruses or ERVs.

About 8% of human DNA is actually old virus DNA. Your genome has a record of some 30,000 endogenous retroviruses inherited from past infections of your ancient ancestors. Where they are in your genome can be used to trace an evolutionary ancestry by comparing your genetic sequence to that of other people, and even other animals. At least a few of your ERVs are an exact match for those found in apes and Old World monkeys in both type and location, indicating a common ancestor with this specific infection. It is possible to genetically trace human ancestry, both by mitochondrial DNA and by sequence analysis of the genome. Even with a small amount of horizontal gene transfer, the eukaryote tree is still a tree. However, the "tree" of life seems to rise out of a network of simpler organisms that is more like a web.

Creationists often complain that science supposedly says there was only one universal ancestor of all living things, and that along the way we evolved both into and then from bacteria, but that doesn't seem to be the case. Twenty-first-century revelations in genomic research now imply that the origins of evolution came quite a while after the origin of life. There are now indications that at the root of each of the largest possible taxonomic divisions, there was a point when "descent" as it is currently understood was not yet occurring, at least not in any determinable lineage. Instead, there was a sort of horizontal gene transfer going on, coupled with endosymbiosis—neither of which could be considered part of the evolutionary process.

By definition, evolution requires inherited genetic frequencies, but the corequirement of descent with modification only allows for one series of ancestors rather than multiple lines of largely unrelated ones being inexplicably blended together. While taxonomy still points to a single common ancestor for all eukaryotes, that ancestor seems to be one of two or maybe three cellular siblings in a soup of similar sisters who evidently did not all descend from any sort of shared conventional parent. At the point where an actual evolutionary phylogeny began to take over more or less exclusively, the domain Eukarya had evidently already emerged separately and quite distinctly from either of the prokaryote lineages.

Creationists, of course, accept none of this. As intelligent-design theorist Jonathan Wells attempts to argue,"The branching tree pattern of Darwin's theory is actually not seen anywhere in the fossil record, unless we impose it with our own minds."

Wrong!

If you're going to categorize any collective, you must first define the grouping by the total tally of traits held in common by every member already universally accepted within that set—without making special exceptions for certain ones—before we can determine whether some new addition truly belongs there. The only way to objectively categorize all sorts of life is by their common characters—that is, those features shared by every member of that collective and only by them. This is how their traits become diagnostic and directly indicative of unique groups.

Let us also remember that the first man to attempt to classify all living things was a convinced Christian creationist who knew of no other option, as he had never heard of evolution nor even conceived of common ancestry, and therefore certainly wasn't trying to defend or promote either one. However, the system he originally devised determines that everything that is truly alive can be divided into two main branches (prokaryotes and eukaryotes), which each then continue diverging in an ongoing series of subdivisions emerging within parental sets, henceforth known as clades.

Even though Linnaean taxonomy didn't allow for evolution, it indicated a pattern that only evolution could explain. That meant there was a problem with our old method of classification and it had to be fixed. Over the last 250 years, we kept patching up the old seven-layer system by adding a suborder, infraorder, superfamily, subgenus, and so on, 'til we can't even tell how to rank the labels anymore. That's when we figured out there are no ranks, so we dropped the labels and found a new system that isn't so arbitrary. The system has been restructured to account for an evolutionary ancestry and to combine morphological and molecular analyses. The new revised version—called cladistic phylogenetics—is twin-nested; it includes a second nested hierarchy of genetic markers, and the old taxonomic ranks have been replaced by clades, many of which are unnamed.

A clade is any taxonomic grouping that includes all its descendant daughter sets. Every new species that ever evolved was just a modified version of whatever its ancestors were. This means that at every level, evolution is just a matter of incremental superficial changes being slowly compiled atop successive tiers of fundamental similarities, which are how clades are identified.

Creationists want to be apart from nature, not a part of it, so they resist being categorized as "mere" organisms. As we examine each of our named ancestral clades in succession, however, our indicated evolution becomes undeniable; every clade is a new tier of traits of classification, another descendant subset of the parent (ancestral) group, but each new clade is also mirrored by transitions indicated in our biology as well as by transitional fossils in paleontology!

I've written so much about this subject over the years that this one chapter could be an entire book all by itself, and it probably should be. There is a *huge* volume of data to present for each of the taxonomic categories listed here, but much of it is dry and technical, and requires a lot of explanation, so I will try to abbreviate each as much as I can without removing too much of the strength of evidence. As we discuss each clade in sequence, remember that each represents what some of your ancestors were in an evolutionary lineage, but each clade also represents what you still are. Humans still belong to each of the following groups:

Eukarya—metabolic organisms with nucleic cells. I have actually argued with creationists who refuse to accept that they're eukaryotes. These same people also deny that they're organisms, because they think both words refer to single-celled microbes or some other gross minimization of humanity. However, there is no speculation required to determine that humans definitely descend from eukaryotes (cells with a nucleus) and that we are still eukaryotes now, because it is a verifiable fact that every one of our cells (including our blood cells) are at least initially nucleic.

Opisthokonta—whose gamete cells have a single posterior flagellum. The domain Eukarya consists of several subgroups, including plants, algae, and those protists that appear to be collectively related as opisthokonts. The opisthokont subcategory is not a formal taxon, but it is a unique grouping of organisms in which those cells that propel themselves with flagellum have that appendage mounted in the posterior "pushing" position. Most opisthokont cells don't have flagellum except on their reproductive spores or sperm. The flagellate cells of most other organisms either pull themselves with front-mounted flagella or they swim with myriad cilia.

Opisthokonts include microspores and their apparent descendants, fungi, as well as choanoflagellates and all their apparent descendants in the animal kingdom. At this level of taxonomic division, opisthokonts are a single lineage, outnumbered by at least a handful of other major groups where everything else is something else.

Now if all life were specially created, you might think that some animals would have bikont reproductive cells, or sperm/spores might crawl on cillia like prokaryote cells do—but so far, not one animal or fungus is like that. They are all opisthokont, as if that trait appeared as an extremely rare and pivotal mutation among certain protists and was then inherited by all the generations that flowered out of that line ever since; that's how phylogenies work. The common ancestry model obviously explains this fact, but to date, no would-be critic of evolution has ever been able to offer any alternate explanation for this or any of the other trends we see in taxonomy.

Opisthokonts are nested within the taxonomic kingdom Protista. This is (I think) the only time that happens in cladistics. Sister clades aren't usually nested in their own siblings. The reason for it in this case is that protists are such an ancient group that they're amazingly diverse, and have all been erroneously lumped into a single clade for ease of classification. Otherwise, there should be more than a dozen different kingdoms of life, and all but a very few of them would be what we currently call protists.

Opisthokonts developed long before the Cambrian explosion, and even before the Vendian period previous to that. The bikont-opisthokont split evidently occurred in the Mesoproterozoic era and represents one of the most significant developments in the history of life on Earth, bowing only to abiogenesis and other pre-Cambrian transitions like the origin of sex and the ascension to multicellular forms.

Metazoa (kingdom Animalia)—multicellular opisthokonts that must ingest other organisms in a digestive tract in order to survive. Creationists howl at the idea that they are animals, but if you have any knowledge at all of what an animal even is, then you know that you are one! This isn't a matter of opinion either; it is a fact, and we can prove it! The biological definition and even the common dictionary definitions both describe humans as belonging to the animal kingdom. The standard biological definition is as follows:

Animal
any organic (Carbon-based) replicative RNA/DNA protein organism:

(a) consisting of multiple diploid cells which each contain a nucleus;
(b) which perform chemical reactions and achieve homeostasis;
(c) whose gammete cells have a single posterior flagella;
(d) which must ingest and digest other organisms in a digestive tract in order to survive.

Merriam-Webster Online Dictionary, meanwhile, defines the word's primary usage thusly:

> **Animal**
> 1 : any of a kingdom (Animalia) of living things including many-celled organisms and often many of the single-celled ones (as protozoan) that typically differ from plants in having cells without cellulose walls, in lacking chlorophyll and the capacity for photosynthesis, in requiring more complex food materials (as proteins), in being organized to a greater degree of complexity, and in having the capacity for spontaneous movement and rapid motor responses to stimulation.

Note that common dictionaries often get biology wrong. There is no such thing as a single-celled animal, as animals are multicellular by definition. So protists are not animals, but people are—and amusingly, even the Bible describes people as animals in Ecclesiastes 3:18–20:

> I said in mine heart concerning the estate of the sons of men, that God might manifest them, and that they might see that they themselves are beasts. For that which befalleth the sons of men befalleth beasts; even one thing befalleth them: as the one dieth, so dieth the other; yea, they have all one breath; so that a man hath no preeminence above a beast: for all is vanity. All go unto one place; all are of the dust, and all turn to dust again.

Obviously, if you and all your ancestors that you know of are animals by definition, then there shouldn't be any argument about whether you descended *from* animals. Nor could you generalize differences *between* humans and animals for the same reason. Instead, you could only cite differences between humans and certain *other* animals.

Those interested in evidence of this type of transition should look at a group of protists called slime moulds, some of which appear to live as fungi and some as animals. Animals and fungi are very closely related. The position of evolutionary theory and the model of common ancestry is that "simple" single-celled organisms are a lot harder to form than it is to build complex multicellular forms out of them. There is substantial evidence in the record of microfossils that various forms of algae and such existed in different parts of the world's oceans as much as a billion years before anything recognizable as a plant or animal.

The slime mould *Dictyostelium discoideum* represents the method by which scientists consider that life made the move from single cells to many cooperative cells, and it is singularly representative of the most profound of all evolutionary stages. This is a communal organism, meaning that millions of individual cells amass themselves to function as a single entity. It travels as a sort of worm, but since it is made of amoebic cells, each cell has to nourish itself independently. So when it encounters food, the worm-unit dissolves itself into its cellular components to osmosize food before reassembling and moving on.

At some point, it seems clear that communal organisms like this stopped dissolving themselves every time they wanted to eat—once the movement of courier cells facilitated some means of internal distribution. These cells of course eventually became dependant on the surrounding body and lost the ability to survive independent of it. Dead cells and waste material (including calcium and keratin) was of course pushed to the outside where they became a covering, like skin, bone, or shell in much later forms. In one particular opisthokont subset, Metazoa (animals), these hard formations appear to have begun as crystalline spicules. Those interested in finding transitional forms here should also look at choanoflagellates as an intermediate form from funguslike protists to the mother of all animals.

In the 1990s, a team of taxonomists, geneticists, and microbiologists working on the Shape of Life project sampled DNA from dozens of animals, and from them identified and sequenced a single gene common to every animal species known to man. This revealed a pattern, a genetic root fingerprint implicating the sponge as the most basal of all animals.

Eumetazoa—all animals more advanced than sponges. Understand that taxonomy is based as much on an organism's reproduction, life cycle, and development as it is on the physiognomy or function of the form itself. The next few subcategories in this sequence should illustrate this.

For example, sponges aren't always sedentary. Their nymph form is mobile and transports itself to the site where it will settle down for the rest of its life. Some of the other primitive animal forms also have a nymph stage nearly identical to that of the sponge, particularly in the case of the hydra. It establishes itself like a sponge does, but never seems to develop beyond that point.

Another of the evident trends of evolution is that the offspring of closely related organisms will look more alike than the adults will. Choanoflagellates very closely resemble the nymph stage of sponges. We are not actually descended from sponges, but from these sorts of nymphs, some of whom lead to sponges while others lead to more advanced animals.

Corals and anemones are also similar in this respect, and are considered closely related. Others may be similar to the sponge nymph too, except that they remain mobile their entire lives. Comb jellies are like that, except a bit sleeker. The last (and most advanced) group is slightly different in that it represents only the earliest phase of development, and begins life with a shape usually more like that of worms. Some of these can develop into beings that are much more complex.

Sponges—the most basal of all animals—are classified in the subkingdom Porifera. All other animals are in a sister subkingdom called Eumetazoa. Most of these develop differently than sponges in adulthood so that their cells are organized into various tissues, and with the exception of the shapeless placozoans, they also have some degree of symmetry in that they are at least tubular with a definite front and rear. These include cnidarians, the first animals to reveal a continuously repeating rhythmic pulse. Scientists believe this led to a heartbeat once such organs began to form.

Bilateria (triploblast animals)—animals that are at some stage of development bilaterally symmetrical. The first hints of the circulatory system appear in one line of a group of Eumetazoans that have achieved a bilateral symmetry. Almost all of the animals that come to mind (including starfish) are bilaterally symmetrical; they can all be divided equally to provide mirror left and right images. But Bilateria have more criteria than just that. All bilaterally symmetrical animals are triploblastic, which means they develop three germ layers:

> *Ectoderm*—Covers the surface of the embryo and forms the outer covering of the animal and also the central nervous system in some phyla.
>
> *Endoderm*—The innermost germ layer that lines the archenteron (primitive gut). It forms lining of the digestive tract, and outpocketing gave rise to the liver and lungs of vertebrates.
>
> *Mesoderm*—Located between the ectoderm and endoderm. It forms the muscles and most organs located between the digestive tract and outer covering of the animal. The circulatory system—and the skeletal system in vertebrates—stem from this. Note that only Bilateria have a mesoderm.

Coelomata—bilaterally symmetrical animals with a tubular internal digestive cavity. There are two subdivisions of Bilateria, based either on the presence of a body cavity or the lack thereof. Acoelomata have no cavity between the digestive tract and the outer body wall. Platyhelminthes (flatworms) belong here. All other animals have some sort of body cavity that joins a mouth of some kind at one end and an anus at the other.

In the pseudocoelomata, a fluid-filled body cavity separates the digestive tract and the outer body wall. This cavity (the pseudocoelom) is not lined or else not completely lined with mesodermal tissue. The so-called aschelminthes (including the Nematoda, Rotifera, and other groups of mostly tiny animals) are pseudocoelomates.

The true coelomata have a body cavity that is completely tissue-lined, resulting in a tube-within-a-tube design. The fluid-filled body cavity, completely lined with mesoderm (the coelom) separates the digestive tract from the outer body wall. Mesenteries connect the inner and outer mesoderm layers and suspend the internal organs in the coelom. All vertebrates and most invertebrates, including arthropods and mollusks, are coelomates.

Deuterostomia—coelomates in which early development of the digestive tract begins with a blastopore where the anal orifice opens before the oral one. There are some bilateral animals that may not look bilateral, as they seem more like mosses or plants, but their bodies can still be divided left-to-right with equal sides. Most bilateral animals look like slugs, cucumbers, or worms. One group typically has multiple legs like caterpillars, and within that group are some with hard shells of keratin-forming exoskeletons. These are the arthropods: insects, crustaceans, arachnids, trilobites, etc. These also have rear-abdominal hearts that (again) appear to have been inherited from an ancestor shared in common.

All the various types of bilateral animals still fall within two basic types, which are each defined by a particular stage of their development. In all these animals, the blastopore (the cavity that will become the digestive tract) develops internally and connects with an opening at one end of the animal and then the other, completing the oral-to-anal passageway. In lophotrochozoans (including mollusks) and ecdysozoans (including arthropods), the mouth opens first and the anus after that, but one line of bilateral animals got it backward. In deuterostomes, the anus is the first to open the connection to the blastopore, and the mouth makes the connection after that. Because this is such an early developmental stage, we could honestly say that at one point we are literally nothing but assholes!

This is a strange thing to have in common with every single "higher" life-form, isn't it? All other life, arthropods, mollusks, bryozoans, brachiopods, etc., develop a connection from the blastopore to the mouth first. This is without exception in any case seen so far. If they were specially created, one might think that any of them could develop by some other means, or in some other order. Maybe snails would develop like mammals and fish like squids—something like that wouldn't always indicate an inherited trait consistent with both the genetics and morphology of common ancestry. But that is never what we see; instead, everything in nature consistently adheres to everything we would expect of a chain of inherited variations carried down through flowering lines of descent. Starfish, urchins, acorn worms, and every single thing that ever had a spinal cord all develop the opening for the anus first. Isn't that odd?

Likewise, vertebrates (Deuterostomes) inherit hemoglobin (red blood), while chelicerates and mollusks (Protostomes) inherit hemocyanin (blue blood). The only exceptions are one family of Antarctic ice fish and one family of snails, both of which have lost those original functions but still have them indicated in their DNA. The common ancestry model obviously explains this fact, but to date no would-be critic of evolution has ever been able to offer any explanation of this, or any of the other trends we see in taxonomy.

Critics of evolution often talk about the Cambrian explosion in an attempt to imply that that all the diversity of life and all the phyla known today appeared simultaneously at that time. However, all the variation mentioned up to this point occurred prior to the Cambrian era, and all remaining developments yet to be discussed from this point on in the sequence occurred during or after the Cambrian era.

Chordata—deuterostomes with a spinal cord. One specific line of deuterostomes is evidently very closely related to another by virtue of two factors; both possess a single central nervous structure, and even pharyngeal gill slits. However, one of these is a worm! This worm with a mouth, anus, heart, and gills (which aren't typical of any other worm) also has a notochord, a sort of not-quite spinal cord. The family name for these worms is Hemichordata, which means "half-chordate."

The fossil record shows examples for the origin of chordates from creatures like this. Tiny swimming worms have been identified from a particular cache of Cambrian and Ordovician fossils in Canada's Burgess Shale. Some of these "worms" have teeth (pretty serious ones too), even though they have as yet no jaws nor bones on which to mount jaws. One particular specimen called Pikaia is considered to be the oldest example of a true chordate. It is a sort of slender slug with an elaborate body-length fin for swimming and a definite indication of a spinal cord, even though it didn't yet have a spine to keep it in.

Every single modern chordate we know of so far appears to be a deuterostome at the same time, so it is reasonable to assume that Pikaia developed that way too, and that it might have lead to other still-invertebrate fish. The fossil record reveals that some of these spineless fossil fish also developed partially calcified internal structures that lead to skeletons.

Craniata—chordates with a brain enclosed inside a skull. Not all chordates have a skull. Those that do are in the subphylum Craniata. Hagfish are craniates in that they have a skull, but they have no jaws or spinal vertebrae to go with it. The lamprey has a skull and vertebrae, but no jaws. Nothing has jaws without a skull or vertebrae, and that includes Chondrichthyes (sharks and rays). The skull and vertebrae are there, but they're now made entirely of cartilage instead of bone, except for the jaw itself. So it appears that the skull developed first, followed by vertebrae, and then the jaw. This is consistent with what we also see in the fossil record, where the first fish to have skulls and backbones and all that were still jawless.

Vertebrata—craniates with a spinal vertebrae descending from the skull. Not all craniates have vertebrae; living and fossil hagfish, for example (as well as some other fossil fish), only have a skull with no spine either of cartilage or bone. Both of these materials form deposits in the body as an animal matures. Sometimes this presents a hindrance, like painful calcium deposits forming inside joints. They can also be advantageous, like the way they evidently developed over time to form around the spinal cord in jointed segments, allowing movement with additional strength.

As with every other aspect of evolution, we have computer and robotic models to explain how it likely occurred, and the sequence is repeated in embryological development. As one might expect, the fossil record also reveals the earliest (and definitely most primitive) transitional skeletal vertebrates from the Cambrian and Ordovician periods, but we have plenty of not-yet vertebrate fish contemporary to, and sometimes even older than these—again implying an evolutionary ancestry.

Remember that we're only following one lineage, and that each of the left or right turns we take cause us to overlook other branches that may be just as hugely diverse as the one we're on, sometimes much more so. Staying on our course, the next fork in the road lies between Gnathostomes, whose skeletons are either cartilaginous or calcified.

Gnathostomata—vertebrates with a jawbone. Acanthodians are the first fish known to have jaws. These placoderms appeared in the early Silurian, near the very dawn of vertebrate evolution. The development of the jaw is in concert with the postvertebrate expansion of the skull, and of course this is consistent with what we see in extant or extinct example species representative of this lineage—both among Chondrychthys and bony fish as well as their jawless, lampreylike predecessors.

The first gnathostomes were the common ancestors of cartilaginous fish and vertebrates with more calcified bones. We have some interesting composite fossils from that era that show traits inherited by a number of subgroups at once. This is where we find the common ancestors and transitional species leading from the earliest and most primitive bony fish to the earliest and most primitive ancestors of sharks and rays.

Osteichthyes—bony vertebrates. Cartilaginous fishes (Chondrichthyes) diverged from Osteichthyes, whose skeletons were heavily calcified into bone. These again divided between teleosts, the ray-finned vertebrates (Actinopterygii) and lobe-finned vertebrates (Sarcopterygii, also known as Crossopterygii). On the issue of cartilaginous skeletons as opposed to those made of bone, I should point out that fossil Sarcopterygiian fish had a significant percentage of cartilage in their bones. This is evident in modern coelacanths, lungfish, gar, and the polypterus.

Polypteriforms (another seemingly "ancient" fish) still have much more cartilage in their skeleton than any modern (teleost) fish. So it appears that gnathostomes (jawed vertebrates) diverged at some point in the late Silurian or early Devonian period, with Chondrychthyes developing progressively more cartilage in their skeletons (except in the jaw of course), while Osteichthyes used progressively more calcium except where these hinge together.

Sarcopterygii—bony vertebrates that have both lungs and legs. Most of the bony fish in the Devonian period had lobed-fins, meaning fins that were attached to limbs complete with bones. Many of these, even some of the most ancient ones, also had lungs. This is evident on every surviving "ancient" species including polypterus. The development of the lung appears to immediately follow the divergence of bony fish from sharks, and that lung was originally nothing more than another sort of birth defect—an asymmetrical distension of the buoyancy bladder common to all bony fish.

Polypteriforms and snake-head fish are obligate air-breathers in their natural environment of warm, shallow, oxygen-depleted waters, and they can even use their short little limbs to amble away from evaporating ponds across dry land to find larger bodies of water. Other fish, even those without actual lungs, use their buoyancy bladders to supplement their oxygen supply when the water is too warm or stagnant to respire with their gills. This is one reason why tropical fish take "drinks" of air.

Cladists (people who really get into this classification system) tend not to use the word "fish" in any scientific application because that word has no consistent definition, so we say "chordates" instead. Similarly, Sarcopterygii doesn't just refer to some ancient order of fish that are now mostly extinct; it refers to them and everything evidently descended from them, including humans. We're still bony vertebrates with lungs and legs.

Stegocephalians—limbed vertebrates with digits on the ends of their appendages. Most modern teleost fish have a pair of pectoral fins on the sides, a pair of pelvic fins and an anal fin underneath, as well as a dorsal and caudal fin on the back, and the tail is usually divided into two flukes with a top and bottom vane. Stegocephalian fish from the Paleozoic had primitive tadpolelike tails with no flukes. They have only the two pectoral and two pelvic fins that have developed into four legs with fingers or toes, even though they may not have a shoulder or pelvis. The development of toes is well documented in the fossil record, starting from a random assemblage of bones that move into an increasingly functional formation.

Proterozoic stegocephalians had internal gills in addition to lungs, and otherwise looked more like salamanders than your typical fish. The bones of some early stegocephalians, like *Livonia multidentada* and *Tiktaalik roseae*, are matched by homology and found to have traits that are half-fish/half-amphibian.

Tetrapoda—gill-less stegocephalians that are skeletally adapted for four limbs. The next important stage in our sequence is Ichthyostega, another tetrapod-fish combo, this time with the tetrapod features dominant and the fish features vestigial. The tail fin-bars are still there, but they've faded significantly and will soon be gone. The gills are no longer internal, although it may have switched to distended external gills similar to those of the axolotl or the nymph stage of some salamanders. Ichthyostega was still not very adept on land, but it may not quite fit the image of a fish out of water either.

Notice that every clade amounts to "all of the ancestral traits, plus this new feature," but it can also be "every accumulated trait so far, plus the loss of this one feature." The loss of internal gills is one of the dividing lines between what are traditionally and informally known as fish, and the next important clade in our discussion.

Tetrapoda means two things, "stegocephalians lacking internal gills" and a "body plan based on four limbs," which these things have obviously already had for some time. The difference is that a true tetrapod also has skeletal adaptations to mount and operate those four limbs more effectively.

The explanation for why all terrestrial vertebrates are based on an originally quadrupedal configuration is that they inherited that, and modified it. The same goes for all those vertebrates that were once four-legged but have since lost limbs, and those that were once terrestrial but that have returned to the water—ichthyosaurs, plesiosaurs, mosasaurs, manatees, whales, etc. Otherwise, frogs, dogs, hogs, snakes and salamanders, bats and birds, turtles and pterosaurs, dinosaurs, lizards, lemurs, llamas, and mice and men are all tetrapods; not by some creative coincidence of magical manufacture, but by an obvious natural inheritance.

Anthracosauria—pendadactyl and postaquatic (or "terrestrial"). Tetrapods are divided into two sibling subgroups: Anthracosauria and Amphibia (modern amphibians). Some anthracosaurs were technically "amphibian" in the sense that they were adapted for both land and water, but they were distinct from the clade where we find frogs and salamanders. In the same sense, crocodiles are carnivorous, but they are not part of the order Carnivora, which is a subset of mammals only. A common argument against monophyletic classification is that we were once fish and then amphibians and we're not either of those things anymore, but true amphibians form a sister clade to ours and are not part of our ancestral lineage, although they were very close and do share some features.

One curiosity about modern amphibians is their lack of scales. I've heard creationist arguments that reptiles and fish both have scales but that amphibians don't, and that fossil amphibians and their ilk can't be considered an intermediate group for that reason. In reality, fish scales are made of dermal bone, and some extant amphibians called caecilians still have them (albeit sparsely) embedded in their skin, which again is just what you would expect from an intermediate grade. They're karyotypic of our anthracosaurian ancestors too. Reptiles on the other hand have either epidermal nodes (which look roughly similar to those on some newts) or follicular scales, which aren't present on any modern amphibian and which are quite different from the scales of fish. The scales of a reptile are made of keratin and are subdermal, like caecilian scales, allowing the skin to be shed without losing a single scale in the process. Another keratinized follicle site is at the ends of the toes—the first claws—which eventually either thickened into hooves on ungulates or thinned into fingernails on monkeys.

One thing that was inherited from preamphibians is the configuration of the fingers. This is where the still fishlike Ichthyostega represents another important upcoming clade. At first, their ancestors had eight fingers emerging amid the rays of their fins; then the front fins lost a digit, and the aft fins followed suit later on. What happened next is interesting. What has now been reduced to a dewclaw on dogs, a wing claw or leg spur on birds, and a thumb on you appears to have begun as a fusion of three digits into one, as seen in the fossils of Ichthyostega. These establish the pentadactyl (five-finger) configuration that you already know like the back of your hand.

An added curiosity is that part of this transition remains in your genome. Even though the genes for pentadactyly (five fingers) is usually expressed, the gene for polydactly (extra fingers) is actually dominant.

Amniota—dry-skinned tetrapods that develop in amniotic fluid and that have keratinized digits (claws, fingernails, hooves). Not all modern amniotes still have five digits with fingernails on each extremity. Some like the three-toed sloth have lost one or more of them. Artiodactyls (deer, cattle, giraffes, etc.) lost their thumb, and their index and pinky fingers have been reduced to nubs forming the "cloven hoof."

The fingernails of all ungulates (perissodactyls and artiodactyls) and afrotherians (elephants, Sirenians, and aardvarks) have thickened into hooves. Perissodactyls (including the rhinoceros) lost their "thumb" and "pinky" toes. Of this group, horses lost all but the middle finger, where the fossil record shows they had more once upon a time. Cetaceans are the only group to lose their fingernails entirely, which the fossil record shows were once hooved. However, they still have all five fingers concealed in their flippers. Sea cows and manatees even have fingernails still visibly present on their flippers. Fossils of snakes reveal they once had complete legs with five toes, each with claws on them, where only a few pythons still have any vestigial claws now, and they're only present on males. Some dinosaurs lost one or two of their fingers, which is why the birds that descended from them still don't have all five digits on their legs or arms, if they still have fingers at all. Most of the earliest dinosaurs (including some of the therapods) had all five. The "thumb" toe on dogs has been reduced to a useless dewclaw, and cats lost their dewclaws altogether. Humans follow what is considered a primitive body plan in that they never lost any of these affectations on either set of their limbs.

Of course the defining characteristic of amniotes is the development of amnion. This represents a significant evolutionary leap in that it is a developmental one, and such fundamental variance is especially rare, but this is owed to the equally dramatic change of moving from aquatic nests to a wholly terrestrial existence.

The eggs of amphibians are dependent upon the water, but are laid along the shore to provide the most light and warmth (and because the parents can no longer breathe water). Just as a peeled grape will develop a kind of skin when it is left to the air, so do the eggs of amphibians, usually at the cost of the contents. Eventually, some lucky generation had a kind of skin preinstalled that withstood accidental exposures to dry air. These obviously survived to pass the benefit of that mutation on. After thousands of generations, subsequent offspring that had tougher and harder skins on their eggs also passed on those genes. Also in the eggs of this particular anthracosaurian subgroup were extraembryonic membranes: the allantois, the chorion, an enlarged yolk sac, and an amnion filled with amniotic fluid.

Anthracosauria included some still very amphibian-like things as well as cotylosaurs, or "stem reptiles," which were the earliest amniotes known. These had a more advanced brain that evolved from modules and paleocircuits of the amphibian brain. This is slightly more advanced than the archipallium "fish brain," which may be a precursor to the mammalian hippocampus.

Solenodonsaurus janenschi is a transitional species between basal anthracosaurs and their apparently nonamphibious descendants. Known from a single incomplete fossil, it shows loss of the lateral line on the head, which was present in amphibians, but still has the single sacral vertebra of previously amphibious tetrapods. Two other specimens known from the early Pennsylvanian period (Hylonomus and Paleothyris) also show the sort of half-amphibian/half-reptile features that antievolutionists keep saying could not exist.

Synapsidae—amniotes with a single temporal fenestra. The first amniotes were captorhinids, which are considered the first reptiles. This is a fine example of the much-debated principle of punctuated equilibrium. The dry land represented a totally new set of environments to be colonized, and among vertebrates, only amniotes had the capability of doing that. So there was an explosion of biodiversity very quickly, where one original group of fossil forms divide into three or four distinct subsets:

> *Synapsids* have only one temporal fenestra, which is a hole in the side of the skull. This is the group that includes mammals and many "mammallike reptiles." The temporal fenestra is closed again on humans, but its outline is still visible as the "temple," a roughly circular crease in the profile of the skull.
>
> *Anapsids* (no temporal fenestra) lead to several interesting forms, some as big as a rhinoceros. Until very recently, scientists thought turtles might be the last surviving members of this group. Although some fossil evidence now suggests turtles may have convergently developed anapsid-type skulls from originally diapsid ancestors; in which case, there are no surviving anapsids.
>
> *Diapsids* (two temporal fenestra) were the "true" reptiles, class Reptilia. They divided into archosaurs on one side, which eventually lead to crocodilians, phytosaurs, pterosaurs, and

dinosaurs (including birds), and lepidosaurs on the other, leading to sphenodons, squamates (lizards and snakes), and numerous other varieties that are now all extinct.

An interesting side note about this division is that archosaurs all have a four-chambered heart where diapsids have a two-chambered heart, and turtles have a sort-of three-and-a-half chambered heart. *Eurapsids* (two temporal fenestra fused into one) began as an offshoot of diapsids and have several transitional species present to document their evolution into platyoposaurs, elasmosaurs, and icthyosaurs. This lineage too is now entirely extinct with no descendants.

Therapsida—synapsids with mammalian skeletal formations. The word "reptile" is another one that has no consistent definition in taxonomy. It used to mean "cold-blooded amniotes with scales and claws," but there are exceptions to those categories, especially when we adhere to the laws of monophyly. There are animals that have descended from "reptiles," and even some that are still universally accepted as reptiles but that either lack claws or scales, or can actually change their body temperature.

Early synapsids of the Carboniferous and Permian periods of the late Palaeozoic era were quite reptilian in the traditional sense. The most interesting group was the pelycosaurs, a paraphyletic group of synapsid "reptiles" that began to display very mammallike traits. From this group came the sphenocodonts; these led to therapsids, which demonstrated a series of transitions in jaw structure, musculature, skeletal adaptations, and the configuration of the inner ear bones, all of which are unique to mammals.

Cynodonta—therapsids with canine teeth. All modern mammals evidently derived from ancestors with canine teeth. Even the ones who eventually lost their canines still belong to this group because some members of these usually fangless mammals still have those teeth. Further, even when something is born without one or more features diagnostic of its parents, it must still be recognized as part of that family.

Some modern groups haven't yet lost this feature entirely; for example, horses and whitetail deer normally have a large and obvious gap from their incisors to their molars, but in some instances there may be an atavism such that small canines still appear. Several other hooved herbivores still have quite pronounced canines, most notably the muntjac, which is one of several saber-toothed deer.

Fossil transitions from the Carboniferous period on show a progressive increase in the neocortex of the brain. Such animals are not only much more intelligent but also capable of emotional feelings—even some degree of compassion, a trait that is mandatory given the mother's need to nurture and defend her young beyond what any mere reptile would ever do.

Some of these forms appeared at about the same time as the first dinosaurs by the early Mesozoic era. After that point, it seems there was little substantial development among mammals until the dominant dinosaurs were finally eliminated at the end of that era, some 150 million years later. Until then, every group of them bore a superficial resemblance to either shrews, rats, or opossums, which all look alike anyway. These are considered the most "primitive" template of mammalian shapes because of that, and also for the fact that their tails still have scales in addition to hair.

Theria (mammals)—endothermic (warm-blooded) therapsids with lactal glands. Modern mammals are defined as warm-blooded, fur-bearing animals with mammaries. It has so far been impossible to determine when any of these traits first appeared since there's little chance of mammary glands leaving any fossil traces, but there are obviously many other details that are particular to mammals alone. These indicate that there were once at least three more distinct mammalian subclasses than there are now, coexisting with the last dinosaurs. Some died out with them; triconodonts, paleorectoids, and multituberculates are known only from fossil fragments. All of the millions of mammals you can think of that exist today are members of only three surviving subclasses, the last remaining monotremes as well as the marsupials and eutherians that both appear to have flowered out of sister stock.

Eutheria—mammals that are born in a placenta and have nipples. There's no way to know how any of these extinct subclasses reproduced. It would be reasonable to assume that most of them likely laid eggs like modern monotremes, but there's no way to be sure if they did. If we didn't have any extant monotremes, we wouldn't even have been able to see how mammals made the move from eggs to live birth. The leathery shells of platypus eggs are so thin and membranous that they're hardly shells at all, and they hatch almost immediately after they've been laid. So it's almost a live birth, almost a placenta rather than a shell, and that's what distinguishes eutherians like yourself from other mammals.

One hypothesis for the evolution of lactation is that the disaccharide lactose evolved from a sort of postnatal "sweating" of lysozyme, an ancient protein widely distributed in the animal kingdom. As the young cling to the mother, often with their little semireptilian mouths, they would naturally ingest some of her sweat, and the development of actual lactose synthase glands would have a preferential selective pressure as a result.

Euarchontoglires—placental mammals with modern but still-generalized skeletal features. After the Cretaceous-Tertiary (K-T) comet or asteroid impact, which marks the beginning of the Cenozoic era and the end of the age of dinosaurs, mammals radiated into many different forms. Many of these were so specialized that they lost some of their original versatility. For example, horses lost all but one toe on each foot, and can now run up to 40 mph, but they can't do much else. Dogs did the same; their ancient ancestors couldn't run as fast or as far as dogs can, because they still had grasping hands that could do other things. So too with dolphins, who are masters of the sea but can't come out of the water anymore. That's what it means to have "generalized" features. It means that we still have all five digits on all four appendages, and we can do many different things that some of the specialists can't do anymore, even though they do what they do much better than we can.

Archonta—placental mammals with fully enclosed optical orbits and binocular vision. The skull of a dog shows the eyes sitting in curved recesses, but the eyes aren't protected from the outside. This is typical of carnivores. If you compare the skull of an oriental civet to an Asian palm civet, then to a yellow or grey mongoose, then to an Egyptian mongoose, you'll see a progression of closely related animals moving east to west wherein each stage shows bony protrusions extending around the eyes until they eventually connect, joining the zygomatic arch and auditory bulla. Then if you look at the skull of the African meerkat (another closely related carnivore), you'll see that arch has thickened to better protect the eyes in a fully enclosed arch of bone. Much the same thing happened with our ancestors.

Whether humans belong to each of the preceding "ancient" ancestral clades is not as hotly contested by creationists as whether we also belong to the following more recent clades.

Primata (primates)—hind-leg dominant archontids with opposable thumbs. It doesn't seem to matter that we actually have a collection of intermediate fossils to accompany every taxonomic clade, or the order of progression we can show in each case for the development of manual dexterity, dramatically increased intelligence, or the ability to walk upright; what's more important is how we can objectively verify that every member of all these groups still belongs to every parent clade already listed, and that we can do the same for every subdivision from this point on.

For example, primates are collectively defined as any gill-less, organic RNA/DNA, protein-based, metabolic, metazoic, nucleic, diploid, bilaterally symmetrical, endothermic, digestive, tryploblast, opisthokont, deuterostome coelemate with a spinal cord and twelve cranial nerves connecting to a limbic system in an enlarged cerebral cortex with a reduced olfactory region inside a jawed-skull with specialized teeth including canines and premolars, forward-oriented fully enclosed optical orbits, and a single temporal fenestra attached to a vertebrate hind-leg dominant tetrapoidal skeleton with a sacral pelvis, clavical, and wrist and ankle bones; having lungs, tear ducts, body-wide hair follicles, lactal mammaries, opposable thumbs, and keratinized dermis and nails on all five digits on all four extremities; in addition to an embryonic development in amniotic fluid leading to a placental birth and highly social lifestyle.

When evangelical minister Mike Huckabee ran for president in 2008, he said, "You know, if anyone wants to believe that they're the descendants of a primate, they're certainly welcome to do it."

But we don't believe this because we want to! Why would we want to? We believe it because we can prove it really is true, and that applies to everyone whether you want to believe it or not. We're not just saying you've descended *from* primates either; we're saying you *are* a primate! Humans have been classified as primates since the 1700s when a Christian creationist scientist figured out what a primate was and prompted other scientists to figure out why that applied to us.

Haplorhini—"dry-nosed" primates. Primates are divided into two main groups: simians and prosimians. Lemurs and lorises are prosimians (suborder Strepsirhini), meaning that they maintain very "primitive" features and are a karyotype of the most basal of the primate order. In fact, they're often used to fill in the missing details we can't get from fossil forms. Like rodents, bats, and even carnivores, prosimians have wet noses, muzzles, moveable ears, and even whiskers, and their fingers still have claws. They're barely primates at all in fact. Some of them even have six mammae, just like a raccoon or some other "lesser" mammal might have. Some lemurs, oddly enough, have only two nipples, and they're located over the pectoral muscles where we have them, and most other mammals don't. This kind of variance within any single lesser taxon is unusual and could imply that prosimians really do represent the basal origin of more advanced primate subgroups. Tarsiers were originally thought to be prosimians but are now known to be haplorhines. Consequently, they've been reclassified into a sister clade alongside anthropoids.

Anthropoidea—also known as Simiiformes, also known as monkeys. This is where the argument turns into a fight. Anthropoids (humanlike primates) are more advanced simians, adding a few more details to the previous list of diagnostic traits. For one thing, they have only two actual breasts, pectoral mammae that are not like those of any other mammal. Unlike lemurs, most monkeys have completely lost the ability to move their ears, although some individuals can still flex them a little bit. Their ears are also uniquely distinctive, unlike any other animal. Look at your ears. Every monkey has ears like that, and only monkeys have ears like that. Monkeys also lack the specialized sensory whiskers that so many "lower" animals have. Among the other obvious external differences are that male monkeys' genitalia are naked and pendulous; they're not in a furry sheath tethered to their abdomens anymore, as they are on most other eutherians. Monkeys have a well-developed caecum, which is a sort of distension in the digestive tract that is mildly advantageous. Monkeys usually have a flatter face than lemurs, and a tendency toward bipedalism. They all have (at least) trichromatic vision, meaning that monkeys can see in color where lemurs and most other mammals can't. Of course monkeys also have larger brains and are typically more intelligent than most other mammals.

Monkeys have detrimental mutations too. They've lost the ability to synthesize either vitamin D3 or vitamin C, and need to supplement both of these in their diet, or else they'll succumb to a condition of malnutrition commonly known as "scurvy."

Now, it should be noted that it isn't just science-denying creationists who object to this classification; many educated science advocates mistakenly balk at this too. One reason why is that we are told all our lives—by respected authorities—that we didn't come from monkeys, and here I am saying we did. However, I'm just repeating what the experts in this field have already determined. Bahinia, Eosimias, and Amphipithecus are each protoanthropoids from the Eocene, and Apidium is popularly considered to be both a parapithecid "monkey" and a potential ancestor of all other monkeys as well as apes and humans. As an online tutorial about early primate evolution maintained by Dennis O'Neil of Palomar College states,

> Monkeys evolved during the early Oligocene or possibly near the end of the Eocene. Their ancestors were most likely prosimians. These monkeys were the first species of our infraorder—the Anthropoidea.

At first I argued against a professional systematist that humans were apes, but not monkeys. After a long and heated exchange, I ultimately lost that argument. So I switched sides. Since then, I've had a few long and brutal arguments with a number of specialists over this, and eventually prevailed in each case. I will not hash out every argument to prove the point, because there are too many, but I will list a couple compelling highlights as we proceed:

Catarrhini—Old World monkeys. After our parapithecid predecessors, the clan of monkeys divides into Old World and New World varieties, classified by a suite of traits, but primarily by the position of the nostrils. New World Platyrrhines have nostrils splayed sideways in opposite directions, while Old World Catarrhines have them pointing downward, the way ours do.

In some respects, New World monkeys are actually older and more primitive than Old World monkeys. Some New World monkeys still have claws; Old World monkeys don't, instead just possessing flat fingernails. New World monkeys have strong prehensile tails, but in Old World monkeys, the tail is weak, diminished, or absent altogether. Some Japanese and Spanish macaques have no tails at all, and neither do apes.

Hominoidea (apes)—Old World monkeys capable of brachiation. Apes are distinguished from *other* monkeys by their ability to brachiate—that is, to fully rotate their shoulders, allowing them to swing through the trees.

By definition, apes are a subset of Old World monkeys, but it has been argued that there are differences between monkeys and apes. For example, all the Old World monkeys universally accepted as such, and which still exist, are Cercopithecidae, treated as a sister clade to Hominoidea. Cercopiths have "butt-pads" called *ischial callosities*, but some of the "lesser" apes do too. Apes also usually have a broader chest than other monkeys do, but some primitive apes had the same thoracic dimensions as any other monkey. Not all monkeys have long canine teeth either. Some are reduced, just like ours. Traits that are "usually" present still aren't definitive unless they are shared by all members, and if they're shared with the other group, then they're not distinct from that group.

Importantly, while cercopithecids are a sister clade to apes, and are not ancestral to them, there is a third Catarrhine group that is little known because its members are all extinct, but that is nonetheless also basal to both modern groups. As *Encarta*'s section on "Human Evolution" stated:

> A primate group known as Propliopithecus, one lineage of which is sometimes called Aegyptopithecus, had primitive catarrhine features.

> Scientists believe, therefore, that Propliopithecus resembles the common ancestor of all later Old World monkeys and apes.

Professor Eric Delson is considered the foremost expert on paleoprimatology, and he lists Propliopithecus as directly ancestral to both Hominoidea and Cercopithecinae. Thus, the progression moves from Aegyptopithecus, an "apelike monkey," to Proconsul, a "monkeylike ape." As *Wikipedia* notes, "[Proconsul] had a mixture of Old World monkey and ape characteristics, so their placement in the ape superfamily Hominoidea is tentative, with some scientists placing Proconsul outside it, before the split of the apes and Old World monkeys."

Finally, there is no trait held by all monkeys but not by any ape, and it is impossible to describe every character common to all monkeys without describing humans. The same thing goes if you describe all the diagnostic traits of Old World monkeys specifically. You'll end up describing people again. So regardless of the outraged objections often raised, it is a fact that whether classified by features or phylogeny, humans and the other apes are uneschewably monkeys, both by definition and derivation.

Hominidae—"great apes." Hominids aren't just "large apes"; there are more distinguishing features than just their size. For one thing, although all of them, including humans, have the same number of hair follicles all over their bodies, they are described as having more sparse hair than "lesser apes" or nonhominid monkeys. Among all other terrestrial animals, they also have an unprecedented brain-size-to-body-mass ratio of any other terrestrial animal. Consequently, they also have the most complex socio-emotional relationships of any other "social" animal. All apes also share a unique dentition of two incisors, one canine, two premolars, and three molars in each quarter of the mouth. In most cases, the canine is reduced, almost useless as a cynodont tooth, and the molars are also unique in all the animal kingdom. Every ape has them, and only apes have them. Each molar comes to five points interrupted by a Y-shaped crevasse. If you want to know what that looks like, go to a mirror and open your mouth real wide.

When you do that, also look at the dome-shape in the roof of your mouth. Therein lies a significant difference between humans and the other apes that are still alive. That shape, that bit of space above the tongue that the other apes don't have, is what allows us to form words, which is why they can't speak. This trait first occurs in *Homo erectus* just over a million years or so ago.

None of the other great apes can speak because of that little extra space in the mouth that only we possess. But other apes are all capable of complex communication, either by their own body language and by grunts and squeals with particular meanings, or by learned forms of sign language or symbols taught to them. Koko the gorilla is the most famous of the articulate signing nonhuman apes. Washoe the chimpanzee and Chantek the orangutan are other famous examples, and of course there are many more around the world who are not so well-known. These beings definitely aren't just mimicking their owners, nor are they "trained" to appear as if they really know what they're saying. This is no stupid pet trick. All of them have been repeatedly proven to possess sufficient cognitive skills and to comprehend the meanings of their words. They may know more than a thousand words, and they all use them to express personal feelings, love, compassion, and even grief for their loved ones who have died.

Koko's life mate, Michael, even remembered his mother being killed on the day he was captured. Upon Michael's death, one of Koko's human visitors told her that he was now an angel, to which the grieving gorilla responded "*Imagine.*" I don't know how to interpret what she meant, but I'd like to think she was saying, "You're imagining that."

The other apes are not as intelligent as we are, because our frontal lobes are larger than any of theirs, but they are definitely intelligent, cognizant beings. Even among the smartest creatures on earth, we are still the smartest by far, but we are not so unique, nor so far above them that we are not still one of them.

When Linnaeus classified humans as apes, the pre-Darwinian creationist community of the eighteenth century reacted harshly. They deliberately restructured taxonomy so that chimpanzees, gorillas, and orangutans would appear to be in their own genus (Pongo), separate from men. For more than two hundred years, extant nonhuman "great apes" were collectively considered pongids, and the word "hominid" had to be devised to explain the fossil "humanoids" we started digging up in the interim.

Hominini—exclusively bipedal humanish apes. The pongid classification strained to keep "apes" and humans separate in the presence of all these new discoveries of fossilized "ape-men," and particularly fossilized nonhumanoid apes, especially since "lesser apes" made "great apes" a contradiction, if their own extinct intermediate ancestors weren't considered apes as well. Just in the last decade, molecular evidence has finally vindicated Linnaeus' case and cleared up all this confusion by discrediting the old notions of men and apes, by recognizing that men *are* apes. Consequently, the Linnaean taxon Hominidae has been elevated from a genus to a family in all publications that are both current and that still favor the Linnaean system. Pongidae is now one of several genera within that family and includes only orangutans and their extinct relatives like sivapithecus and gigantopithecus. The seven surviving species of gorilla are now their own genus and are believed by some to have begun with *Sahalanthropus tchadensis* some six million years ago. Chimpanzees (bonobos and troglodytes) compose another hominid genus called Pan. They're still not human, but they're all humanoid; they're all hominids.

There has been another recent transition in nomenclature. I met an anthropology teacher a decade or so ago who put a misleading question on one of his tests: *"Explain the difference between an ape and a hominid."* I had to explain to him that "hominid" *means* "great ape," and that the word we used to apply to more humanoid things like australopiths is no longer hominid, but hominine, as in tribe hominini. He was having trouble with that and the concept of monophyletic classification. I encouraged him to look this up and gave him some recent sources, but he was very old and lamented that he didn't want to unlearn and relearn everything just to keep up. It's not really that difficult. We just realized there's one more category than we used to consider; that's all.

Homo (man)—bipedal human-apes with enormous brains. All fossil species within the genus "homo" are human, including the apelike *Homo habilis* and all the different varieties of *Homo erectus*, and even *Homo* *neanderthalensis* and *Homo floresiensis*, also known as the "hobbits" of Indonesia.

This brings us to sapiens—the "wise man." We call ourselves wise, yet we invented apologetics in order to defend and promote beliefs that we already know are plainly false.

These taxonomic clades wouldn't be this way if different "kinds" of life had been magically created unrelated to anything else—not unless God wanted to trick us into believing everything had evolved—because the phylogenetic tree of life is plainly evident from the bottom up to any objective observer who dares compare the anatomy of different sets of collective life forms. It can be just as objectively confirmed from the top down when reexamined genetically. This is why it is referred to as a "twin-nested hierarchy," but there's still more than that, because the evident development of physiology and morphology can be confirmed biochemically, as well as chronologically in geology, and developmentally in embryology. Why should that be? How do creationists explain why it is that every living thing fits into all of these daughter sets within parent groups, each being derived according to apparently inherited traits? They don't even try to explain any of that, or anything else. They won't because they can't, because evolution is the only explanation that accounts for any of this, and it explains it all.

THE 11TH FOUNDATIONAL FALSEHOOD OF CREATIONISM

"Macroevolution Has Never Been Observed"

The peregrine falcon may be the fastest bird in a powered dive, but ducks are actually the fastest birds in sustained level flight. That is of course except for the Pekin duck. It is larger and heavier than its wild ancestors and its wings aren't as big, so it can't fly at all. It doesn't have to either, because it's entirely domesticated; it doesn't migrate anymore because we've kept them cooped up for thousands of years and provided for them all year round. Then we allowed preferential breeding of the birds that were most like what we wanted them to be, until we got flightless white ducks that any predator could spot in an instant and that can't escape from anything. That works to our advantage, so that's how they turned out.

More recently, a German breeder has produced a lineage of giant rabbits up to four feet long that are consequently slower and more vulnerable to predators. They're raised in cages to protect them, but only long enough to be harvested since they were bred to be eaten (although they apparently also make good pets).

People have intentionally or inadvertently cultivated hundreds of different recognized breeds of rabbits, dogs, cats, cattle, barnyard birds, and other domestic livestock this way. Whether they were aware of it or not, they used the same evolutionary mechanisms of random mutations and environmental factors of reproductive success as in natural selection; it's still evolution either way.

However, artificial selection is significantly different from natural selection in two ways. First, we're doing it on purpose with a goal in mind. That is a more direct and therefore a much faster process than waiting for population mechanics to eventually get around to highlighting any particular trait. Second, we also tend to breed animals based on whatever traits are in our best interests rather than theirs. This is often the very opposite of the pressures applied by natural selection. What we consider "prize" livestock are usually judged according to criteria that natural selection would have weeded out. Take sheep for example. We've made them dependant on our practice of regular shearing, because their overgrown wool is unmanageable in the wild and would naturally have been selected against. The same goes for certain breeds of dogs that would never survive even one generation in any wilderness setting.

In a sense, artificial selection turns the tables on natural selection because we've bred animals that are disadvantaged, being slow, weak, dependent, and/or tasty—criteria that nature wouldn't have selected. The only value to these animals' existence is a potential that isn't reached until after they've been killed so that they may then be skinned or eaten. The irony in that is thinking that nature is cruel while we are "humane." At least natural selection has the effect of improving that organism's ability to survive and compete against adversaries or rivals.

Naturally selective pressures tend to give preference to descendant traits that offer some inherent advantage to that organism in a given environment according to population mechanics. Traits that create a hindrance for that animal's survival or reproduction are obviously eliminated relatively quickly, but those defects are sometimes the very traits we find arbitrarily appealing. Artificial selection means the offspring can even be silly in some cases, like Shar Pei puppies having too much skin for their bodies.

For another example, in the 1800s, fancifully styled sculptures of lions were popular in Chinese art, even though they bore little resemblance to real lions. The empress Tzu Hsi reportedly found a puppy with a birth defect that she thought made it look like an Oriental depiction of a lion, so she bred that dog with normal females until she eventually got more puppies with the same type of birth defect. They continued breeding those puppies with other dogs that looked the most like them until every generation produced the same weird-looking Chinese "lions." There are two breeds like that: the Pekinese (named for the city of Peking, which is today more commonly referred to as Beijing) and the Shih Tzu, which were named after the empress herself. Obviously, neither of these breeds looks much like an actual lion. Since Chinese artists didn't have any real lions to look at back then, their renderings weren't very accurate, but their lion sculptures do coincidentally look a bit like really big Pekinese dogs.

Our practice of cultivation has had a similar impact on agriculture. We didn't just find corn and learn to grow it as we found it; we made it ourselves out of something else. There was a series of genetic tests and hybridization experiments done to confirm that corn was derived from a type of bushy grass called teosinte. Roughly ten thousand years ago, Mesoamerican farmers sorted their seeds and only planted the ones most like what they wanted. They chose the ones that grew best and were the most robust, that had the longest cobs, with the largest, most edible, and best-tasting kernels, and so on. These variations were favored over others, and that led to a series of maize and then eventually to modern sweet corn as it is known and mass marketed today.

Creationists may argue that the transition of mere grass seeds to a leaf-sheathed ear of corn requires new "information," although they've never identified any metric by which we could measure that. Now that we know that is what happened, they rarely argue that anymore; now they say it's just changing the information that is already there, as if that weren't the case in every example of evolution at this scale. We can still be sure that it is a construct of selective breeding and human-guided evolution. We know this not just from archaeology and the fact that we still have transitional intermediates, but also from genetics, because the two morphologically distinct species at either extreme are still genetically compatible. The differences between corn and teosinte have been identified as a modification in only five genes, each of which were brought about or cultivated by artificial selection. So corn is one of the first GMOs (genetically modified organisms).

Another more amusing example is the banana. Ray Comfort of Living Waters Ministries famously described the banana as being a nightmare for atheists, because of what he said were clues of its intelligent design. Using what appears to be a store-bought, commercially produced yellow banana as a prop, he shared his thoughts in a video:

> Behold the atheist's nightmare. Now if you study a well-made banana, you'll find on the far side, there are three ridges, on the close side, two ridges. If you get your hand ready to grip a banana, you'll find on the far side there are three grooves, on the close side, two grooves. The banana and the hand are perfectly made, one for the other.
>
> You'll find the maker of the banana, almighty God has made it with a non-slip surface; it has outward indicators of inward contents. If it's green, too early; yellow, just right; black, too late.
>
> Now if you go to the top of the banana, you'll find that as the soda can makers placed a tab at the top, so God has placed a tab at the top. When you pull the tab, the contents don't squirt in your face. You'll find a wrapper which is biodegradable, has perforations. Notice how gracefully it sits over the human hand. Notice it has a point at the top for ease of entry, it's just the right shape for the human mouth. It's chewy, easy to digest, it's even curved toward the face to make the whole process so much easier.

Even if Ray was correct about the banana, that wouldn't explain why coconuts, citrus fruits, pecans, pomegranates, and most other fruits and nuts are so hard to get at and eat. It also brings into question of what sort of god would create the durian fruit of Southeast Asia? Durian is an ugly mass of sharp spikes concealing a slimy pusslike interior. It reportedly tastes okay if you're accustomed to it, and some of the natives in that region really like it, even though it smells like a corpse rotting in a sewer.

Ray says the whole of "creation" testifies to God. If so, most of it doesn't give a very good report. On the surface, Ray's banana seems to be a rare exception. But when he made the above statement, Ray obviously didn't know that much about bananas; as always, there is a "big picture" image that is out of focus in the creationist worldview. There are more gods in more religions than they know about. There are more stars in more galaxies than they know about. There are more extinct and extant species than they know about. There are more subspecies or breeds that we've cultivated than they know about, and there are a lot more types of bananas than Ray Comfort knows about.

One thing he should have noticed was that the banana he was eating didn't have any recognizable seeds. That should have been his first clue. Bananas are a type of elongated berry produced by a few different species within one genus of huge herbs. The only time any fruit doesn't bear seeds is when it has suffered a mutation that renders it unable to reproduce. That's when natural selection shuts it down, but artificial selection picks it up. What happened to the yellow Dwarf Cavendish banana is the same thing that happened to the navel orange: we took advantage of that mutation and kept cuttings of that original mutant alive, in order to market the seedless variety.

There are a few different banana cultivars available in different sizes. When they're ripe, they're either brown, yellow, red, purple, or always green. The original food banana is a green Asian plantain, which is crammed full of large hard seeds amid a hardcore matrix that is virtually inedible unless it's first softened by cooking. There is another more common wild-type banana that has noxious seeds and is so rigid inside that we still can't chew it, and wouldn't want to even if it was cooked. What does that do to Ray Comfort's argument? Neither of these wild-type bananas are anything like the way Ray Comfort described bananas. They don't even have the five ridges he talked about; the original plantains have only four ridges.

People have apparently been cultivating bananas for 7,000 years (a millennium more than Ray thinks the world has even existed), but the greatest advances in that crop have only come very recently. The mutation for soft-tissued sweet yellow bananas first occurred in Jamaica in 1836. Since then, many other cultivars have been developed from them, so that we now have the five-sided seedless variety too. Ironically, this proves that Ray Comfort's banana did have an intelligent designer, but the designers were men taking advantage of how evolution applies to agriculture.

In every form of life, small subtle mutations usually occur almost regularly and fairly constantly, and these tend to build up over time, resulting in slight chemical or physical fluctuations. If a particular difference is isolated and encouraged in subsequent generations, either by conditions or intent, that difference will be further exaggerated such that it and those like it can be distinguished from the original parent group.

The further back you trace any two visibly related lineages, the more their recessed ancestors will resemble each other, because at one time they were the same thing and had each grown their own way since. The same pattern is reflected in the aging process. This is why the young of two closely related species look more alike than the adults do.

Remember also that you can't look back on your own life and determine the hour when you had no more childish traits but were a fully mature adult. For the same reason, even if you could view all your consecutive ancestors over a million generations, it would still be impossible to positively identify any one individual as the first human. All you would find throughout that time are communities where individuals show only subtle variations from each other. You would not be able to recognize the year when those once unique traits came to dominate the whole tribe, nor could you ever find one generation in which no one could still interbreed with anyone from the generation before.

You could find a collective of chemically interfertile contemporaries whose genetics have become so unique as to become incompatible with some of those who were a thousand generations distant, because life is not as static as folks imagine. Living designs are a fluid dynamic. When a population is divided in complete isolation, unique distinctions will arise in each group, and the two will grow increasingly apart. The longer this goes on, the less likely it will be to bridge the widening gap and produce viable offspring.

To illustrate this, you share more traits in common with your siblings than you do with your cousins due to the recent ancestors you share with them, your parents. Deeper down, you share more in common with those in your extended family than you do with neighbors and classmates, etc.—people you don't recognize as part of your biological family. But you must realize that on some level you're still related. Deeper down, one could likely recognize subtle indications of cultural demes, which most people will still agree all descend from one common ancestral lineage despite their current apparent diversity and unfamiliar ways.

Deeper down, we've seen that new breeds of barnyard birds, domestic pets, livestock, corn, even bananas have to some degree been engineered by human intervention via artificial selection, and new subspecies have occurred in the wild via natural selection. In both cases, these stem from common ancestry, be that hundreds of breeds of dogs coming from one strain of wolves, or dozens of commercial bovines being derived from the now-extinct European aurochs.

All these reveal that life is a fluid dynamic producing new and subtle variation with every descendant. But evolution only occurs when new alleles are spread throughout a given community, at the population level. This is where selection comes into play, because the parent gene pool actually does more to inhibit new aberrations than to promote them. A smaller gene pool is much easier to influence, so what you usually get are more significant changes emerging in smaller colonies that have been genetically isolated from the main population.

Eventually, they may get to the point where the two groups are distinct—where a trait now held in common by every member of one group is not shared by any member of the other group. This is known either as a subspecies or a breed, depending on whether it was naturally evolved or artificially bred. If the two groups resume interbreeding, then all that may meld together again as if it had never been. But if they're isolated long enough, they will continue to drift further apart both physically and genetically until it becomes difficult to interbreed at all anymore. Eventually, they'll only be able to sire infertile hybrids, if they can still produce anything living. At the point when two sexually reproductive populations can no longer interbreed with viable offspring, then they have become two different species.

This is the most significant level in the whole of phylogenetics, when the daughter strain is now unrestrained by the once-dominant parent gene pool and is therefore free to express an even greater variance, thus continually widening the gap between them. This is a practical parallel of what creationists call the addition of new "information." Thus, "species" is the only taxonomic division that is genetically significant, and it is also the only one of the old Linnaean ranks with an applicable definition. Consequently, it is the only level of taxonomy that can be objectively determined, and the act of speciation is the only event in the contested scale of evolutionary history that could ever be observed, even if we had a time machine. So it is the only possible point of division between the largely unnecessary distinctions of macro- and microevolution.

Once upon a time, creationists openly dismissed both macro- and microevolution and argued accordingly. Today, such zealots tend to accept microevolution while confidently proclaiming that macroevolution is a matter of faith because (they say) no one has ever seen a new species evolve. When they don't know any better, they feel safe dismissing speciation as impossible. As just one representative example of this tendency, consider the following from the website of Exchanged Life Discipleship:

> The difference between micro- and macro-evolution is a major point of confusion between the Christian worldview and the Darwinian evolution worldview in today's culture. Microevolution is the adaptations and changes within a species while macro-evolution is the addition of new

> traits or a transition to a new species. Microevolution is a fact that is plainly observable throughout nature. Macro-evolution is a theory that has never been observed in science. Evolutionists usually argue that those who believe in creation are ignoring the facts. However, there is nothing that the evolutionists observe in science that creationist or Christians as a whole disagree with.

Religious people also typically employ erroneous and misleading usage of words so that they can never be proven wrong. They don't like to define their terms and won't accept your definitions either, so we have to glean their meaning from the implications of whatever their people say. They like to fall back on the "No True Scotsman" logical fallacy too, so if you're talking with creationists and you try to quote other creationists to establish or compare what they mean, it had better be someone they consider to be a like-minded representative authority.

For example, Phillip E. Johnson is a born-again Christian lawyer who cofounded the Discovery Institute and is considered the father of the intelligent design movement. On April 30, 1994, Johnson clarified the distinction between macro- and microevolution during a debate with evolutionary biologist William Provine at Stanford University:

> The theory is perfectly valid at that level [of microevolution]: minor changes that do not produce new kinds of organisms, and that above all do not add to the genetic information. . . . breeders are able to produce change only within boundaries. Even those dogs are all members of a single biological species, which are chemically interfertile. We don't get dogs getting bigger and bigger indefinitely—as big as elephants, or whales, much less changing into elephants or whales. . . . and the claim that if selective breeding hasn't produced the kind of macro changes, the kinds of new forms of life, new biolo—complex organs that are needed, that's only because there hasn't been enough time.

Amusingly, even though that was decades ago, he also predicted the imminent collapse of macroevolutionary science due to what he said was a lack of evidence. Creationists have been faithfully repeating that prophecy for more than a century now, but exactly the opposite is all that keeps happening.

Notice that Johnson identifies breeds as "members of a single biological species" according to whether they are chemically interfertile. He also identified the chemical fertility of species as his boundaries, and he contrasts that with "macro changes," which he identified as "new forms of life." Clearly, that meant "new species" to him, so in that debate, he was using the terms *microevolution* and *macroevolution* correctly.

Johnson also equated biological "species" with the creationist concept of "kinds," being whatever original organisms creationists think God magically created in the beginning. Then notice that Johnson expressed that evolution meant that dogs could or should *"turn into elephants or whales."* This is a key failure in the creationist concept. They throw out all evident ancestry and everything we know about the gradual accumulation of distinctive mutations, and instead expect that evolution is supposed to abruptly cause drastic changes *between* evolutionary branches rather than accepting that branching tree patterns gradually emerge within any given clade. They've created an impossible criterion of an absurd expectation that descendant species should be so different from their ancestral lineage that they're not even recognizably related to their parents.

Creationists insist that macroevolution has never been observed, and the excuse they use to deny that it has requires the addition of a bogus condition that simply does not apply. They say that we've never seen one kind of thing turn into a completely different kind of thing. They'll argue that evolution can only occur within "definite limits," and then only to subtle variance within an organism's "kind." They say new diversity is limited to rare and unviable hybrids between those kinds, and they usually say that the emergence of new species is impossible.

In the evolutionary perspective, any single ancestral species can diverge into two or more daughter species, each becoming so distinct that eventually either of the new species would be unable to interbreed with the ancestral and/or sister species anymore. Thus, one species becomes two, then four, eight, and so on—yet they are all obviously related, and still belong to all the ancestral clades as their parents did. They can create an additional clade, a new one their ancestors don't belong to, but it will still be part of the same sequence of clades as their parents were.

For the purpose of illustration, imagine some archaic long-fanged panther such as Dinofelis, one whose descendants include both lions and tigers. There are plenty of these sorts of cats in the fossil record, predating any of the big cats we have today. We have plenty of genetic evidence as well as fossils to indicate that all modern panther-types were derived from a prehistoric predecessor, similar to the Asian clouded leopard, which has the longest fangs of any living cat. Likewise, a pliohippus' descendants should logically include the many types of horses, zebras, and asses we have today.

By ignoring fossil forms, the creationist's perspective has that completely backwards, insisting instead on an illusory sequence of separately conjured "kinds" which (like "information") must remain forever undefined. Then they straw man the science by saying that that evolutionary diversity can only occur by mixing these "kinds" in hybridization. So in the creationist perspective, you start with lions and tigers, horses and zebras, ignoring or rejecting all of their ancestral clades. The lion and tiger produce a tigon or a liger, while horses and donkeys produce a mule or a hinny. None of these hybrids are fertile, although there has been at least one documented exception to this. Thus two species becomes three, for one generation only, and that's as far as it goes if it even gets that far.

This isn't the only way creationists argue their contention of course. Others argue that hybrids are only possible from species of the same "kind." This argument holds that horses and donkeys are the equine kind, while lions and tigers are the cat kind. In this view, God magicked up a cat and a horse—one species each—and all the subsequent subsets of those two kinds that we have today were derivations of those two original creations.

There are enormous problems with this. One of the worst ones for young-earth creationists is that they have to imagine all prehistoric forms from every geologic period must have lived at the same time just a few thousand years ago. The problem is worse than they know, because there are thousands more fossil species than they could possibly be aware of. The family trees of extinct ancestors for so many different taxons can be conveniently eliminated simply through the blindness of ignorance. This is another reason they reject transitional fossils, and instead demand some blend only between current "kinds." Their perspective has no depth.

Since most young-earth creationists also believe the end is nigh and that we're living in the last days, they deny both our past and our future. They may refuse to acknowledge geologic time scales and cannot admit that any new organism might be unable to interbreed with the stock whence it came because their sacred fables say they were *"created each after their own kind."* But of course, they can't say what a "kind" is either. Once again, the Bible fails to provide any clear explanation. Take, for example, Genesis 1:24–25:

> And God said, Let the earth bring forth the living creature after his kind, cattle, and creeping thing, and beast of the earth after his kind: and it was so. And God made the beast of the earth after his kind, and cattle after their kind, and every thing that creepeth upon the earth after his kind: and God saw that it was good.

At first, it seems as if the Bible identifies "cattle" as a kind, but it says the same about beasts. How could they be different kinds? What's the difference? Then we realize that it also said that "living creature" is a kind too. So they're really all just undeterminable subcategories.

Importantly though, the passage above, and even the rest of that chapter, allows that God could have permitted his creatures to evolve naturally instead of magically conjuring different kinds separate from each other. The essential difference between creationists and theistic evolutionists is in how they interpret Genesis 7:14: "They and every beast after its kind, and all the cattle after their kind, and every creeping thing that creeps on the earth after its kind, and every bird after its kind, all sorts of birds."

Ignoring for the moment that the Bible seems to provide a taxonomic category for "creeping things," it also seems to name birds as a kind, but then subdivides them into all "sorts" of birds. How many kinds of birds are there? Why aren't they all collectively referred to as the bird kind? Why are there multiple kinds within one kind? Not only do we still not know what a "kind" is, but we also don't know what a "sort" is either. Are they the same thing? What about feathered dinosaurs that are virtually indistinguishable from birds? Which "kind" are they?

This is the problem with unscientific thinking. It's inherently inaccurate because it's deliberately vague, just like a horoscope. One may as well learn to count: *One, a couple, a few, many, a lot.* It's like when I used to visit my grandmother's house out in the country; I'd ask when her pies would be done, and her answer could be "a while," "a spell," or "a jiffy." In each case, the answers given imply that the real answer is "I don't know."

Before they knew any better, creationists confidently claimed that no new species had ever evolved. Whatever examples I might present were dismissed as being something that was discovered only recently but might have been there all along, so they demanded that I show examples of new species evolving under direct observation. At first it seemed pretty fair odds against that, but eventually they realized that actually could happen, and it might have *already* happened. So they needed to change their story so that they could still accept that it did happen. As a result, they changed the definition of created "kinds" from species to some higher taxonomic classification. This change also gave them a way to explain how Noah managed to cram millions of species into a box the size of a football field and still float. Accepting speciation meant they wouldn't have to account for so many species on board.

They couldn't just move the definition of "kinds" up to the genus level, however, because that had already been confirmed too. There are two species of camels—dromedaries with one hump, and bactrians with two. No one argues that these are obviously related, even though they can't interbreed. They can only procreate "after their own kind." That should make these camels *different* kinds, but of course creationists won't accept that.

There was a species called Poebrotherium, an early generalized camelid known only from the fossil record. Chronologically and morphologically, it appears to be the karyotypic common ancestor of every camelid we have today. There are also intermediate species. Procamelus could be the granddaddy of llamas, alpacas, and vicuña in South America, and Oxydactylus could be the common ancestor of dromedaries and bactrian camels in Asia. That's if we select just three of the fifty camelids known in the fossil record. But that's just if you believe the "evolutionists," which of course the fundamentalists don't.

Camels clearly are closely related to llamas. This is too obvious to deny, even though camels have "new information" that llamas don't have, being their definitive humps. Despite this, creationists would rather that camels and llamas be two different "kinds." After all, they've grown far enough apart that they neither could nor would interbreed even if they were still on the same continent. They're not just different species, and according to the old Linnaean system, they're not even the same genus! Yet, they are closely related.

Although species remains the only defined taxon, I and others have suggested that a genus should be a collective of species who are related close enough to produce living (albeit sterile) hybrids. If their hybrids are fertile, they're the same species. If they're not, they're different species of the same genus. If they can't produce living hybrids at all, then they're too different genera or some higher level of division. In this case, taxonomists recognized the same level of classification as I would have according to a different set of criteria. Otherwise, a taxonomic genus is not defined the way I suggest it should be. It's not defined at all.

If it *were* defined the way I say, then what if hybridization is still possible but only through artificial insemination and laboratory assistance? Would that be a family-level distinction?

British biologist Julian "Lulu" Skidmore went to Dubai determined to prove a familial relationship between llamas and camels. She didn't just adhere to fossil similarities, definitive traits, or genomic sequencing—things that creationists habitually misunderstand and reject; she found more conclusive proof than that. In 1998, she managed to cross-breed a 990-pound camel dad with a 165-pound momma llama to produce Rama the "cama." These two lineages have been genetically isolated for roughly 30 million years. After extensive research, she devised a means of using laboratory assistance to breed viable hybrids that could neither reproduce nor even exist naturally. This is proof of an unobserved speciation event.

This may be one of the reasons why Ken Ham of Answers in Genesis has redefined "kinds" at the family level, ignoring all the obvious implications of that, and the fact that we're in the family of Hominidae, also known as the "great apes." Ham admits there are inconsistencies in his assertion, and that he arbitrarily places different "kinds" at different taxonomic levels as required by his interpretation of his favored folklore. One of the many problems with that is that each of these "kinds" still belong to a sequence of parent clades that are also the parents of several other "kinds" too. That can't be if they were all specially created, so creationists devised "barimology," an attempt to appropriate some acceptable percentage of cladistic taxonomy while trying to excuse the higher commonalities and to portray each of these "baramins," uniquely created "kinds." Although they still won't admit it, this attempt fails for several reasons.

First of all, creationists believe that australopithecines shouldn't be classified as homoines (humans) despite our nearly identical morphology and obvious evidence of evolution, owing to some imagined genetic barrier, the last ditch of the desperate in denial. Of course they say this having to ignore the March 2011 *PLOS Genetics* paper, "A Molecular Phylogeny of Living Primates," by Polina Perelman et al., which exactly confirms our position among the great apes and our genetic proximity to chimpanzees, which we had already determined by other means.

We're so close to chimpanzees that it might still be possible to produce living hybrids. After all, using laboratory techniques, we've interbred more distantly related species than that, and no one has conclusively determined whether we're chemically interfertile with chimps. In truth, no one really wants to prove whether a human-chimp hybrid is possible either; because most of us don't even want to know the answer to that question for obvious reasons. So it is still safe to call chimpanzees a different species regardless of whether we could interbreed with them or not. But all this implies that fertility between much more humanlike australopiths and primitive men was a virtual certainty. This reminds me of Dr. Zira in the original 1968 movie, *Planet of the Apes*, when she begrudgingly accepted a kiss from a human astronaut: "Alright, but you're just so damned ugly."

The other problem with the classification of created kinds is an even bigger one. In fact it is the biggest problem creation "scientists" have. I call it the phylogeny challenge, which I'll explain. Creationists usually accept that taxonomy is superficially accurate, but they'll only concede that to a degree, because they insist that their god miraculously conjured a series of definitely different kinds of animals, which were each specially created separate from one another. Creationists allow that each of these kinds have since diversified—but only within mysterious limits that they refuse to rigidly define—and they say that no lineage can be traced beyond their alleged original archetypes. However, they're unable to identify what those kinds are, how many there are, or how they could be recognized.

I would challenge them to show me their mystic divisions among the following taxa.

- Are mallards related to pochards, wood ducks, and muscovies?
- Are all ducks also related to geese and all other anseriformes?
- Are anseriformes related to galliformes and other neognathes?
- Are neognathes related to paleognathes?
- Are any extant birds related to hesperornis, ichthyornis, enantiornis, or other euornithes?
- Are euorniths related to confuciusornis or archaeopteryx?
- Are all early aves related to microraptor, velociraptor, or other nonavian dinosaurs?

- Are dinosaurs related to pterosaurs, phytosaurs, and other archosaurs?

If evolution from common ancestry is not true and some flavor of special creation of as-yet unidentified kinds *is* true, then there would be some surface levels in a cladogram where you would accept an actual evolutionary ancestry, but there must also be subsequent levels in that twin-nested hierarchy where life-forms would no longer be the same kind and wouldn't be biologically related anymore. At that point, they would be magically created separate kinds, and distinctly unique from those listed around it as well as those apparently ancestral to it. So . . .

- Are Bengal tigers related to Burmese tigers and all other tiger species?
- Are all known species of tiger related to each other and all other panthers?
- Are all panthers related to felines and scimitar cats?
- Are all felids related to nimvarids or viverrids? And how could we tell?
- Are all of Feloidea related to any or all other members of the order Carnivora?

Those who promote creationism's bewildering inanity should be able to show exactly where and why uniquely created kinds could not be grouped together with any parent clades that would otherwise only imply an evolutionary ancestry. Throw away any other argument you might be thinking about; none of them compare to this! If creationism is true of *anything* more than a single ancestor of all animal forms (if not the entire eukaryote collective), or if the concept of common ancestry is fundamentally mistaken, then there *must* be a point in the tree where taxonomy falls apart—where what we thought was related to everything is really unrelated to anything else; and unless you're a scientologist or a Raelian, that criteria must apply to other animals besides ourselves. So . . .

- Is the short-tailed goanna related to the perentie and all other Australian goannas?
- Are all Australian goannas related to each other and the African and Indonesian monitors?

- Are today's terrestrial varanids related to Cretaceous mosasaurs?
- Are varanids related to any other anguimorphs including snakes?
- Are anguimorphs also related to scincomorphs and geckos?
- Are all scleroglossa related to iguanids and other squamates?
- Are all of squamata related to each other and all other lepidosaurs?
- Are lepidosaurs related to placodonts and plesiosaurs?
- Are lepidosauromorphs related to archosaurs and other diapsids?
- Are all diapsids related to anapsids or synapsid "reptiles" like dimetrodon?
- Are all reptiles related to each other and all other amniotes?
- Are all amniotes related to each other and all other tetrapods?
- Are all tetrapods related to each other and all other vertebrates?

And so on. Which of these are related? Which of these are created?

Remember, if there is any validity to creationism whatsoever, or if there is some critical flaw in the overall theory of evolution from common ancestry, that flaw must be found *here* or it simply can't be anywhere else. That is the phylogeny challenge. This challenge has been unanswered for more than ten years for the simple reason that there is no such thing as a "kind." There is no point where any collection of animals appears to be original baramins. Baraminology is without basis because it is impossible to identify any point in taxonomy where everything that ever lived isn't evidently related to everything else.

Every animal, including humans, belong to a sequence of parent clades ultimately culminating in one encompassing category for all of them. Worse for creationists, that sequence continues from there, such that it also includes fungi, plants, and every other kind of eukaryote into a single category, so that at some level we're all the *same* "kind." To quote pioneering environmental activist John Muir, "When we try to pick up anything by itself, we find it attached to everything else in the Universe."

Professional creationists have to deny macroevolution for the same reason they have to deny transitional species: not because these combined realities can only indicate an animal ancestry, but because either one alone proves that such is at least possible, and religious extremists are not permitted to admit even that. If it is possible to walk twenty feet, it's possible to walk twenty miles, so these pseudoscientists insist there must be some definite boundary blocking the evolution of new "kinds." But they won't say where or what that boundary is.

Committed creationists habitually misdefine their terms, if they can be forced to use definitions at all, because they will not be accountable. They can't be, because they've decided in advance never to change their minds even if they're proven wrong. If they were to find out that macroevolution was ever actually seen and proven to have happened for certain, their cultish faith would still forbid them to admit it. Instead, they'd have to redefine their terms, to move the goalposts from whatever evolution can show to something that would still be impossible no matter how true evolution is, just to be safe.

For example, as a young man, I didn't expect scientists to actually witness the evolution of a new species under controlled conditions. Such an event, I thought, would be too fine a gradation to identify, and too rare an occurrence to expect that it would happen while anyone was looking. So the first time I read about that happening with a new species of fruit fly that evolved under direct observation in a lab, I pointed that out to one of my creationist friends. Of course that wasn't good enough. Nothing ever is or could be. *"That's not macroevolution,"* he said, *"because it's still a fly; it didn't turn into something else."*

What is a new species of fruit fly supposed to turn into? We started with one species and now we have two. They're living in sealed containers with no chance of contamination. They didn't just carelessly let in other flies and fool themselves into thinking they appeared there. These populations were artificially selected, purposely cultivated, and carefully separated by professional scientists performing a sensitive experiment. This is a whole new population, and a verified strain of the parent clade, genetically distinct from it and chemically incompatible with it. This is a new species—that's macroevolution!

Creationists often complain that evolution isn't scientific because it can't be repeated in laboratory experimentation, but it can be and it has been. The first time that happened, it was repeated at the hand of pioneer geneticist Theodosius Dobzhansky back in the 1960s, but it has happened a number of times since then too. We know how to make that happen now. As *New*

Scientist described in a June 9, 2008 article titled "Bacteria Make Major Evolutionary Shift in the Lab,"

> A major evolutionary innovation has unfurled right in front of researchers' eyes. It's the first time evolution has been caught in the act of making such a rare and complex new trait. And because the species in question is a bacterium, scientists have been able to replay history to show how this evolutionary novelty grew from the accumulation of unpredictable, chance events. Twenty years ago, evolutionary biologist Richard Lenski of Michigan State University in East Lansing, US, took a single *Escherichia coli* bacterium and used its descendants to found 12 laboratory populations. The 12 have been growing ever since, gradually accumulating mutations and evolving for more than 44,000 generations, while Lenski watches what happens. . . . All 12 evolved larger cells, for example, as well as faster growth rates on the glucose they were fed, and lower peak population densities. But sometime around the 31,500th generation, something dramatic happened in just one of the populations—the bacteria suddenly acquired the ability to metabolize citrate, a second nutrient in their culture medium that *E. coli* normally cannot use.

If you Google the above article, you'll find systematic attempted rebuttals on creationist websites, because they have to find some way to discredit or dismiss every example of macroevolution actually happening. One of those failed rebuttal attempts resulted in an amusing confrontational email exchange between Richard Lenski, the evolutionary biologist who performed the described experiment, and Andrew Schlafly, a creationist homeschooler with Conservapedia. I won't detail that exchange because a full account of it can be found on RationalWiki, and Richard Dawkins already addressed it in his book *The Greatest Show on Earth*. I mention it just to note what macroevolution really is—to show that it can be repeated under controlled conditions and that it also produces new "information," even though creationists keep insisting that it can't. I mention this also to highlight the fact that creationists have a compulsion to reject and discredit any and all evidence that ever stands against them.

In 2012, a young high school activist named Zack Kopplin admirably organized a campaign to repeal a law passed by the Louisiana Legislature that allowed the teaching of creationism and the censorship of evolution in the state's public high school science classrooms. During testimony before the Louisiana State Senate, one of the science experts enlisted by Kopplin described the above experiment with *E. coli*. Upon hearing that evolution had been observed, State Senator Mike Walsworth interrupted, asking whether this bacterium had evolved into a person. That's how jaw-droppingly ignorant creationism is! When Kopplin's team heard that, they sat stunned and stared for a moment, because how could anyone be so stupid and yet be in charge of anything? They naively thought that these politically powerful right-wing religious ideologues were adequately educated and somehow sincere and willing to reconsider an indefensible position honestly. I've made that same assumption myself many times, but I'm usually similarly disappointed by the reactions of those defending belief in complete nonsense who clearly do not want to understand what is really real.

It is of course much easier to observe evolution when we're talking about organisms with rapid reproduction and short generations. Even then, remember that the *E. coli* evolution experiment ran for a quarter of a century! If it typically takes tens of thousands of generations to see any measurable significance, then we might expect to only see this with microbes and insects with a life span of a few days. But we have actually seen it with vertebrates too, again when we're talking about short generations and rapid reproduction.

In 1419, Portuguese sailors discovered a few volcanic islands west of the strait of Gibraltar. They settled around the area and accidentally imported a few European mice to each of these settlements. Five centuries later, French biologist Janice Britton-Davidian laid out hundreds of mousetraps in forty locations around the semitropical island of Madeira. She discovered that these very small and isolated populations of mice have evolved into six different genetically distinct species. They all look about the same, but the ancestral European mice have 40 chromosomes, and the Madeiran mice range from 22 to 30 chromosomes. They didn't lose DNA; rather, some of their chromosomes fused, much like chromosome 2 in humans, packing more DNA into fewer units. This is a different type of speciation, because such a chromosomal variance could qualify them as distinct species even if they can still interbreed.

About this discovery, biologist Scott Edwards said, "What is surprising is how fast this has taken place, but this is an interesting case because it may prove to be an extreme case of rapid speciation." Just to reiterate: showing a population-level chromosomal variance in just 1,500 to 2,000 generations is "extremely rapid." Even so, as other studies suggest, mice in the Faroe Islands, which have a similar origin story, may have evolved twice as fast. Evidence of such rapid evolution is by no means unique to mice. As a April 18, 2008 article in *Science Daily* titled "Lizards Undergo Rapid Evolution after Introduction to a New Home" reports,

> In 1971, biologists moved five adult pairs of Italian wall lizards from their home island of Pod Kopiste, in the South Adriatic Sea, to the neighboring island of Pod Mrcaru. Now, an international team of researchers has shown that introducing these small, green-backed lizards, Podarcis sicula, to a new environment caused them to undergo rapid and large-scale evolutionary changes.
>
> "Striking differences in head size and shape, increased bite strength and the development of new structures in the lizard's digestive tracts were noted after only 36 years, which is an extremely short time scale," says Duncan Irschick, a professor of biology at the University of Massachusetts Amherst. "These physical changes have occurred side-by-side with dramatic changes in population density and social structure."

Creationists think there should be abrupt dramatic complete transformations in single generations, and usually with individuals instead of whole populations. That's what happens when all you know about evolution is what you saw in *X-Men*. We need to understand what evolution *really* is and what we can really expect from it.

Consider how pathetic prosthetics and robotics technology still is today in comparison to the type of technology conceived in television shows from the 1970s like *The Six Million Dollar Man* and *The Bionic Woman*. If we were to build an autonomous robot in 2016 that can correct its balance if it stumbles and that can read signs and recognize people's faces, we should be impressed because that's progress. It's certainly better than we could do a half-century ago. But back then, sci-fi fans thought that a human-shaped composition of transistors should be able to do everything we can do better than we can, and that it should act exactly like a human, apart from being curiously unable to understand our emotions. To expect droids to be like that even now is an unrealistic expectation, but to expect it back then is to think like a creationist.

There are no fruit flies turning into something that isn't a fly anymore, no dogs turning into elephants, or bacteria turning into a human, and there's no mammals suddenly growing birds' wings either; nothing like that. Creationists know that if they ask for the sort of evidence we should really expect if evolution is really real, we can provide it because evolution *is* really real. But they don't want to know that, so instead they demand to see things that no scientist could ever produce, and that would actually go against what evolution really is. Evolution has to adhere to certain biological laws and one of those is that no lineage can jump to a different branch on the tree of life.

We've seen new species evolve. We could say we've even seen a new genus evolve, depending on what you mean by that. So science denialists don't ask for that anymore. That's not good enough. Demanding to see family-level evolution is too risky because we might produce that too. So they jumped a couple spaces and now challenge science to show them direct observation of the evolution of new phyla instead.

There is one distinct difference between the old Linnaean ranks of classification and systematic cladistic plylogenetics. We still use the names of the old Linnaean ranks like "mammal" and "chordate" and so on, but we only use them as sign posts to keep from being lost in a network of unnamed clades. Otherwise, there is no real significance to any of the ranks themselves because they're undefined. Now we know there really is no level above species, because every other rank in taxonomy is more or less arbitrarily assigned as a construct of human convenience. The Linnaean ranks of family, genus, order, and phyla are all factually illustrative, but otherwise virtually meaningless because every new taxa that ever evolved began with speciation (the emergence of a distinctly new species), but in such a way that the daughter species was still just a modified version of whatever its parents were, and whose eventual descendants will always belong to whatever categories their ancestors did no matter how much they may change as time goes on.

The only reason creationists cling to these micro and macro distinctions is so they can have some excuse to accept "small-scale" evolution (which they begrudgingly admit cannot be denied even with the greatest faith), while still denying "large-scale" evolution, where their exact parameter of how large must remain elusive to prevent it ever being disproved. Of course that means "large-scale" evolution can mean whatever they want it to at any given moment.

Frank Sherwin from the Institute of Creation Research defined macroevolution in one discussion as the origin of every kind of animal, and later on in the same discussion, he changed his definition to the origin of all life. He knows he's using the terms incorrectly; he simply doesn't care! But the fact is he doesn't get to conveniently redefine what these terms have always meant to the scientists who invented those words in the first place.

The different breeds of dogs are an example of microevolution, while the different species of wolves and wild dogs (as well as panthers and felines, horses and zebras, and llamas and camels) are all examples of macroevolution. Each set is definitely biologically closely related, but they're each different species, not different "kinds" nor the same kind.

The terms macroevolution and microevolution were first coined in 1927 by Russian entomologist Yuri Filipchenko. It was an evolutionary biologist who invented these words, so science is the authority on what they mean. According to universities actually teaching this subject, microevolution is variation *within* species, and macroevolution is variation *between* species. That means the emergence of new breeds or subspecies is microevolution, but the emergence of new species is macroevolution. Creationists refuse to admit that's what it means.

But if you don't believe me, look it up.

> "Macroevolution is evolution on a scale of separated gene pools. Macroevolutionary studies focus on change that occurs at or above the level of species, in contrast with microevolution, which refers to smaller evolutionary changes (typically described as changes in allele frequencies) within a species or population."
>
> —*Wikipedia*

> "[M]ajor evolutionary transition from one type of organism to another occurring at the level of the species and higher taxa."
>
> —*Dictionary.com*

> "[E]volution that results in relatively large and complex changes (as in species formation)"
>
> —*Merriam-Webster Online Dictionary*

Even though each of the above is reasonably accurate, you can't always trust common dictionaries for laypeople when they're trying to talk science. So sticking with scientific sources, let's start with the University of California–Berkeley's online primer called "Evolution 101": *"Macroevolution refers to evolution of groups larger than an individual species."*

We get the same definition from Duke University: *"Evolutionary patterns and processes at and above the species level."*

And from University of South Carolina–Beaufort: *"Macroevolutionists study the processes that cause the origination and extinction of species."*

And from Stanford University: *"Microevolution is defined as changes within a species that aren't drastic enough to create an entirely new species. Changes that result in a new species are part of macroevolution."*

Now let's look at the reference website Biology Online: *"Macroevolution involves variation of allele frequencies at or above the level of a species."*

So macroevolution is properly defined as the emergence of new taxa at or above the species level. The only time creationists will use the proper definition is when they are as-yet unaware of the fact that speciation has already been directly observed and documented dozens of times—both in the lab and in naturally controlled conditions in the field. A couple different lists of observed speciation events and observed evolutionary changes are available online, each with specific scientific citations. As they're usually not talking about sexually reproductive animals, they identify speciation by novel traits, genetic distinction, and the emergent process.

In addition to the documented development of new enzymes and chromosomes, novel synthesis abilities, denovo genes, and retroviral resistance, notable examples include the evolution of a new multicellular species arising from unicellular algae under direct observation in the lab. That's one example of many different types of speciation. In fact, they've seen this so many times they've had to categorize recurrent types of macroevolution they've seen so often repeated.

Allopatric speciation is what I've mostly described so far, where a given population is simply divided into genetically separate groups, usually by geographic barriers, and experience genetic drift. For example, compare the North American raccoon to the South American coatamundi, or the American cougar and the African caracal.

Peripatric speciation is when a tiny group is completely isolated and experiences a genetic bottleneck (founder's effect), often producing much faster or more dramatic evolution. The mice of Madeira are a good example of this.

Parapatric speciation is similar to peripatric, except that it's not a complete isolation, but a limited connection to a sister group. Over time, genetic dissimilarity leads to behaviors or mechanisms that prevent interbreeding. Tigers and Indian lions are a good example of this, as are wolves and dogs.

Sympatric speciation is when a new species emerges within or among the parent group, often by exploring a different food source in the same area. The best example of this are the cichlid fish of the Rift Valley. These are also often good examples of how vertebrate species can switch back and forth from sexual to asexual reproduction, again under direct observation, as a reaction to a dynamic environment.

Once creationists find out about all this, their first reaction is to use the excuse that some newly evolved species of fruit fly or fish somehow doesn't count because it's "still" a fly or it's "still" a fish. Well of course it is! Evolution couldn't permit them to be anything else, according to the evolutionary law of monophyletic variation.

Critics of evolution demand that the new species be so different from their parents that one can't even tell they're related. The irony there is that evolutionary theory never suggests that one "kind" of thing ever turned into another, fundamentally different "kind" of anything, not unless you ignore all the intermediate stages—which of course they do. That is not all they ignore. Consider this telling exchange from the Scopes "Monkey Trial" of 1925:

William Jennings Bryan:
I do not think about things I don't think about.

Clarence Darrow:
Do you think about things that you do think about?

William Jennings Bryan:
Sometimes.

(laughter in courtroom)

To comprehend evolutionary theory, one must first understand that it's only ever a matter of changing proportions, altering or enhancing existing features to build on what is already there. Developmental biology, genetics, and comparative morphology combine to confirm many of these taxonomic stages such that organs do not seem to have appeared abruptly or fully formed as if out of nowhere, because there is an implied evolutionary origin evident in every case.

Comparative morphology has shown us how lungs evolved, from an originally asymmetric distension in the buoyancy bladder of primitive fish. The fossil record has shown us how wings evolved in the ancestors of birds, from feathered arms that insulated a larger clutch of eggs. A concerted view of paleontology and embryology showed us how feathers evolved, with every incremental stage of that development genetically identified and indicated in biological development of living birds as well as recapitulated in sequential fossils in a line of maniraptoran dinosaurs.

We know how all of that happened. But if we lived back then in the midst of these developments, and didn't know in advance what they would become, creationists would still dismiss each of those features as simply being distortions of whatever is already there. They don't want to know and don't want to understand what they don't want to believe.

The fact is that even the transition of fish to tetrapods, dinosaurs to birds, or apes to men are each just a matter of incremental, superficial changes being slowly compiled atop successive tiers of fundamental similarities. These represent monophyletic clades that will forever encompass all the descendants of that clade. This is why birds are still dinosaurs, and humans are still apes, and both are still stegocephalian chordates.

No matter how much you or your heirs may change, you obviously can't outgrow heredity. The very concept of common ancestry is a multitiered and intertwined complex phylogenetic system that shows why there can't be any distinctly separate "kinds" to begin with!

At the same time, the act of speciation splits the population presenting an eventually impassable boundary between them. We often see this demonstrated live in the form of ring species, where different evolutionary stages exist all at once in a geographic rather than chronological distribution. There are a few examples of this, usually various species of gulls with home regions circling the North Pole. Each is a chemically interfertile breed with its neighbors to the immediate east or west, but is no longer fertile with those on opposite ends.

A favorite example is the Ensatina salamanders living all around

California's San Joaquin Valley. As they slowly spread along that circular mountain range, they developed as a series of subspecies along the way, which were all fertile with each other. But once they closed the loop, they were too genetically different to interbreed with those on the original point. As you circle the ring, Subspecies (A) may breed with Subspecies (B), and (B) may breed with (C), and (C) with (D), but (A) and (D) cannot interbreed because by the time their territories overlap again, closing that geographic loop, they've grown too distant genetically and can't come back. This is when we see the formation of new features, organs, or skeletal structures, each examples of new genetic "information." Collectively, what all of this shows is that even though a new species of perching bird (for example) is "still" a finch, it is now a different "kind" of finch, a distinctly different and genetically distant descendant species, where each and every such case proves there is no "boundary" against macroevolution.

THE 12TH FOUNDATIONAL FALSEHOOD OF CREATIONISM

Creation "Science"

I confess, I am fascinated by evolutionary history—whether the classification of life forms and their interconnected lineages, or the idea of a pristine, almost alien world during the Carboniferous period, where ammonites and trilobites were commonplace but where there was no hint of humanity and not a trace of litter. All the different environments and types of life that once existed are of sincere interest to me. But more than all of that, I like to read the works of early scientists, to get in their heads and see such confusion when what they've been taught and what everyone around them believes doesn't match the evidence at hand.

I know that some 2,500 years ago, early attempts at science predicted that living things were related somehow, though they couldn't possibly have understood that at the time. Still, it's hard to imagine how it took until Darwin's day for the world to realize that all birds must have been related to one original population that did not look like any bird living today. How was there ever a time when that wasn't obvious? The reason that is (and the reason I find it so interesting) is that throughout the Enlightenment and even into the nineteenth century, superstition, supposition, and folklore (their equivalent to urban legends) had culturally conditioned and permeated everything such that not even the greatest minds of that time were immune. It didn't matter how clever they were; they were wading through a vast network of nonsense universally imposed with no way to confirm or refute most of it, so they had no idea how to shake that stupor that everyone around them shared.

I've heard people say this conflict has been raging since the days of the ancient Greeks, "between science like Democritus and faith like Aristotle," to quote a YouTube music video on that subject titled "Beware the Believers." But it is seldom true that scientists reject all pseudoscience woo, and most of the leading religious advocates of history are touted as respected scholars even by some nonbelievers. We're all just people essentially, and products of our respective environments. Everyone thinks they're rational, yet only a few of us are able to keep from wading in bullshit deeper than the height of our boots.

Paleontology in its early days was a difficult discipline. Nineteenth-century pioneers had to devise a way to test and confirm the claims that were made because they were often confused by their own preconceptions and made simple mistakes that went undiscovered for decades. Unwary professors were sometimes made the victims of pranks, and not every respected scientist was entirely honest. For example, in 1726, Johann Bartholomew Adam Beringer was senior professor and Dean of the Faculty of Medicine at the University of Würzburg, Germany. He paid some boys to go on a fossil dig, and they returned with what clearly would be the most extraordinary items of his collection. The excavated items weren't exactly lifelike; they were somewhat stylized, and not all of them were even of organisms. Some of them depicted the sun and planetary objects, and a few of them even included the name of God inscribed in Hebrew, Latin, and Arabic. Unbeknownst to Beringer, colleagues of his had themselves carved the sandstone reliefs and planted them in a maliciously intended prank.

Remember that no one yet knew for certain that fossils were the impressions left by once-living organisms—but that hypothesis was regularly debated, including by Beringer. But these most recent finds were different. He initially thought they might have been carvings by ancient pagans, except that they bore the name of God, which those pagans would not have known. He decided they were the "capricious fabrications of God," thinking God had actually chosen to sign some of his playful doodling!

Beringer wrote a book based on his exhaustive analysis of these stones before he finally learned that they were fraudulent. However, he had invested so much of himself in that study that he couldn't bear the thought of having wasted that much time on a fool's errand. He acknowledged suspicion of the hoax when it was first reported, but then convinced himself his fossils were real. He published his book *Lithographiæ Wirceburgensis* in 1726, even after his colleagues warned him that they'd faked the entire collection just to make a fool of him. Now throughout history, he's known as the fool who believed in "the lying stones."

American paleontologists around the turn of the twentieth century weren't always reputable scholars either—many were egotistical adventurers. The fictional archaeologist Indiana Jones is actually based on real-life paleontologist Roy Chapman Andrews (1884–1960), director of the American Museum of Natural History. His predecessors were less like Indiana Jones and more like Yosemite Sam; Edward Drinker Cope (1840–1897) of the Academy of Natural Sciences in Philadelphia and Othniel Charles Marsh (1831–1899) of the Peabody Museum of Natural History at Yale were highly competitive underhanded rivals who famously resorted to bribery, theft, vandalism, and libel against each other. Because of their rivalry, the great dinosaur rush of the late 1800s is now better known as the "bone wars."

In an unintended collaboration now made famous, O. C. Marsh discovered a large sauropod dinosaur called Apatosaurus in 1877, but he couldn't find the skull. In the same year, E. D. Cope came into possession of the skull of another large sauropod called Camarasaurus. In 1879, Marsh claimed to have discovered a third species of large sauropod, which he named Brontosaurus. When it was mounted in the American Museum of Natural History, it wore the sculpted skull of Camarasaurus, a fact which was not publicized. Brontosaurus became the world's most famous dinosaur. Until the last few decades of the twentieth century, more people could remember that name than even Tyrannosaurus.

Yet ever since it was named, experts have argued that there is no such thing as Brontosaurus—that Marsh had only found a second apatosaur and tried to sell it as a different genus and species. Amusingly, in 2015, a computerized analysis of every diplodocid dinosaur ever unearthed revealed that Marsh's famous composite exhibit was not actually an apatosaur with a fake camarasaur head; it actually was a different species, and we have the proper skull for it now too. There is no need to name this species because we already know what it is. For more than a century, those of us in the know were confident that brontosaurus was a fraud. But now we have to say, "Whoops, never mind. Yes Virginia, there is a Brontosaurus."

Before Darwin was famous, England was already a leader in paleontological studies, largely due to an intrepid fossil hunter named Mary Anning. Sadly, she didn't get much recognition at the time, partly because she was too poor to associate with academic aristocracy and partly because she was labeled a religious dissenter, but she was primarily prohibited from participation simply because of her gender. Once upon a time, England was very religious and—not coincidentally—also very racist and sexist. It was an arrogant, ethnocentric empire, much like the United States is today. Back then the Geological Society of London was an old boys' club with no girls allowed. Experts consulted with Anning, but her name was not included on any of their publications. Such was the patriarchy of the time. Mary Anning died in poverty and relative obscurity in 1847, and the significance of her contribution to science wasn't formally recognized by the Royal Society until 2010, 163 years after she died.

Two hundred years ago when she was just twelve years old, she codiscovered the first ichthyosaur fossil ever found. She was also first to discover a plesiosaur, and she found the second pterosaur, Dimorphosaurus. (A Pterodactylus had already been discovered in Germany.) Neither of these was very big, so they weren't that different to Victorian eyes from bats or birds.

Another English paleontologist named Gideon Mantell noted having been inspired by Mary Anning, and he made a few impressive discoveries himself. In 1822 either he or his wife happened across fossilized teeth like those of an iguana, except that they were twenty times larger. His discovery was disputed by other paleontologists, both as to how old they were and what they came from. William Buckland said they were the teeth of a fish. Georges Cuvier initially dismissed them as the incisors of a rhinoceros, though he changed his mind soon after. On further inspection at a later date, Cuvier identified the teeth as belonging to a giant herbivorous reptile.

Even after being vindicated by Cuvier, Mantell was still challenged by Sir Richard Owen. Mantell soon discovered other skeletal fragments of the animal he named Iguanodon. At first, he envisioned a giant monster, similar to the one shown in the 1966 movie *One Million Years BC*, an impossibly huge iguana with a rhinolike spike on its nose. Owen agreed with that, albeit at a smaller scale, but he still erroneously dismissed several of the vertebrae as coming from various different species—probably because they didn't look like what he expected.

Other curious fossils came to the British Geological Survey representing yet another giant reptile unlike anything alive in the modern day. Owen was a renowned anatomist of the Royal Society when he described the few bones then known to belong to Megalosaurus. The Latin name essentially means "big-ass lizard," although Owen somehow imagined it having a much more bearlike appearance. Then in 1832, Mantell discovered another distinctive species, Hylaeosaurus—a type of ankylosaur. Such a creature had no analogue in the modern bestiary.

Owen compared Megalosaurus to Iguanodon and Hylaeosaurus and recognized traits shared in common by these giants, from the fused sacral vertebrae in the pelvis to the fact that the legs don't sprawl like lizards or crocodiles. They were more like mammals in that the legs supported the weight directly like columns. In 1842, Owen classified these specimens together under the new taxonomic name "dinosaur," a word meaning "terrible lizard." That wasn't the way Owen meant to describe them. He wanted them to be "fearfully great" lizards, perfect creations in their original form—and yes, he thought they were "created."

In an earlier chapter, I mentioned how Sir Richard Owen was a celebrated biologist and the foremost authority on paleofauna in the world in his time. He was credited with the establishment of the British Museum of Natural History and inventing the word "dinosaur," but he built his reputation by suppressing the work of other scientists and taking credit for their discoveries himself. Owen was not well-liked, and was often described as dishonest, malicious, and hateful. He was devoutly religious, but he was also a leading anatomist and zoologist, respected and unrivaled in each of these fields, and he was both Darwin's superior and fiercest adversary. Unlike Darwin, Owen believed religion should guide and even override scientific research.

Decades before Darwin ever published his theory of natural selection, many scientists already suspected that evolution happened; they just didn't know how to explain it. Almost all of these scientists were Christian, but they weren't dogmatic literalists and weren't inclined to ignore or conceal evidence. Owen however was a creationist, albeit not like we have today. There was as yet no theory of evolution for him to deny, but he did object to the earlier notion of transmutation, that one species could *somehow* become another.

A typical creationist today is blissfully ignorant that there even is a fossil record, much less as to all that's in it, and they don't know what taxonomy is either. Owen didn't have that excuse. Once Cuvier died in 1832, Richard Owen became the world's leading authority on paleontology and taxonomy. He knew better than anyone about the progressive stages shown in the fossil record, and this caused him to reconcile things in an unusual way. Where today's creationists talk about the baraminology of created "kinds," Owen believed in archetypes of divine design. A lot of people who have read a substantial amount of his work have commented that it's hard to make out exactly what he's talking about. He's not always consistent and he contradicts himself in confusing ways, but it seems as if he had the idea that God would occasionally release new and improved models to replace the old series once they're worn out, such that Megalosaurus and Iguanodon were replaced by newer mammalian predators and prey. That's why we had ichthyosaurs in the past and dolphins now. That was his explanation for why the fossil record for pterosaurs ends before the fossil record for birds begins. At least, that's how it seemed in his day. As he wrote in 1841, in the second part of his *Report on British Fossil Reptiles* for the British Association for the Advancement of Science,

> If therefore, the extinct species, in which the Reptilian organization culminated, were on the march of development to a higher type, the Megalosaurus ought to have given origin to the carnivorous mammalia, and the herbivorous should have been derived from the Iguanodon. But where is the trace of such mammalia in the strata immediately succeeding those in which we lose sight of the relics of the great Dinosaurian Reptiles?
>
> The evidence . . . permits of no other conclusion than that the different species of reptiles were suddenly introduced upon the earth's surface, . . . negative the notion that the progression has been the result of self-developing energies adequate to a transmutation of specific characters; but on the contrary, support the conclusion that the modifications of osteological structure which characterize the extinct reptiles, were impressed upon them at their creation, and have been neither derived from improvement of a lower, nor lost by progressive development into a higher type.

Mammals obviously didn't evolve from dinosaurs. Even if they did, we wouldn't have meat-eating mammals evolving from one dinosaur and plant-eating mammals evolving from another dinosaur. Remember the law of monophy, where all mammalian lines branch out of one common ancestor. Thus, the above quotation is proof that even when a creationist is a world-renowned expert on taxonomy and paleontology, they still have a fundamental misunderstanding of evolution. To be fair, though, this sort of confusion was common throughout academia at that time. Darwin's explanation of hereditary biodiversity published a couple of decades later was the first and only one with consistent clarity, which isn't saying much when you look at the befuddlement of his contemporaries.

Like many creationists, Owen had the false impression that evolution meant a ladder of increasing complexity and progressive advances to achieve higher forms. But he recognized that dinosaurs were bigger and better than any modern reptile, and to him that meant they couldn't have evolved. Instead, he said that these magnificent monsters were much more mammallike originally but that they must have "degenerated" over time to become the sluggish, cold-blooded, barely functional reptiles we have today.

Owen only had partial skeletons or a few scattered bones for each of these dinosaurs, not enough to know much of anything about them. But he wanted to show them off, because who wouldn't? However, he wanted to do it with something more than just illustrations. The average person couldn't look at a femur and tell what kind of animal it was, nor could they extrapolate how big the rest of the animal must have been, and the most impressive thing about dinosaurs is that they got *big*—bigger than anything in human experience! So in 1854, Owen had life-size sculptures made out of a combination of concrete, stonework, and lead, all in sculpted detail, and he had them arranged in a natural setting in London's Crystal Palace Park where they can still be seen today. These included not only dinosaurs, but also other prehistoric animals such as ichthyosaurs. (Owen depicted them as looking like crocodiles, where we now know they looked more like dolphins.)

Amusingly as more fossils were found and these forms were fleshed out, it turned out that dinosaurs were even more advanced than Owen originally imagined. This presented some problems for his position because they might be even more advanced than modern mammals, and that didn't suit Owen's preferred belief. The first problem was when Mantell discovered that the front legs were much smaller than the back legs. This is true of all dinosaurs, but no one knew what that might mean. The limbs were also more gracile than Owen's depiction of them. Owen's vision of Iguanodon now looked more like a reptilian rhino, but with a longer tail. Also their posture indicates slower, less efficient beasts than modern mammals, in keeping with Owen's interpretation of a perpetually improving succession of archetypes.

For decades, Iguanodon was known only from a few unconnected bones. Then in 1878, the mostly complete and still-assembled fossils of thirty-eight individuals were discovered together, more than a thousand feet deep in a coal mine. They showed that the horn on its nose was wrong, that it was actually a thumb spike. There were also clear indications that Iguanodon could run bipedally. As more fossils of Megalosaurus emerged, it turned out they were definitely bipedal too.

Another anatomist named Thomas Huxley, famously known as "Darwin's bulldog," saw Owen's Megalosaurus and immediately drew a connection with flightless ratite birds like ostriches and emus. Huxley argued that birds were essentially reptiles, and he remarked on the obvious similarities of their hind legs. However, the rest of the scientific community was not yet able to imagine how an ostrich could evolve from Iguanodon.

There was also a good deal of argument over which traits were inherited and which traits were convergent. Harry G. Seeley (a rival of Owen's and Huxley's) argued that the congruence between the hind limbs of birds and dinosaurs could be attributed to a shared mode of life and not a family relationship; walking bipedally simply forced the legs of dinosaurs and birds to take a similar form. Huxley turned the argument from convergence against his opponents. The traits supposedly shared between birds and pterosaurs had to do with flight, and given that both lineages had become adapted to flying, common traits in their skeletons were to be expected. On the other hand, the diagnostic traits in the hips, legs, and feet of dinosaurs were found in *all* birds, not just ground-dwelling ones. This meant that these characters marked a true family relationship and not just a shared way of life. He also suggested that common blackbirds were in fact dinosaurs in disguise.

These arguments were carried out in public literature—you published your argument and then pissed off your peers, who then published their counterarguments against you. It wasn't much different from today's blogosphere, just a lot slower.

Louis Dollo somehow never noticed the avian structure of dinosaur legs. His vision of Iguanodon looked more like a giant kangaroo with an erect posture, balancing on its tail. But many of these dinosaurs have fused caudal vertebrae. That's not just an accident of fossilization; those bones were fused in life, too. The tail wasn't dragged along the ground like a crocodile, or arched tripodlike as with kangaroos. Dinosaur tails were often held aloft as a rigid counterbalance that got in the way if they stood up straight. The tail couldn't bend the way Dollo wanted, so the only way to mount that skeleton with an erect posture was to break the fossil where they wanted it to bend. So they did, rather than consider that they might have gotten it wrong.

Nowadays our understanding of Iguanodon is that they were mostly quadrupedal, especially as they got older, but that they could also run on two legs. So Owen's and Dollo's interpretations of Iguanodon were both wrong. Owen's vision of Megalosaurus was especially bad, though; it looked a bit like a gigantic wolverine, but now we know they looked more like flightless eagles, albeit with long tails, arms, and teeth.

Now as soon as everyone realized that a theropod foot looked just like a bird's foot, they pointed to trackways that had previously been attributed to giant birds of enormous proportions and guessed that these must have been the tracks of dinosaurs. Owen was determined to divide birds from dinosaurs. In fact that is one of the reasons why he named them terrible *lizards*—he couldn't accept a "transmutation" into birds.

He had successfully silenced a lot of these claims when he reconstructed the skeleton of a Moa from New Zealand. Here was a gigantic bird capable of making many of those tracks. That ratite birds and dinosaurs had exactly the same shape and structure to their legs was (according to Owen) a "borrowed" trait, essentially the argument of common design indicating a common designer—except that in this case, it was a designer who could freely swap parts from other creations like a manufacturer of machines.

Owen was trained in Linnaean taxonomy, which presented fish, amphibians, reptiles, birds, and mammals as separate categories, where it was impossible to derive one from another. But these were usually rooted in fundamental commonality, which Owen didn't always adhere to. Owen's idea of archetypes had him classify animals by what he perceived as their "unity of type." He held that reptiles and amphibians were essentially the same thing, and he placed birds and mammals side by side at a higher level due largely to their warm-bloodedness. At one point, he thought dinosaurs were warm-blooded too, but he later argued against that when he didn't want anyone to see that the categories of reptiles and birds might overlap.

Owen compared birds to mammals regarding their similar intelligence and behaviors like parental care. These he thought were diagnostic traits that distinguished warm-blooded animals from cold-blooded ones. He also said, "The particulars in which Birds differ from all Mammals and agree with Reptiles are comparatively unimportant ones of the skeleton."

So according to Owen, the underlying foundational structure didn't matter, and only the coincidental surface features are definitive. One may as well establish relationships based on whether two species are the same color. It seems absurd to us now, but at that time, there were notable scientists who actually thought like that.

As I previously explained, evolution leads to a series of incremental changes being slowly compiled atop successive tiers of fundamental similarities, and these tiers of similarity establish taxonomic clades, meaning that the basic commonality matters more than any subsequent surface attributes. I would trace an evolutionary lineage according to that rule, and embryology would bear that out too, but the people of the eighteenth and nineteenth centuries were surprisingly superficial, more concerned with the facade than the foundation. When attempting to determine the origin of birds, for example, some people actually allied them to turtles; partly because sea turtles swim by flapping their forelimbs, but mostly because turtles have beaks!

If your contemporaries are able to believe something as ridiculous as that, then imagine their reaction once pterosaurs were discovered. Folks readily assumed that birds must have evolved from pterosaurs despite how different their fundamental structure was. Remember Europe had some old traditional folklore that if you adopt the clothing and behavior of a thing, you could become that thing. Maybe that's where these ideas came from.

Darwin however was unusually progressive for his time, well above the typical prejudice of judging a book by its cover. He examined life as it develops—from the inside out rather than the other way around. He predicted that if his theory was true, there should be a bird found in the fossil record with unfused wing fingers. His prediction was vindicated just a couple years later when they found the first Archaeopteryx. Imagine the reaction to that. Paleontologists didn't know what to make of it.

Huxley argued that the giant flightless ratite birds like ostriches and emus may have evolved directly from dinosaurs, and that we see their wings as they're developing. He imagined that ostriches might fly one day, but he decided that the birds that could already fly must have evolved from Archaeopteryx.

Harry G. Seeley tried to cite Archaeopteryx as a transitional species between birds and pterosaurs, as if the ability to fly could have been inherited from beings with a completely different type of wing.

Andreas Wagner (who promoted a kind of evolution, but not common ancestry) wanted to name it Griphosaurus, thinking it to be a pterosaur with feathers—as if that makes any sense.

All these men eventually recognized how and why they were wrong and changed their minds accordingly, but not Owen. He had a reputation for never admitting his mistakes. He seized on this collective confusion to devise another wedge between birds and reptiles. He largely ignored Huxley's minority opinion that birds had evolved from dinosaurs and instead tried to strengthen the link to pterosaurs that had been suggested by Wagner and Seeley.

In 1874, Owen wrote a detailed monograph of *Mesozoic Fossil Reptiles* in which he compared Archaeopteryx to Pterodactylus. He conceded that "every bone in the bird [Archaeopteryx] was antecedently present in the framework of the pterodactyle."

The way he phrased it, it seemed that if birds had evolved from anything, pterosaurs had to be the closest link. Having built on the illusion already endorsed by the majority of his contemporaries, Owen then went on to describe how other details made it impossible to derive the structure of birds out of the template provided by pterosaurs.

He was completely correct in dismissing any possible evolution of birds from pterosaurs for precisely the reasons he gave, but he did this in an attempt to discredit Darwin's theory, and he knew that wasn't Darwin's argument. Owen deliberately misrepresented Darwin's position.

Owen hadn't yet succeeded in keeping birds and reptiles separate. There was still the issue of Archaeopteryx's long tail and grasping wing fingers, which Darwin had predicted. Owen could not honestly accept or admit a transitional species, so he examined Archaeopteryx and found a subtle feature in the ankle that was still uniquely avian, not known to be shared by other dinosaurs. Thus he described Archaeopteryx as an unusual bird, but still just a bird nonetheless. At the same time, he applied a double-standard by dismissing Darwin's prediction of unfused wing fingers as "relatively unimportant."

So all that matters is whatever supports Owen's argument, and when it doesn't it can be dismissed without consideration. Owen had a reputation for spinning information to make it look like it supported him, and he wasn't above making things up to suit his position either. Huxley had called him out on that before, and again publicly exposed Owen's deceptive analysis when he published his own counterexamination.

These men famously viewed each other with contempt. When they posted their studies, they included criticisms against each other, like a flame war. Can you imagine professional scientists in the modern day berating each other in public media? I've seen it happen.

Owen made a prediction of his own, hoping to counter Darwin's. The first Archaeopteryx fossil ever found was missing its head. Archaeologist Sir John Evans thought that he had discovered a portion of its toothed mouth on the same slab as the rest of Archaeopteryx and he thought they belonged together, but because it had teeth, other scientists dismissed the jaws as belonging to a fish. So Owen predicted that later finds would show that Archaeopteryx had a normal toothless beak like all other birds known to that time. Of course, later finds proved him wrong. First of all, fossil birds like Hesperornis and Ichthyornis turned out to have teeth in their beaks, just like some pterosaurs did. But worse than that, it turned out that Archaeopteryx didn't have a beak at all, just a reptilian maw full of teeth. Owen's prediction backfired. However, before anyone knew that, another dinosaur—a nonavian dinosaur—had already ruined Owen's prediction, and his position.

As I said, scientists of the eighteenth and nineteenth centuries were superficial, judging affectations over foundation. Here's a demonstration of that: not all of the Archaeopteryx fossils bore feather impressions. Two of them didn't, and for that reason, they were both assumed to be something else, another dinosaur that looks like Archaeopteryx otherwise. Having no detectable feathers to confuse anyone, Compsognathus was classified as an unambiguous dinosaur using all the same criteria that Owen himself had already listed for that taxon. Archaeopteryx has longer arms and a wishbone, but otherwise their skeletons were virtually identical, and were confused with each other for some time.

Carl Gegenbaur studied the tiny, chicken-sized Compsognathus and discovered that it even had the same ankle as Archaeopteryx, so that trait wasn't uniquely avian anymore. In fact, the only trait that was uniquely avian was where the collar bone had fused into a furcula, or wishbone. Owen argued that mere cold-blooded animals did not have a wishbone because wishbones were unique to birds. However, wishbones have since been found on Dinonychus and Allosaurus and Tyrannosaurus and a host of others. In fact, no part of Archaeopteryx's skeleton is uniquely avian. That is obvious once you compare it to Compsognathus and then compare it to any unambiguous bird skeleton like that of a chicken. For this reason, many (including myself) have argued that Archaeopteryx should not be classified as a bird, but that it is definitely a dinosaur. In 1870, Thomas Henry Huxley wrote an essay titled "Further Evidence of the Affinity between Dinosaurian Reptiles and Birds" in the *Quarterly Journal of the Geological Society of London*,

> I find that the tibia and the astragalus of a Dorking fowl remain readily separable at the time at which these birds are usually brought to the table. The enemial ephphysis is also easily detached at this time. If the tibia without that epiphysis and the astragalus were found in the fossil state, I know not by what test they could be distinguished from the bones of a Dinosaurian. And if the whole hindquarters, from the ilium to the toes, of a half-hatched chicken could be suddenly enlarged, ossified, and fossilized as they are, they would furnish us with the last step of the transition between Birds and Reptiles; for there would be nothing in their characters to prevent us from referring them to the Dinosauria.

By then even Seeley was convinced, having become something of an expert on dinosaurs and pterosaurs. Soon he found himself bashing Owen too, because Owen cast pterosaurs as cold-blooded sluggish gliders. Owen admitted that pterosaurs had hollow bones just like birds and dinosaurs, but again he tried to suppress the importance of that and the implication that these animals were warm-blooded, highly energetic, and very strong fliers.

Seeley knew better than Owen. Seeley's popular 1901 book *Pterosaurs, Dragons of the Air* found that birds and pterosaurs are closely parallel. His belief that they had a common origin has been proved, for both are archosaurs—just not as close as he thought. He upset Owen's characterization of the pterosaurs as cold-blooded, sluggish gliders and recognized them as warm-blooded active fliers. But because Owen was perceived as an authority, his word held sway. So that a hundred years later, scientists were still arguing over whether these were all cold-blooded reptiles. That's how all the books depicted them when I was a kid.

In 1887, Seeley swept aside Owen's classification of dinosaurs and argued that the ruling reptiles actually fell into two distinct groups: the lizard-hipped Saurischia and the bird-hipped Ornithischia. Of course he thought that birds must have evolved from "bird-hipped" dinosaurs. But as we move down the lineage of the lizard-hipped variety, the fossil sequence shows the pubis bone moving into the avian position. Scientists also find fossil impressions apparently revealing four-chambered hearts, as well as every other attribute once believed to belong to birds alone, including gizzards and gastroliths. They've found quite a few dinosaurs that are very close to birds, and some are so close they can't tell whether they're birds or not. They've also found a few dinosaurs that are definitely not birds, yet they still had feathers. We've learned more about the paleoworld in the last two or three decades than in the past couple centuries prior.

Once Darwin published his landmark theory in 1859, Owen's resistance to transmutations became an objection to evolution, a working theory with substantial support. That forced a more desperate attempt to rebut it. Owen promised his religious fellows that he would succeed where Linnaeus had failed in finding some physical trait by which to distinguish man from apes, regardless whether that was really the case or not. First he presented similarities as differences, and when he couldn't find any legitimate differences, he made up entirely fictitious ones. As if his authority would always remain unquestioned, he proclaimed amid other scientists in peer review that the hippocampus minor was a uniquely human lobe of the brain and absent among apes. Such an expert as he couldn't have made such an obvious mistake, and his curious inability to concede any error except by way of evasive maneuver finally allowed Huxley to indict him for perjury. Lecturing at the Royal Institution in London, Huxley stated,

> Now I am quite sure that if we had these three creatures fossilized or preserved in spirits for comparison and were quite unprejudiced judges, we should at once admit that there is very little greater interval as animals between the gorilla and the man than exists between the gorilla and the baboon.

Although Owen gave many lectures, wrote hundreds of scientific papers, and received honors, lands, and titles from the crown, he was also accused throughout his career of deliberate deceit, of "lying for God and for malice," and even of writing anonymous letters to the press praising himself in third person while raving disdain against his colleagues. Finally, Sir Richard Owen was dismissed from the Royal Society's Zoological Council amid myriad charges of plagiarism spanning the entirety of his career.

Throughout history, there have been many scientists who believed the universe was "created" in the same sense that Christian proponents of natural sciences still believe today. I have often seen evangelical websites listing scientists who were Christian, as if that made them "Christian scientists" who believed in biblical inerrancy as opposed to just scientists who happened to be Christian. ("Christian Science" is an American fundamentalist religious denomination famous for refusing medical treatment in favor of prayer.)

The fact that Konrad von Gesner (1516–1565) or Robert Boyle (1627–1691) believed in the Christian god does not mean that creationism has any scientific support, nor that each of these men would have denied evidence against a literal interpretation of Christian mythology. Listing the names of scientists who lived before Darwin cannot imply that they would have rejected Darwin's explanation or the evidence he cited; only a few like Sir Isaac Newton were so fanatic that we could reasonably expect that. Other names that creationists often list as being among their own don't belong there at all.

- Galileo Galilei (1564–1642) challenged the church's belief that the sun orbited the earth. He was imprisoned for life by the Inquisition, who forced him to recant this truth. Three hundred and fifty years later, in 1992, the Church finally formally dismissed the charges against him, admitting Galileo's views were correct after all.
- Johannes Kepler (1571–1630) was an astrologer turned astronomer. His failure to identify "platonic solids" (indicative of divine design in astronomy) became a breakthrough in celestial physics.
- Carolus Linnaeus (1707–1778) attempted to distinguish "created kinds," but instead he discovered a multilevel descendant hierarchy of taxonomic groups, inconsistent with creation, which compelled him to classify humans as apes.

- James Hutton (1726–1797) discovered geologic gradualism and evidence of deep time, both of which creationists would like to refute.
- Sir William Herschel (1738–1822) announced having seen stars whose light he could prove to have taken two million years to reach Earth.
- Georges Cuvier (1769–1832) discovered and confirmed the extinction of species, a concept that was considered heretical according to creation proponents of that time.
- Rev. William Smith (1769–1839) was a clergyman, one of several scientists who were Christian credited with the discovery of the geologic column that young-earth creationists deny even exists.
- Charles Lyell (1797–1875) established the principle of uniformitarianism, that "the present is the key to the past," to which every young-earth creationist objects.
- Louis Pasteur (1822–1895) was at least a Lamarckian evolutionist who didn't believe in creation. He used methodological naturalism to prove the church wrong again when he provided convincing evidence for the germ theory of disease. The fathers of Protestant Christianity had said diseases were caused by demons.

For some reason, even though Richard Owen was definitely a creationist and a foremost expert and a leading authority, none of the creationist websites I've seen list him among their "Christian scientists." I wonder why?

Those men who believed in God and made historic contributions to science still relied on necessarily natural methodology, because that is the only way science can progress. In many cases, they found natural explanations for things previously believed to be miraculous, and they only succeeded when they did not allow religious convictions to subvert or inhibit their inquiry. None of them were able to vindicate the Bible stories, and their efforts to do so only ever indicated another origin. They discovered a lot about evolution, but the only thing they discovered about creationism is that a lot of it just couldn't be true. Thus, these men wouldn't have supported creationism as we know it today, and many of them wouldn't have been creationists if they knew anything about evolution.

But Richard Owen was a creation scientist—both in the sense that he preferred magical manifestations to material mechanisms and because he deliberately misrepresented evidence in an attempt to mislead others into believing as he did. According to an informational essay about Owen on the University of California Museum of Paleontology website,

> Owen was not well disposed to Darwin's theory of evolution by natural selection when Darwin published Origin of Species in 1859. However, his pronouncements on the subject of evolution were puzzling and contradictory; in later years he alternately denied its validity, professed ignorance on the matter, and claimed to have come up with the idea himself almost ten years before Darwin.

Owen believed in common archetypes rather than a common ancestor, and his conduct presents an archetype of the modern creation scientists, except that they submit to peer review rarely (if ever) and none of them are experts in anything. They've never produced any research indicative of their position. They cannot substantiate any of their assertions, and they've never successfully refuted anyone else's hypotheses either. But every argument or claim of evidence they've ever made in favor of creation has been refuted immediately and repeatedly. All they've ever been able to do was criticize real science, and even then the absolute best arguments they've ever come up with are disproved time and again across all scientific fields, with mountains of evidence standing against their every allegation. Yet creationists still use those same ridiculous rationalizations, because they will never accept where their beliefs are in error! The only notable strength creationism has and the only impressive aspect of the creationist position is how anyone can be so consistently proven to be absolutely wrong about absolutely everything 100% of the time for such a long time, and still believe that theirs is the absolute truth.

More amazing still is how often they will actually lie in defense of their alleged truth. For one of many examples, geologist Andrew Snelling documented an erosional period of more than a hundred million years on a rock formation that he himself determined to be billions of years old. He detailed all the evidence and methods leading to this conclusion and submitted these findings to scientific journals where they would be peer reviewed. At this point, he was living a double life, because at the same time he was also writing articles for the Creation Science Foundation, which would become Answers in Genesis. Therein he pretended to refute his own current observations—minus the evidence and methodology, of course. In those articles he wrote that the earth was only 6,000 years old simply because the Bible said so, and that nearly everything in geology should somehow be blamed on a global flood—one that his alternate self apparently didn't believe in. Clearly, he was not being honest with himself or anyone else.

Every publication promoting creation over any avenue of actual science contains misquotes, misdefinitions, and misrepresented misinformation, while creationists' every appeal to reason is based entirely on erroneous assumptions and logical fallacies. There is a madness to their method, but it is naught but propaganda.

What are the fruits of creation science? What are its predictions? Advances? Discoveries? What are its applications? Can it even make any contributions? Has there ever been an occasion where a supernatural explanation turned out to be verifiably correct? Can you cite one time in the history of science when supernatural assumptions have ever improved our understanding of anything? The answers are obvious, yet this doesn't stop Ken Ham from saying with a straight face: "We don't believe in evolution. Evolution is an idea some people have to explain life without God."

No, evolution is science, and as such it's a way of explaining life without magic; there's a difference. Even at the kindergarten level, science is defined as a way of learning about the natural world—"natural" meaning "in accordance with the laws of nature." Nature is further defined as the sum of all forces or phenomena in the entirety of perceptible reality. Everything that really exists has properties, and anything that can be objectively indicated, measured, or tested is therefore natural. The supernatural is contrasted with this, being defined as that which is beyond the material universe, outside our reality, a transcendent dreamlike dimension indistinguishable from illusions or the products of imagination—independent of, and even defiant of, physical laws, and thus neither detectable nor describable by science. The evocation of inexplicable paranormal forces or supernatural entities to influence natural events or phenomena are usually described as "miracles" but are also clearly "magic" by definition, as discussed in chapter 1.

In brief, miracles = magic. Don't believe me? Look it up.

Yet creationists contend that they don't believe in magic. But speaking anything into existence is an incantation, and the Bible is full of spells of one sort or another. Animating golems, or conjuring interdependent systems, and causing complex organisms to appear out of thin air are each logically implausible and physically impossible according to everything we know about anything at all. Yet this is exactly what religiously motivated pseudoscientists actually promote!

How do we test these ideas? How can we tell them apart from any of the thousands of fables men have concocted for the ghosts and gods of other religions? How can we tell whether any of this is even real, and not something someone just made up? Despite anyone's assertions of personal conviction, it is impossible to distinguish miracles from subjective impressions imagined out of nothing. In the realm of fantasy, it's easy to demonstrate psionic talents, astral entities, and magical manifestations. Until they do that in reality too, then science has nothing but nature to work with.

Those who know the necessity of naturalism can list millions of practical advantages that continue to come from that, so science requires a way to weed fiction from function through independent verification and the process of elimination. That's why natural science works and creation science doesn't. It's also why faith healers don't work in hospitals.

The National Academy of Sciences defines science as the "systematic enterprise of gathering knowledge about the world and organizing and condensing that knowledge into testable laws and theories"—a definition further endorsed by the Academic Press Dictionary of Science & Technology.

Fraudulent evangelical charlatans often say that creationism is scientific, but there's a total lack of determinably accurate evidence behind any of their assertions and no way to construct any hypotheses to explain any of their claims, because no experiments could possibly support them. Furthermore, faith prohibits believers from ever admitting when their notions would be falsified. Creationists are therefore unable to add to the sum of knowledge, instead only offering excuses trying to actually reduce what we already know. They've no way to recognize their own flaws and won't correct them, so they can neither confirm nor improve their accuracy. This runs counter to everything real science is and does.

Consequently, since the dawn of rational thought, the advancement of science has been retarded by the minions of mysticism, and profound revelations have often been opposed or suppressed by the greater part of the dominant religion because dogmatic faith is not based on reason and zealots will not be reasoned with.

A classic example of that is the fate of Socrates. The famed philosopher of the fifth century BCE dared to say that lightning had a natural cause, and that it wasn't a god hurling magic javelins. When he refused to recant, he was convicted of impiety, the crime of refusing to acknowledge the existence of gods. So the intolerant (and insecure) Hellenist theocracy sentenced him to death by poisoning.

The earth was found to be round about 300 BCE, but sacred scriptures written after that still describe the world as flat. Some believed that until the time of Columbus; some still do today! Galileo never expected the Spanish Inquisition to imprison him for life just for showing us that the earth was a planet that moved around the sun. Even today, there are many creationists who believe that the sun orbits the earth, and that the earth is somehow at the center of the entire universe. It seems to be a matter of whether one has a choice to believe what is evidently true, or to believe something else instead.

Before Answers in Genesis, the Institute for Creation Research was the most prominent source of antievolution propaganda. This was the organization most obviously trying to promote "creation science." Several years ago, when Ben Stein released his intelligent design movie *Expelled: No Intelligence Allowed*, some people in my area held a seminar of sorts to discuss the film. The theater was packed. There were only three people on the panel. One of them (if I remember correctly) was there to give some token defense of science. Another was a film critic who admitted having no relevant importance to the evening's discussion at all. The last panelist was Frank Sherwin, who was then referred to as the bulldog of the Institute for Creation Research. (This was another attempt at creating an illusion of equivalence, since Thomas Huxley was famously known as Darwin's bulldog.) To give you an idea of Sherwin's sophistry, he exclaimed at one point: "So this is a war of the worldviews, and all science is creation science. Would you agree with that this evening? All science is . . . ? Creation science. All truth is . . . ? God's truth! And certainly science is the search for . . . ? Truth!"

At that time, I was just getting involved enough that I began to consider myself an activist. Almost none of the people I knew in real life were able to understand why I cared enough to do what I do. My then fourteen-year-old son, Connor, was one of my worst critics back then. So I took him with me, just so that he could see what this controversial conflict was really all about.

It was an eye-opening experience for him. Every time Sherwin spoke, it was of course utter nonsense that any reasonable person should be able to disprove in a moment, but Connor looked around to notice a full audience of naive patrons eagerly swallowing it all up. He saw little old ladies who clearly knew nothing about science and were hopelessly gullible, nodding obediently as Sherwin fed them lies. Deliberately deceiving innocents is something neither I nor my son will tolerate. The boy didn't speak, but he was so obviously irritated that he was vibrating.

At the end of the presentation when everyone had a chance to get up and talk, I wanted to confront Sherwin, but I was targeted and immediately surrounded by a crowd of emotionally charged believers, each trying to challenge me with embarrassingly ignorant questions like, "If you think we came from apes, where's your tail?" I was detained by the mob, having to explain what an ape is, and the fact apes don't have tails. Connor was small enough then that he managed to slip through the crowd like a barracuda. I could see him clearly berating Sherwin, a man who was at least as tall as me. I couldn't hear what was being said and I couldn't break away either, but I kept glancing at my son shaking an accusing finger at Sherwin and criticizing him hotly while a separate crowd gathered around them.

Moments later, Sherwin was gone. No one else but my son got to speak to him. Once my own crowd dispersed, my son's crowd gave me the report: the boy bluntly and fearlessly accused Sherwin of lying on a series of listed points. Sherwin's first defense was an attempt to confuse Connor with a nonsensical word salad, a tactic my boy saw through immediately. So Sherwin demanded that Connor give him a transitional species. Connor fired back, naming Hesperocyon. Sherwin didn't know what that even is and didn't dare discuss it. There can't be a situation where he would honestly consider evidence against his position, and he doesn't know what to do when confronting someone who knows better than he does. So he changed the subject, and demanded that a fourteen-year-old kid give him an example of a beneficial mutation. Again Connor's response was instantaneous: "LRP5 function for hyper bone density; surely unbreakable bones is a beneficial mutation, isn't it?!"

Again Sherwin had no response to Connor's contemptuous tone. People were patting me on the back, bragging about my son. They said he had Sherwin visibly sweating and confused until he literally fled the room! So much for ICR's "bulldog." Since then, Connor has attended many secular conferences with me, and I'm proud to say that he fits in very well with that group.

Yes, Frank, science is a search for truth—whatever the truth may turn out to be, even if it's evidently not what we wanted to believe it was. In science, it doesn't matter what you believe; all that matters is why you believe it. This is why real science disallows faith, promising instead to remain objective, to follow wherever the evidence leads, and to either correct or reject any and all errors along the way, even if it challenges whatever we think we know now, or would rather believe. Carl Sagan stated exactly this in his keynote address at the 1987 Committee for the Scientific Investigation of Claims of the Paranormal conference,

> In science, it often happens that scientists say, "You know, that's a really good argument; my position is mistaken," and then they would actually change their minds, and you would never hear that old view from them again. They really do it. It doesn't happen as often as it should because scientists are human and change is sometimes painful. But it happens every day. I cannot recall the last time something like that happened in politics or religion.

But creationist organizations post written declarations of their unwavering obligation to uphold and defend their preconceived notions, declaring in advance their refusal to ever to let their minds be changed by any amount of evidence that is ever revealed. Antiscience evangelists display their statement of faith on their own forums, as if it were something to be proud of—as if admitting to a closed and dishonest mind wasn't something to be ashamed of or to beg forgiveness for. Every creationist organization posts a statement of faith wherein they admit that they assume their conclusions at the onset for reasons that have nothing to do with factual information, and more importantly, where they refuse to ever admit when they're wrong. They even admit in advance that they refuse to be reasoned with!

> "By definition, *no* apparent, perceived, or claimed *evidence* in any field, including history and chronology, *can be valid if it contradicts the Scriptural record*."
>
> —Answersingenesis.org

> "[V]erbal inspiration guarantees that these writings, as originally and miraculously given, are *infallible and completely authoritative*

on all matters with which they deal, *free from error of any sort*, scientific [sic] and historical as well as moral and theological."

—Institute for Creation Research

"[This school] . . . stresses the Word of God as the *only source of truth in our world*."

—Canyon Creek Christian Academy

"[T]he autographs of the 66 canonical books of the Bible are objectively inspired, *infallible and the inerrant Word of God* in all of their parts and in all matters of which they speak (history, theology, science, etc.)."

—Mark Cadwallader's Creation Moments

"The Bible is the divinely inspired written Word of God. Because it is inspired throughout, it is *completely free from error*—scientifically, historically, theologically, and morally. Thus it is the *absolute authority in all matters* of truth, faith, and conduct. *The final guide to the interpretation of the Bible is the Bible itself*. God's world must always agree with God's Word, because the Creator of the one is the Author of the other. Thus, where *physical evidences* from the creation may be used to confirm the Bible, these evidences *must never* be used to correct or interpret *the Bible*. The written Word *must take priority in the event of any apparent conflict*."

—Mark Ramsey's Greater Houston Creation Association

"Revealed truth: That which is revealed in Scripture, whether or not man has scientifically proved it. *If it is in the Bible, it is already true without requiring additional proof*.

". . . *Fallacy: that which contradicts God's revealed truth*, no matter how scientific, how commonly believed, or how apparently workable or logical it may seem."

—Bob Jones University, *Biology Student Text*, vol. 2 (3rd ed.)

[All emphasis added]

Each of these organizations announces that they will automatically and thoughtlessly reject without consideration any and all evidence that might ever arise should it conflict with their interpretation of Bronze Age folklore, so they have already rejected all the evidence there could ever be, no matter what we may discover in the future, or how conclusive that is. In other words, no matter how true the truth really is, no amount of proof will ever change their minds.

No scientific organization would participate in anything so Orwellian, so grossly biased, and so blatantly dishonest as this. If truth mattered more than whatever you prefer to believe, this wouldn't be necessary and it wouldn't be permitted. But among creationists, truth is irrelevant.

There are plenty of facts that I can show to be true of evolution, but there are no facts pertaining to, concordant with, or supportive of creationism. There is no mechanism to creationism, no indicative evidence whatsoever, and not one law applicable to it or that is not broken by it. It can't be indicated or vindicated, verified or falsified by any means creationists would accept, and yet it has been utterly disproved in every testable claim that it makes.

They don't want to do science, they want to *undo* science! They try to segregate experimental science from historical science, ignoring the fact that both are based on empirical observations and both can be checked with testable hypotheses. Worse, they want to redefine science in general so that subjective impressions of faith and excuses of magic can supplant the scientific method whenever necessary in defense of their beliefs. They're only open to critical inquiry so long as it does not challenge their sacred scriptures nor vindicate any of the fields of study to which they're already opposed. In short, everything science stands for, or hopes to achieve, is threatened by the political agenda of these superstitious subversives.

Once upon a time, our ancestors believed that thunder, lightning, and volcanoes were gods in action, that comets were an omen, that the stars and planets had human characteristics, that sickness was a curse of witchcraft, and that epilepsy was demonic possession—all because that's what religion would have us believe. In each case, the real truth might never have been discovered had we been satisfied by those lies. And in each case, the reality was a revelation of whole new fields of study previously unimagined, and vastly more complex than the simple excuses we made up in our ignorance. No doubt that pattern will continue, such that if we ever do discover the cause of the Big Bang, or some better explanation for the origin of life, the universe, and everything, it too will be a wealth of new information with practical application, and so advanced that it will render our previous belief in gods, ghosts, and magic just as laughably silly as every other field of study so far has already shown.

Science relies on rationalism, allowing that many things *might* be true, but that one can never claim anything definitely *is* true without substantial, positively indicative and verifiably accurate physical evidence that can be examined and critically analyzed by experts in peer review, and adequately explained in a theoretical construct. You can believe whatever you like. As long as you admit that it is a belief, you don't have to defend it. But if you assert your belief as a statement of fact, then you *do* have to defend it! Stating anything as definitely true when there is insufficient evidence to back it is dishonest. Making such positive proclamations without any evidence at all is a matter of faith, and promising in advance to forever defend an unsupportable *a priori* preference even against an avalanche of evidence against it is apologetics, which is all creation "science" really is.

THE 13TH FOUNDATIONAL FALSEHOOD OF CREATIONISM

"Evolution Is a Fraud"

In order to convincingly demonstrate just how dishonest creationism is, I offer a two-part challenge for believers hoping to defend their position: (1) name one evolutionary scientist who lied in the act of promoting evolution over creationism, and (2) name a professional creationist who did *not* lie when trying to defend creationism or condemn evolution.

Remember that a lie is misinformation, or information misrepresented with deliberate intent to deceive. To satisfy the first part of this challenge, you should identify the person and quote their lie verbatim, then show how we know their statement was false, and finally how we can tell whether the scientist in question knew it was false when they said it. You'll only bear the burden of proof for the first part of this challenge as it pertains to evolutionary biologists, by which I mean geneticists, embryologists, geologists, zoologists, etc., specialists in fields directly relating to evolution. Of course, that person should also accept the mainstream scientific position, and not be on the payroll of Answers in Genesis or some similarly biased organization. Also remember we're only talking about those who lied in their critique of creationism or as an attempt to promote evolution by comparison. I'm not saying there are no dishonest scientists—there are, and I can name a few of them off the top of my head. But they don't lie about this. There is no reason to.

I am saying that there are no honest creationists though, not among the professionals (by which I mean anyone with an income from preaching this nonsense). I don't mean that there are no honest people who happen to believe in creationism; it's possible. But once sincere believers begin to investigate the evidence and arguments on either side of this alleged controversy, they will very quickly face a life-altering choice: whether to remain honest, or whether to remain creationist, because it is no longer possible to be both. Either they can honestly concede what the definition of macroevolution really is and thus be forced to admit that it has been observed, or they can lie and say there's never been any transitional species or beneficial mutations. It is not possible to defend creationism honestly. Professional creationists are theists who have made a career out of objecting to evolution, so they know full well what lies they have to tell at each cue and that there's no way around that.

For the second part of the challenge pertaining to professional creationists, I will bear the burden of proof, meaning that once you provide the name, I'll be the one having to present statements made by that person that are demonstrably wrong, and I'll have to show how we can tell whether that person knew they were wrong when they said it. As I've already indicated in each of the previous chapters, every professional creationist I've ever looked up is guilty of willful deceit or they wouldn't be able to play their game, and it doesn't take long to produce citations to prove that.

I've been posting this challenge for decades now, and no one has ever been able to answer either part of it yet; though right away, they all think they have the perfect answer to the first part. Most can only think of one person, and their first indictment is always the same name.

Ernst Haeckel (1834–1919) was a pioneer biologist, physician, and philosopher. At home in Germany, he was a professor of comparative anatomy for half a century, but he was also a field naturalist. In the course of his travels to different continents, he had personally discovered, described, and named thousands of new species. He was a pioneer taxonomist, the first to produce a visual map of the genealogical tree of life. He published dozens of scientific works, many of which were wonderfully illustrated because he was also an artist. The man was an award-winning science communicator, the *Cosmos* host of his time. He even has mountains named in his honor, both in the United States and in New Zealand. But his numerous contributions to biology go largely unnoticed compared to a couple rather embarrassing errors.

As I mentioned before, and as Karl Ernst von Baer originally noted in 1828, the more closely related any two species are, the more similar their development. He also observed that vertebrate animals in their embryonic stages seem to have a common design, whereas adult forms show difference. Arm buds are virtually indistinguishable at first formation but might become a wing, an arm, a leg, or a flipper. At another level of this same pattern, the young of two closely related species will look more alike than the adults do. An excellent example of this is found in crustaceans. Although adult crabs look quite a bit different from lobsters or shrimp in that they don't have tails, we see that crab nymphs (or zoea) actually *do* have tails, and that these fold into the carapace of mature crabs, leaving only a vestigial crease indicating where the tail was. Of course, this trend continues into embryology too, and the implications of that confused naturalists and anatomists of the nineteenth century.

For example, Haeckel studied dozens of embryos under a microscope and interpreted them to promote his idea that "ontogeny recapitulates phylogeny," suggesting that embryonic development reflects the organism's evolutionary ancestry. Haeckel's "biogenetic law" of embryological parallelism was one of many notions of biological transcendentalism first proposed by Lamarckian naturalist Etienne Geoffroy Saint-Hillaire in about 1760. Haeckel believed that the developing fetus mirrored its evolution such that it might pass through phases of becoming a fish, then an amphibian, then a reptile each in succession before becoming a mammal. To illustrate this, he produced about a hundred drawings of embryos at various stages. But he later admitted that about a half-dozen of them were either false or speculative due to a lack of visual references. In the first edition of his best-selling book *Naturliche Schopfungsgeschichte* (Natural History of Creation), Haeckel used the same image to represent the embryos of dogs, chickens, and turtles. When a reviewer alerted him to this, he said that no one could tell the difference at that stage, which was probably true given the instrumentation of the time. While this is not necessaruily a lie, the fact that any of his drawings were admittedly without reference has disgraced Haeckel's name in the annals of science despite the fact that these were corrected in each of the later editions.

Creationists now insist that Haeckel was reportedly convicted of fraud by a German court, though that doesn't seem to be the case. His creationist contemporaries, including Rudolf Virchow and Louis Agassiz, did accuse

Haeckel of deception, but the charge of fraud didn't emerge until 1997 with the research of embryologist Michael Robertson, as detailed in an article in *Science* by Elizabeth Pennisi titled "Haeckel's Embryos: Fraud Rediscovered." No one argues that a scientist shouldn't promote speculation as fact; leave that to religion. But the drawings Haeckel didn't have references for in his first edition aren't the reason for the recent charge of fraud. It was a computerized analysis of his artwork as compared to microphotographs of the same species at the same stages of development. The charge was that he embellished these drawings to imply more resemblance than there was based on a critique of his artistic skill. However, it has been shown that this same analysis would also indict Haeckel's enemy contemporaries on the same charge, as well as modern embryologists too.

A later paper titled "Haeckel's Embryos: Fraud Not Proven," written by Robert Richards and published in the journal *Biology & Philosophy* in 2009, offers a compelling rebuttal to the charges detailed in the 1997 *Science* article, "The historical and biological evidence, however, shows the charge against Haeckel to be logically mischievous, historically naive, and founded on highly misleading photography."

The images under scrutiny were taken from Haeckel's hastily assembled first edition of *Anthropogenie*. However, each of the subsequent editions had the advantage of better instrumentation, and the accuracy of the drawings improved. But there was nothing wrong with those images to begin with. The damning microphotographs published by Michael Robertson in 1997 showed these embryos with yolk and other maternal material that made them look very different. That, and the chicken was photographed at a different angle with a different lens effect than the others, while the salamander was a different size. Haeckel clearly indicated that his drawings were only of the embryos, omitting things like yolk, and that he made them all the same size and oriented the same way for ease of comparison, so there's no foul to fault.

Robertson, the very researcher who indicted Haeckel in 1997, seems to have softened his view since then, perhaps after his own errors in the indictment itself were brought to light. In a November 2002 paper published in the *Biological Reviews of the Cambridge Philosophical Society* titled "Haeckel's ABC of Evolution and Development," Robertson (with G. Keuck) writes,

> Haeckel's much criticized embryo drawings are important as phylogenetic hypotheses, teaching aids, and evidence for evolution. While some criticisms of the drawings are legitimate, others are more tendentious.

What is especially odd about the criticism against Haeckel's "embellishments" (if such were ever intended) is that they're unnecessary. Creationists adamantly complain that textbooks referred to his admittedly inaccurate drawings for so long, but for some reason, they continue to accuse those authors of fraud even when the textbooks replace the drawings with microphotographs! Do they think the photographs are fraudulent too? Or do they object only because those images still indicate the same evolutionary parallels that Haeckel saw?

Darwin wrote that embryology contained compelling evidence of evolution, and correctly so. In his letter to J. D. Hooker, for example, he stated, "Hardly any point gave me so much satisfaction when I was at work on the Origin as the explanation of the wide difference in many classes between the embryo and the adult animal, and the close resemblance of embryos within the same class." Creationists dismiss this on the assumption that Darwin's theory was inspired by Haeckel's allegedly fraudulent drawings, and that consequently, evolution is a fraud. But of course the truth is the other way around. Darwin referred to real embryos; Haeckel's drawings didn't even exist until years after Darwin's final publication. Haeckel eventually befriended Darwin and even convinced him of his biogenetic law at one point, but Darwin had already published his definitive work, and Haeckel had no influence over that.

In his landmark publication, Darwin had accurately depicted the resemblance of closely related embryos and never suggested that they progress through the adult stages of their evolutionary lineage. Modern biology does recognize numerous connections between ontogeny and phylogeny, and explains them using evolutionary theory without recourse to Haeckel's specific view. Haeckel's original assumption that embryonic development would indicate adult species in an organism's ancestral history was proven false by 1910. Recapitulation has since been replaced by a more accurate study of the parallels between embryological and evolutionary development, colloquially known as "evo devo." Among other discoveries, this field revealed the evolutionary origin of the feather as implied by transitional stages in the fossil record being recapitulated in the stages of embryological development of chickens. But the fact Darwin recognized, that embryology does provide testable confirmations and predictions of phylogeny, were already evident before Haeckel ever picked up his pencil.

It is no hoax that mammalian embryos temporarily have pharyngeal pouches, which are morphologically indistinguishable from the gill slits in modern fish embryos, and that the divergence of development from there matches what is indicated in the fossil record. This is fact, not fraud, and none of these facts should be true unless evolution were true also. Why else do whale embryos have four limbs? Why do glass snake embryos have feet complete with toes? Why do chicken embryos have three-fingered hands? Why do human and bird embryos have tails?

So Haeckel didn't lie, so far as we know. At least we know he didn't lie when he could have, and that's a good indication. Even if he had lied, it wasn't while promoting evolution over creationism. Haeckel wasn't trying to convince evolutionists of evolution; he was promoting recapitulation, an alternate notion of fetal development. His posit failed against Karl Ernst von Baer, the leading authority who conceived the laws of embryology.

The second most common attempted answer to the first part of my challenge to name an evolutionary scientist who lied in the act of promoting evolution over creationism is Piltdown Man. Seriously, several people have answered the question this way. I have to explain to them that Piltdown Man was not the name of an evolutionary scientist.

A hundred years ago, the only human fossils yet known were a few Neanderthals, Cro-Magnon, and *Homo erectus*. Then Charles Dawson, an English attorney and amateur archaeologist, presented bones and associated artifacts of what appeared to be an as-yet unidentified species. It was reportedly unearthed by unnamed quarrymen in the same area of Sussex where the first recorded dinosaur discovery was made a century earlier.

British Imperialists were generally accepting of the news, because they liked the idea of man originating in their country. But French and American scientists were skeptical, doubting that the skull and jaw even belonged together. The British Museum touted the Piltdown Man as authentic, but the American Museum of Natural History displayed it only as a "mixture of ape and man fossils," which is what it eventually turned out to be.

E. Ray Lankester, director of the British Museum, initially thought the piece was an important find, but he doubted that the jaw and skull were from the same individual. He would not comment on its age and did not believe it responsible for the tools found around it. Arthur Kent of the Geological Association was another British skeptic. Henry Fairfield Osborn, then president of the American Museum of Natural History, was one of the skeptics of the initial find. However, every time some doubt arose about the authenticity of the bones and every time there was a challenge to the idea that they came from the same animal, an astonishingly well-timed discovery would put these doubts and challenges to rest. Eventually, Osborn was convinced by subsequent "discoveries," even though he doubted the union of the skull and jaw, and said Piltdown was unrelated to any other species of man in the fossil record.

There was no way to adequately examine such things back in 1915. Chemical tests common today didn't yet exist, and we didn't yet have a practical understanding of radiation. Before the first australopiths were discovered, we didn't know exactly what to expect of the links that were then still missing between humans and the other apes known at that time. But as we began filling in the gaps in human evolution with thousands of legitimate fossils, a pattern emerged that left Piltdown an increasingly obvious anomaly.

Even Arthur Smith Woodward, credited as one of the codiscoverers of Piltdown, began to question the find, saying, "there appears to be no place for a stage resembling that of any adult existing ape. It is difficult even to understand how Eoanthropus can be one of the series." F. H. Edmonds questioned the source of the Piltdown animal fossils. He and Alvan T. Marston also attacked Dawson's dating methods, insisting that the Piltdown collection was much younger than suggested. Marston and Ashley Montague also suspected eoanthropus of being a composite piece, which would eventually prove to be correct. Consequently, it was taken off display and stored away almost continuously for decades. It lost importance in most discussions because in light of everything else we discovered over the next few decades, it just never fit, and was eventually dismissed from the list of potential human ancestors for that reason.

As the years wore on, criticism arose against everyone who ever promoted the Piltdown collection because there seemed to be so much wrong with it. Finally, in the 1950s, it was taken back out of the box and scrutinized via more modern means. First, fluorine dating revealed that it was much too recent, and it was shown to have been chemically treated to give a false impression of its age and mineral composition. Then it was finally determined that the jaw must have come from an orangutan, and that it had been deliberately reshaped with modern tools in a well-crafted and deliberate forgery.

No one knows who did it either, and more importantly, why. Errors were already known and previously reported, but few ever suspected fraud because what would be the motive?

The actual culprit may be unknown, but suspects have been identified. Museum volunteer Martin Hinton apparently had issues with the authorities directing the British Museum. Recently, a trunk belonging to Hinton was found to contain several unfossilized bones that had been chemically stained and treated in the same way the Piltdown bones had been. Charles Dawson, the primary suspect at the time, was considered ambitious and unscrupulous even for a lawyer. He was credited with each of the more significant finds and there was no more evidence of Piltdown Man to be found after he died in 1916. He is also known to have forged other archaeological finds. Notably, Sir Arthur Conan Doyle, author of the Sherlock Holmes mysteries, happened to be Dawson's neighbor and was present at the dig sites. He had also written *The Lost World*, which spoke of faking fossils as practical jokes. It was published in the same year as the Piltdown "discoveries."

Doyle had all the means and even a motive, a religious one. As a member of a number of paranormal organizations, he objected to the "naturalism" of "Darwinists." Doyle could have perpetrated the Piltdown hoax for the same reason as Hinton, to humiliate the scholarship of the British Museum. But Doyle's supernatural beliefs caused him to be the fool of a hoax himself. In his later years, he became a devoted proponent of mysticism, spiritualism, and psychic phenomena, including a strong belief in fairies! His conversion cost him both friends and credibility, especially when he was a duped by a pair of little girls using trick photography. The Cottingley Fairy fraud convinced many believers for more than sixty years. Only in 1983 did the then-elderly girls finally confess that most (but not all) of their fanciful photos were faked. Maybe that was the motive; maybe Piltdown Man was just a practical joke that had gone too far. But no one was laughing, and they weren't going to let it happen again.

Although such a hoax would count as a lie, we can't answer the challenge if we can't name the guilty party. Even if we knew the answer, it still wouldn't be a lie against creationism, nor an attempt to promote evolution either, because the motive of most of the suspects would have been to *discredit* scientists. In Dawson's case, even if he had no obvious motive to discredit scientists, his very action would have risked discrediting them.

In my top-four list of failed attempts to answer my challenge for an evolutionary scientist lying against creationism, the third most common answer is itself based on a lie. I sometimes hear that Darwin himself lied, that he didn't believe his own theory, and that he'd recanted evolution on his deathbed. All three of these claims are themselves fibs, mostly from an evangelist called Lady Hope. She claimed to have visited Darwin just before he died and that he confessed his doubts to her. Everyone in Darwin's family, however, denies this. His son Francis said it was "quite untrue" and publicly accused Lady Hope of lying. Darwin's daughter Henrietta didn't believe that Lady Hope had ever seen her father. "I was present at his deathbed. Lady Hope was not present during his last illness, or any illness. . . . He never recanted any of his scientific views, either then or earlier. We think the story of his conversion was fabricated in the U.S.A." His son Leonard called Lady Hope's story a "purely fictitious hallucination."

But we're not talking about the lies of creationists; I asked for a lying "evolutionist." The closest we get to satisfying that challenge in the top few attempted answers I hear is Nebraska Man. Again, it's not a name, not an evolutionary scientist, not a promotion of evolution as opposed to creationism, and not a lie or a hoax either. Rather, as occurs often in science, it was a mistake that was later retracted.

Even before the Piltdown hoax was officially exposed, an American paleontologist earned himself a lifetime of embarrassment when he applied the label Hesperopithecus (ape of the Western world) to a tooth from an extinct piglike animal discovered in Nebraska. The cheek teeth of pigs and peccaries are sort of similar to ape molars, and this one was badly worn such that it was hard to distinguish them. Upon examining the find, Henry Fairfield Osborn believed it to be human, which is strange because he never believed that about Piltdown Man. But the real embarrassment came in June 1922 when the story was published in the *Illustrated London News*, just weeks after publishing to *Science* for peer review. As I mentioned earlier, the way popular media sensationalizes stories does a disservice to science.

Creationists like to say that scientists were as duped by Nebraska Man as they were by Piltdown Man, but they weren't.

"The occurrence of a man-like ape among fossils in North America seems so unlikely that good evidence is needed to make it credible."

—Arthur Smith Woodward (1922)

"[The Nebraska tooth] combines characters seen in the molars of the chimpanzee, of Pithecanthropus, and of man, but . . . it is hardly safe to affirm more than that Hesperopithecus was structurally related to all three. . . . the prevailing resemblance of the Hesperopithecus type are with the gorilla-chimpanzee group."

—Gregory and Hellman (1923)

"In 1920, Osborn described two molars from the Pliocene of Nebraska; he attributed these to an anthropoid primate to which he has given the name, Hesperopithecus. The teeth are not well preserved, so that the validity of Osborn's determination has not yet been generally accepted."

—Prof. George Grant MacCurdy (1924)

"The suggestion that the Nebraska tooth (Hesperopithecus) may possibly indicate the existence of Mankind in Early Pliocene times is, as I have explained . . . still wholly tentative. The claim that real men were in existence in Pliocene and Miocene times must be regarded as a mere hypothesis unsupported as yet by any adequate evidence."

—Grafton Elliot Smith (1927)

Everyone who saw the fossil agreed that it did look like an ape's tooth. But with only a couple tentative exceptions, the entire contemporary scientific community either immediately rejected the accuracy of Osborn's assertions or they demanded more substantial evidence to back them.

Osborn obviously couldn't provide that evidence, even after another five years of searching. Eventually, he came to the sad realization that his fossil probably wasn't really human after all. When a man admits his own mistake on his own accord upon realization of it, and before anyone else even knows of the error, that is honorable and can hardly be called a hoax.

One of Osborn's former students, the zoologist William K. Gregory, then published a formal retraction in *Science*: "The scientific world, however, was far from accepting without further evidence the validity of Professor Osborn's conclusion that the fossil tooth from Nebraska represented either a human or an anthropoid tooth."

Creationists often accuse scientists of contriving the illustration of Nebraska Man and of conjuring a whole skeleton and facial construct out of a single tooth that was never even human in the first place. But the fact is that the popular magazine commissioned its own artist's impression and scientists of the day (including Osborn himself) immediately reacted with harsh criticism. As a result, the article was never reprinted. Even when Osborn himself still believed in his "Hesperopithecus" early on, he criticized that illustration: "Such a drawing or 'reconstruction' would doubtless be only a figment of the imagination, of no scientific value, and undoubtedly inaccurate. . . . One of my friends, Prof. G. Elliot Smith, has perhaps shown too great optimism in his most interesting newspaper and magazine articles on Hesperopithecus."

Now even though Piltdown Man was eventually exposed by evolutionary science itself, and even though Nebraska Man was simple stupidity, honestly and voluntarily admitted, and even though there were no other such examples in the history of paleoanthropology, creationists still portray both of these events and many others that were never in question—as if they were all part of some ridiculous unified international conspiracy intended to fool the world into believing evolution over creation *ex nihilo*. These paranoid propagandists also commonly contend, based only on these exceptions and other imagined ones, that each of the thousands of fossil hominids we've found and confirmed before and since were all proven to be fakes too, even when the alleged authorities making these claims are themselves convicted frauds. I'm referring to Mr. Kent Hovind, the charlatan who calls himself "Dr. Dino" and who bills himself as an expert on dinosaurs and the Bible. However, being no more than a virtually illiterate flim-flam man and thus unable to participate in academia or even write a book, Hovind teamed up with Chick Publications (another creationist organization recognized by the Southern Poverty Law Center as a hate group) to make ridiculous religious tracts in miniature comic book format.

The most famous of these excessively ignorant, bigoted, and hateful tracts was *Big Daddy*. I have torn this stupid tract apart on Talk.Origins, on a live podcast of the Bible Reloaded, and in my keynote speech to Broward College on Darwin Day. The worst part of it for me is a centerfold panel that shows a few badly drawn hominines arranged like Rudolph Zallinger's famous 1965 painting *March of Progress* depicting human evolution. Hovind captioned the hominines with grossly inaccurate distortions of fossil humans.

> **Lucy**: *Nearly all experts agree Lucy was just a 3 foot tall chimpanzee.*

None of the experts believe that. All of the experts agree that australopithes are *not* chimpanzees, but that they lie between humans and modern apes, or that they're simply a basal human form.

> **Heidelberg Man**: *Built from a jawbone that was conceded by many to be quite human.*

Homo heidelbergensis was "quite human" because he *was* human, just not the same paleontological species we are. He was a descendant of *Homo erectus* and apparently ancestral to both *Homo sapiens* and *Homo neanderthalensis*. He wasn't known from a single jawbone either, but from more than four thousand bones, including multiple skulls representing nearly thirty individuals found in one site alone, with dozens more found elsewhere.

> **Nebraska Man**: *Scientifically built up from one tooth, later found to be the tooth of an extinct pig.*

. . . and rejected by the entire scientific community.

> **Piltdown Man**: *The jawbone turned out to belong to a modern ape.*

A fraud perpetrated against evolution, not by it.

> **Peking Man**: *Supposedly 500,000 years old, but all evidence has disappeared.*

Some forty examples of Peking Man were lost in World War II, but casts of those fossils remain. What was once called *Anthropopithecus erectus* and then *Sinanthropus pekinensis* is now better known as *Homo erectus*, which is known from the fossils of hundreds of individuals found all over Africa, Asia, and southern Europe. So it's not as if all evidence of them has disappeared.

> **Neanderthal Man**: *At the Int'l Congress of Zoology (1858) Dr. A.J.E. Cave said his examination showed that this famous skeleton found in Germany over 50 years ago is that of an old man who suffered from arthritis.*

That first fossil was diseased, but it was still Neanderthal, and the physical signs of that disease had no relationship to nor impact on that classification. We've also found hundreds of Neanderthal men, women, and children, and even their DNA, which (in addition to their uniquely definitive skeletons) has provided proof that they were a different subset distinct from our species!

> **New Guinea Man**: *Dates way back to 1970. This species has been found in the region just north of Australia.*

I couldn't find any references to this. There is no correlation with any actual hominine species. Hovind seems to have pulled this one out of his ass.

> **Cro-Magnon Man**: *One of the earliest and best established fossils is at least equal in physique and brain capacity to modern man, so what's the difference?*

There isn't a difference. Cro-Magnon isn't a taxon; they were just the first *Homo sapiens* known in Europe, displacing the sons of Heidelberg Man.

> **Modern Man**: *This genius thinks we came from a monkey. "Professing themselves to be wise, they became fools." (Romans 1:22)*

Modern man didn't just come *from* a monkey, but as a member of the infraorder Catarrhini, he *is* a monkey by definition! Revealing inexcusable ignorance of a whole field of study is not an argument against it. Hovind clearly has no idea what he's talking about, and why is it always only people who have already proven themselves to be fools who cite Romans 1:22?

It's perfectly fine to be as skeptical as you like and tell someone you don't believe them. However, in my view, accusations of dishonesty shouldn't be treated lightly or issued casually. I accuse creationists of lying only because I know I can prove it—even in a court of law, if need be. You shouldn't accuse someone of lying unless you can already prove it right then and there, but be careful. If you can show that someone said something wrong, it's often hard to show whether they knew it was wrong or were trying to deceive anyone. Hanlon's Razor says to never ascribe to malice what can be adequately explained by incompetence. But if you accuse someone of lying when you can't even show that what they said is wrong, then you're the liar—especially when the claim you're challenging has already been vindicated, and you therefore have no grounds for criticism. That's why the following accusation of fraud on the website of the Northwest Creation Network is itself fraudulent: it is a lie.

> In an attempt to further their careers and justify the claims that evolution is a legitimate theory, many scientists have fraudulently deceived the world by planting or reconstructing fossils which they would claim to be authentic finds.

Before the advent of the Internet, these disingenuous shysters either printed moronic comics or made low-quality videotapes that were distributed in churches to undereducated patsies who would likely never know they were being duped. That's where we find Marvin Lubenow, author of *Bones of Contention*, the next loser in the league of liars for the Lord, who stated in one such video: "Homo habilis was made up of at least two, if not more, different groups that did not belong together. They're an assemblage of several different types of animals put together and made into one."

Lubenow acts like only one fossil exists and that it was a doctored assembly like Piltdown Man. But they've found the remains of dozens of *Homo habilis* individuals, and about a half-dozen *Homo rudolfensis* too. These were once thought to be distinct hominine species, but they're so similar that many paleoanthropologists now consider them to be two variations of one species. It is hard to imagine how Lubenow could be so wrong about so much and still not know it. He can't even show any indication that any of these were composite fossils. His accusation is therefore utterly empty. It is a lie.

Later in the same video:

> *Homo erectus* or "Java Man" isn't a half-man, half-ape either. The man who discovered it admitted before he died that it was a fraud. He confessed that he had found an ape's skull about fifty feet away from a human leg and two human skulls, and had mixed-and-matched to create a fictitious creature.

As I previously showed, it's impossible to have a half-man/half-ape for the same reason that it's impossible to have a half-duck/half-bird. In either case, they would be 100% both. Despite the many lies repeated by Lubenow and other creationists, "Java Man" (also known as *Pithecanthropus erectus*) turned out to be another one of hundreds of *Homo erectus* fossils found so far. That means the "ape skull" he mentioned was definitely human, and not what Lubenow would recognize as an ape. All modern ape skulls are typified by an obvious sagittal crest along the top. One of the distinguishing features of extinct hominine skulls is they don't have that. Their cranium is similar to ours, so if there had been a mix-and-match, it wouldn't have made much difference. But there was never any indication of any mixing and matching anyway. Lubenow apparently made that up. The two modern skulls he mentioned weren't fifty feet away either; they were found in a mountain cave over sixty *miles* away! Eugène Dubois, the man who discovered Java Man, promoted it as a missing link because he thought the fragments he found were more gibbonlike than hominid. He turned out to be wrong, but that doesn't qualify as a fraud, and there was never any admission of fraud either. That's another of Lubenow's lies. So all of these creationist's accusations of evolutionist frauds are themselves fraudulent fabrications.

Duane Gish proved that himself when he declared the skull of Java Man to be that of an ape and not a human, because he also declared Turkana boy to be "fully human" even though the two crania are an exact match! They're the same thing!

Also creationists say that *Homo floresiensis* were just normal people who suffered from microcephaly—that is, undersized brains. But there was a whole community of them. At least nine hobbit-sized adults are known thus far. Their wrists and ankles are more apelike even than *Homo erectus*, including bones that australopiths have but that we don't. They were also fully bipedal toolmakers, and CT scans reveal that their tiny (417cc) brains had uniquely overdeveloped frontal lobes. There was no question but that they were a distinctly different species of people. Hobbits were evidently a sibling group to *Homo erectus*, rather than descendants of them as we were.

So they were a more distant human relation, but they were clearly human nonetheless.

Similarly, John Morris of the Institute for Creation Research said that Lucy (the first *Australopithecus afarensis* ever found) was assembled from unrelated pieces found in different places. He said that the strongest evidence of her bipedality was a knee joint that was found over a mile away and hundreds of feet deeper in the strata. But those were different individuals who each bore their own independent evidence of strict bipedality. All of the bones shown in photographs of Lucy were found at a single location. Also, the same geologic layer may be shallow, exposed, or absent in some areas and deeper in others due to erosion, sedimentation, upheavals, etc.

Henry Morris, the founder of the Institute for Creation Research, said that all known fossils of ancient humans would fit on a pool table. Well, not anymore. Now you'll need a whole pool!

While they like to accuse "evolutionists" of fraud, it turns out that creationists are the ones perpetuating frauds. For example, young-earth creationist archaeologist Ron Wyatt claimed to have found the Ark of the Covenant directly beneath what he said was the actual site of Jesus' crucifixion, which was marked by an "earthquake crack" that no one else noticed. In addition, Wyatt claimed to have found Jesus' DNA and he said that blood tests confirmed that Jesus was born of a virgin. Wyatt also sold artifacts that he said were obtained from the site of Noah's Ark, which he also discovered. Associates who once believed him now insist that what he found was just a vaguely boat-shaped mound of dirt, and they're angry that his videos depict them as if they're endorsing him.

In the creationist tradition, if you can't find any actual evidence, you create your own. For example, in the 1930s, a number of human footprints had been carved out of sandstone by George Adams of Glen Rose, Texas. Some of them even combined a human track with a dinosaur print, and he sold them as Cretaceous tracks. These fraudulent tracks were used to convince people that the Cretaceous period happened only a few thousand years ago. These tracks were shown by a number of fundamentalist groups until the 1970s, when Adams' nephew explained how they were made. Adams used a hammer and chisel, softened the edges with muriatic acid, and artificially aged the finds by soaking them in manure. This practice seems to have continued into the modern day, since all the evidence of creation is always buried in bullshit.

Likewise, if you can't find any authorities to support you, you create those too, and if those alleged experts can't get appropriate credentials, you can also create those. For example "Dr." Carl Baugh is one of the few professional creationists still trying to defend the Paluxy River "man tracks." He also has the dubious dishonor of having been called out as a fraud even by other creationists. Like Kent Hovind, Mr. Carl Baugh is a charlatan, someone who calls himself a doctor but who doesn't have a legitimate PhD. He claims three doctorates, one of which was only an honorary degree given by an unaccredited school that he never actually attended. Another is from a school that didn't offer the degree he claimed. When called out on that, he said he got the degree from an extension of that school located in Australia. But no such extension exists. His third doctorate is another honorary degree from another unaccredited school, which means it can't give degrees, honorary or otherwise; and he should have known that, because he was the president of that school when he gave the degree to himself. But he got away with it. He now runs a creationist museum and hosts a weekly TV show where he pretends to refute evolution.

Even though there have now been innumerable examples of natural selection acting under direct observation and a multitude of experiments gauging these, creationists are still trying to deny even the first of these observances, the peppered moths of industrialized England. Jonathan Wells, author of *Icons of Evolution*, said that was a fraud too because the photos had to be staged—not for the normal convenience of photography, but because he claimed that peppered moths don't rest on tree bark. But a thirty year study at Cambridge University by Michael Majerus on *industrial melanism* revealed that in fact most of them do! As Majerus stated in an email to the entomologist Donald Frack,

> The suggestion that [Bernard] Kettlewell ever "faked" a result is offensive to his memory. He was an honorable, good scientist who reported his findings with honesty and integrity. . . . The case of melanism in the peppered moth IS ONE OF THE BEST EXAMPLES OF EVOLUTION IN ACTION BY DARWIN'S PROCESS OF NATURAL SELECTION that we have. In general it is based on good science and it is sound. [Emphasis in original]

In one of these videotaped seminars, we see a rare moment of ironic honesty from Kent Hovind, though we weren't intended to realize he was unintentionally referring to himself.

> To some people in this world, money is more important than truth; and if they have to lie to you to keep their paycheck coming in, they will lie to you.

It's rare that Hovind says anything that is actually true. But this is, and there are a lot of evangelical pastors and Catholic bishops and such making tens of millions of dollars selling dogmatic nonsense in exchange for tithe. Let's not forget that Hovind himself made millions evangelizing until he went to prison for tax fraud.

But it isn't just about the money, because most religious extremists are very poor. Rather, it is a need to believe, to pretend that you have the ear of the most powerful being imaginable and the commander of the universe. That makes you seem more important than you really are. You can also pretend you're immortal, and that you can even request that two plus two becomes five if you ever need to. You just have to convince yourself of that, and that's easier to do if other people believe it too. Then it will at least *seem* real, and you won't be the only weirdo who thinks that the earth is the flat center of the universe where snakes can talk. Because you know that if you tie a towel around your neck like a cape and call yourself Superman, rational people are just going to think there's something wrong with you. It's not going to do you any good at all unless you can convince others to respect your delusion, especially if they really want to pretend like you're Superman too. I've seen it happen.

Importantly though, science does *not* work like that, and it is designed such that it never will work like that. There have been plenty of times when powerful corporations have paid huge amounts to individual scientists in order to get them to misrepresent the amount or effects of lead spreading through the environment from our gasoline, or to say that global warming is a hoax, or to deny any connection between cigarettes and what we now know to be a whole slough of smoking-related illnesses. In each case, an individual scientist could be swayed, but the required elements of the practice of science meant that other scientists, an overwhelming majority of them, will expose such liars even among their own ranks. In the realm of science, accuracy and accountability are paramount, as is credibility. Not everyone who promotes science is honest, but for some people in this world, truth matters more than any amount of money. Those people accept science for that reason and must necessarily reject lies like creationism.

Despite raving fundamentalist accusations, it is unlikely that any hoax artifact could ever fool science anymore. No forged or faked fossil could ever survive the battery of analysis that is the peer-review process. Proper scientists neither want nor need fraudulent fakery the way creationists do because they're not committed to defending an *a priori* position like creationists are. Evolution never needed any artificial artifacts because its actual evidence was always ample and increasingly abundant. That's actually why paranoid apologists are bellowing so much louder now. For a case in point, look no further than Kirk Cameron:

> If you do the research, you'll find that a Chinese farmer glued together the head of a bird and the parts of a reptile, and completely fooled the worldwide scientific community including National Geographic with what they thought was a transitional form. It was called Archaeoraptor.

Archaeoraptor was the name given to a fossil purchased illegally from Chinese smugglers. The man who bought it, Stephen Czerkas, wasn't a scientist; he was a paleoartist, and a really good one, who ran the Dinosaur Museum in Blanding, Utah, with his wife Sylvia. He had devoted much of his life to illustrating the prehistoric world. Just prior to this controversy, I had actually called the Czerkas just to tell them how inspiring their sculptures were to a wannabe paleontologist like me. I first saw their traveling exhibit at the Dallas Museum of Natural History and was very impressed. Most of their work is extremely lifelike and beautifully well done, especially their renderings of Deinonychus, with or without feathers.

Unfortunately, Czerkas was fooled by a cleverly hashed-together fossil that, if genuine, might have been an important contribution to science. He paid $80,000 to acquire it, then had it examined by experts after it was too late. Only a handful of scientists ever saw Archaeoraptor, but every one who did noted that it was a composite piece, and the artistic amateurs who paid for the fossil were repeatedly warned that some parts of it might not even belong to the whole. Paleontologist Phil Currie noticed in the initial examination that the left and right feet mirrored each other perfectly and that the fossil had been completed by using both slab and counterslab with no connection to the tail. Preparator Kevin Aulenback concluded that the fossil was a composite of at least three separate specimens, and up to as many as five. Paleontologist Timothy Rowe ran CT scans indicating that the bottom fragments, the tail, and the lower legs were not part of the larger fossil. He informed the owners there was a chance of it being a fraud. Chinese paleontologist Xu Xing recognized on first inspection that the tail belonged to a different feathered dromaeosaur. He returned to China and eventually found a fairly complete Microraptor, whose tail was the counterslab of the fossil in America. This proved that Archaeoraptor was a fake.

Czerkas could have bought a house for what he paid for that fossil, and he had created one of his lifelike artistic sculptures of it too. Having invested that much into something he really believed in, he didn't want to hear that he had been gypped. He asked the scientists to keep quiet about it, and maybe they did just because they didn't want to humiliate him any further. But the humiliation had only just begun.

Still refusing to accept that he paid eighty grand for a worthless forgery, Czerkas tried to convince himself and others that it was real, and he kept to the original plan of publicizing his find in *National Geographic*. Regardless what Kirk Cameron thinks, *National Geographic* is a popular magazine for regular people in the mainstream, but it is not a scientific journal. They had to submit their paper elsewhere to get it peer reviewed so that they could also do a press conference and have all the desired fanfare. However, the leading journals *Nature* and *Science* both rejected their paper on the grounds that the fossil had been illegally smuggled and had obviously been tampered with to increase its value.

Still undaunted even after all of that, *National Geographic* foolishly ran the story anyway, convinced that the fossil would be peer reviewed eventually, even after it had already been rejected by everyone who knew anything about it. Once the article was published, it was immediately exposed as a fake by multiple experts. So Archaeoraptor had not "completely fooled the worldwide scientific community" the way Kirk Cameron said it did; it couldn't partly fool even one single scientist.

There is more to this story, though. The irony there is that the tail of the alleged Archaeoraptor turned out to belong to the as-yet unidentified or described Microraptor, a four-winged and apparently gliding feathered dinosaur. It was a predicted transition between diminutive dinosaurs like

Compsognathus and Archaopteryx. As such, it was even more compelling proof of avian evolution from dinosaurs than Archaeopteryx was in Darwin's day. This specific evidence proving dinosaur-to-bird evolution was predicted a century ago, in 1915.

An ornithologist named C. William Beebe detected a trait in Archaeopteryx that most others had missed; it had a few flight feathers on its hind legs. Birds have feathers on their legs, but not flight feathers like they have in their wings. These feathers were in the wrong place, and they weren't long enough to be useful. So why were they there? He concluded they must have been vestigial. Beebe had examined a recently hatched dove with rudimentary feather quills attached to its upper leg, implying an atavism—that is, an occasionally recurring ancestral trait. Usually these appear and disappear in embryology, but this was a recently hatched chick.

Archaeopteryx appeared to have long feathers on its legs too, and their wings weren't as big as modern birds either. Thus, Beebe surmised that the ancestors of birds must logically have had even shorter wings early on that were not big enough to fly with. So maybe they also had leg wings that helped balance them out while parachuting out of the trees until their arm wings were broad enough to do the job alone. He said their evolutionary path to flight had gone through a "Tetrapteryx" stage, which is exactly what Microraptor is. It should have been called Tetrapteryx.

That's not the only irony here, either. If Czerkas had bought the complete fossil of Microraptor instead, then he actually would have had a prize worth the money he paid. But then he would have rejected its value, and the reason that he would have done so is mystifying.

The last time I drove through Utah, I called Czerkas and told him I was coming way out of my way to see his museum. I told him when I would arrive and who all was coming with me, and I asked if he could be there because I wanted to meet him. But he declined. He didn't care to meet me, and I couldn't understand why he was being so bland about it. The reason was obvious once I got there and wandered the halls unescorted. Among his magnificent renderings was a wall-sized cladogram, a phylogenetic tree of archosaurs where half of everything listed was in the wrong place. It seemed that part of it was actually backward, implying that birds had not evolved from dinosaurs at all, and as if some (but not all) dinosaurs had evolved from birds instead. It was a polyphyletic tree with birds both outside of dinosaurs and basal to them at the same time. I stared at that jumbled mess on the wall the same way you might stare at an altar to Bast that you accidentally found in your neighbor's dungeon-style basement. That's when I realized the problem: Stephen Czerkas belonged to a group called B.A.N.D. (Birds Are Not Dinosaurs).

For no reason that I have ever heard, scientists John Ruben, Alan Feduccia, and a few others banded together in the 1980s and vowed to deny any and all evidence that would ever imply that birds were descended from dinosaurs. This was back when John Ostrom and his protégé Bob Bakker started publishing lists of avian traits that were being discovered in dinosaurs, as well as a series of increasingly birdlike transitional theropods. I cannot fathom how people of science could ever impose an *a priori* required belief like that. A promise to reject evidence yet to be revealed? I thought only creationists did that. As far as I can tell, Feduccia is the only Christian among the BANDits, and he's not exactly a creationist. So what gives? What possible science-supporting reason could there be in applying a doctrine of dogmatic data denial?

Their last good argument was disproved way back in 2005, but they still persist in repeating it. Anytime any article anywhere claims that birds did not evolve from dinosaurs for whatever reason, check the source. Ruben's or Fedduccia's name will always be connected to it. They're the only scientists saying that, and they're not doing it for any reason, but rather, as far as I can tell, according to a covenant. That I consider dishonest, if not fraudulent, and I know of several paleontologists and other scientists who think so too.

But even that doesn't answer my challenge to name one evolutionary scientist who ever lied while promoting evolution over creationism. Czerkas may have lied, but he wasn't a scientist (although his website says he was). I should note also that as I finished writing this, I learned that Stephen Czerkas had just died. I'm sorry to have to speak ill of the recently deceased.

In my opinion (and in the opinion of others), Ruben and Feduccia are definitely lying, but they're arguing against mainstream evolution, trying to propose an upside-down broken phylogeny instead. Not surprisingly, they're both constantly quoted by creationists. Maybe Czerkas et al. wanted to promote Archaeoraptor as an ancestor to dinosaurs—who knows? The very idea of it causes me to shake my head with a furrowed brow.

So since the first half of my challenge cannot be answered, what about the second half? To name one professional creationist who has *not* lied while promoting creationism or condemning evolution. I've listed a few of the leading liars already, but I haven't mentioned their biggest lie, the one they all tell and they tell it everywhere.

Creationists often comment that Darwinism begat racism, and that Christians accept that every human is one of God's children. They say that as if those who accept evolution cannot also believe in God. They don't know their own history, either; before Darwin ever published his theory and for millennia before he was even born, Christians commonly believed that the different races of men were not even related.

Before Darwin suggested common ancestry, both religious and nonreligious people thought that the various races of men had no common ancestors at all. Some believed that the different races they perceived were convergently derived from different species, or that they spontaneously generated from different sources. Those who believed in God proposed other options. Mormons for example believed that black skin was the mark of the devil. Others believed that the Genesis character Cain got his mystery bride among the apes. Still others believed that God created other kinds of people from other Adams that weren't mentioned in the Bible and who predate anyone the Bible mentions by name.

There was at that time a common belief that God had created multiple races of people, and that he put them on separate continents with the intent that they remain separate. Europeans believed in Adam and Eve, and they believed that Adam and Eve were white. Everyone else was thought to be of some other ancestry, and not actually related. I met people who still believed this in the twentieth century too! The Ku Klux Klan still believes that now! This racist excuse was the best way to explain the paradox of Genesis 4:14–17, where Cain feared being discovered by others, and went out of the presence of God, to take a wife and build a city in a foreign land. The story gives the impression that God's presence is limited to one geographic region, and also that there were other people in the world besides any of Cain's brothers or sisters that the Bible forgot to mention.

In another creation myth, the Titan Prometheus created both people and animals all out of clay, just like the Semitic gods did in Enuma Elish and in Genesis. Then each of these clay figurines had life literally breathed into it. The father-god Zeus said that Prometheus made too many animals and forced him to make some of those animals into men, albeit with bestial souls. This is the first reference to creationists believing that some people are fully human and that some are essentially soulless animals.

But that's just the beginning. This theme was repeated again and again down through history. How could Darwin have instilled such racism into the people who lived before him? It was not Darwin but the authors of antiquity who numbered other races and called them inferior.

Darwin proposed a tree of life where different lineages could be equally evolved, but Aristotle proposed a ladder of life where you have higher and lower life forms, and where one race is superior to all others—not just in specific ways but in general, and it was often the case that one group would refer to another as subhuman animals.

It was so bad that in the sixteenth century Catholic missionaries to the New World held the Valladoid debate where one man argued that Native Americans were fully human. His four opponents, also in the Catholic clergy, argued that these tribes were soulless, thoughtless, talking animals in human guise.

To be fair, it wasn't just the devoutly religious who held such views. In the eighteenth century, this problem was rampant among scientists and secularists too, including David Hume, Voltaire, and the naturalist George Buffon, who said that white people would "degenerate" into dark-skinned people when they moved to the tropics. Buffon also said there were six different species of human alive in his time.

It was in this century, a hundred years before Darwin, that the great scientist Carolus Linneaus classified humans into six different species: European white people, American red people, Asian yellow people, and African black people, in addition to chimpanzees (*Homo troglodyte*) and orangutans (*Homo nocturnis*). Linneaus believed that humans were apes, but he also believed that apes were humans. The existence of gorillas hadn't yet been confirmed, and Linneaus didn't yet know about Australian aborigines either. But once they were both discovered, the Australoid race was declared to be the lowest form of humanity, largely because they and gorillas were the same color. That's how superficial eighteenth-century scientists were, and they weren't any better in Darwin's day either. This was the common language of both spiritualists and naturalists alike from at least Plato's day, thousands of years before Darwin was born. So how was any of this his fault?

In fact, the children of black slaves were called "objects of property" and "domestic animals" according to an American legal statute dated the same year as Darwin's birth. So how did Darwin's theory cause that? Abraham Lincoln was born on the same day in the same year as Darwin, and Lincoln was extremely racist. He said that white people were generally superior, and that blacks should not be voters or jurors, nor were they qualified to hold public office. Yet Lincoln is heralded as the great emancipator, while Darwin is criticized as a racist. How is that fair?

Even in the modern day, the evangelical Oral Roberts University would suspend students for interracial dating and expel them for interracial marriage. Further, the Ku Klux Klan say they are exclusively creationist, so it's a bit hypocritical of Christians to criticize Darwin for being racist.

Creationists have told a lot of lies about Charles Darwin, but the worst of all of them was this: in 2001, Louisiana legislator Sharon Weston-Broome submitted a bill saying that creationists believe that all men are created equal, with inalienable rights including life, liberty, and the pursuit of happiness. She says the writings of Charles Darwin promoted the justification of racism, because the core concepts of Darwinist ideology is that certain races and classes of humans are inherently superior to others. She even went so far to say that Adolf Hitler was a Darwinist who exploited these racist views. She's not the only one to say that either; I've heard this repeated hundreds of times throughout my life, but it is a disgusting misrepresentation of all involved.

For one thing, racist American lynch mobs didn't agree with Darwin about anything. That is the very thing Darwin argued against! That's what happens when religious extremists are in charge of everything. So much for life, liberty, and the pursuit of happiness.

All my life, folks have told me that Hitler was an evolutionist. So I challenged them many times to show me where Hitler ever promoted or accepted evolution. They pointed me to volume II, chapter 4 of *Mein Kampf*, where Hitler mentions evolution and talks about what brought mankind away from the animal world. That does seem like he's promoting Darwin's idea. But in that passage, Hitler specifies that he is only talking about cultural evolution, and he says that man's inventions were the ticket, not any biological process. In that same paragraph, Hitler talks about what "everyone who believes in the higher evolution of living organisms must admit." Again that might seem like he's promoting evolution, but creationists don't recognize their own arguments in his words. Hitler is criticizing a belief he does not share, which is why he questions how it all began.

Hitler did mention the natural law of evolution, and he equated that to the strong dominating the weak, but it is clear from the context that he is not talking about evolution in a Darwinian sense. Hitler talked about cultural, political, industrial, and military evolution, and the only time he mentioned natural evolution, he obviously didn't mean it in a biological sense. For this reason, some translations use the word "development" instead. Even when Hitler mentioned organic evolution, he was talking about management of an organization.

He never mentions Darwin's name, and only once does Hitler refer to Darwinian evolution, in volume I, chapter 11 of the same book. I cited this passage in a couple of videos, where Hitler uses the same collection of arguments that creationists commonly do. He says he accepts only *micro*evolution, saying that evolution can only occur within definite limits, producing only subtle variants within their "kind." He said that new diversity is limited to rare and inviable hybrids between those kinds, and he said the emergence of new species is impossible. Hitler used each of these common creationist arguments, saying things no "Darwinist" would ever say. I've never seen any statement Hitler ever made where he even acknowledges Darwinian evolution, except for this one passage where he rejects it outright. Later in the same chapter, he even said that such evolution "is a sin against the eternal creator." Further, in *Hitler's Table Talk*, we find

> From where do we get the right to believe, that from the very beginning Man was not what he is today? Looking at Nature tells us, that in the realm of plants and animals changes and developments happen. But nowhere inside a kind shows such a development as the breadth of the jump, as Man must supposedly have made, if he has developed from an ape-like state to what he is today.

Creationists of course don't want to admit that Hitler was one of their own, but they have no evidence to excuse him. By their own admission, the same criteria they would use against him would also excuse the majority of mainstream Christians today. We have no way to know what someone's beliefs are beyond what they say; Hitler had always only ever described himself both publicly and privately as a believer in God, specifically as a Christian, and more specifically, as a Catholic. But in *Mein Kampf* and elsewhere, he made plain that he was also a creationist. It doesn't matter that he once made a disparaging comment against Christians as a political force. Modern Christians of every denomination do that too. That doesn't exclude them or him.

As further proof that Hitler rejected Darwinism, he banned all of Darwin's books and ordered they be burnt. Joseph Stalin did so as well, albeit for different reasons. Both men wanted to believe that man's development could be controlled by personal ambition as an act of will, but reality isn't like that. Nature only allows you to play the cards you're dealt. Most importantly, Hitler praised racial purity and superiority. Darwin by contrast taught that purity leads to congenital defects, and that superiority is a variable and determined by the environment. Hitler's prejudice and Darwin's process could not be any more at odds.

I asked those creationists to show me where Darwin ever said anything to imply that he was racist. It turns out that in all of his collective works, there is only one paragraph, in *Descent of Man*, chapter VI, which could be interpreted that way:

> At some future period, not very distant as measured by centuries, the civilised races of man will almost certainly exterminate and replace the savage races throughout the world. At the same time the anthropomorphous apes, as Professor Schaaffhausen has remarked, will no doubt be exterminated. The break between man and his nearest allies will then be wider, for it will intervene between man in a more civilised state, as we may hope, even than the Caucasian, and some ape as low as a baboon, instead of as now between the negro or Australian and the gorilla.

What he said there does sound racist to a twenty-first-century audience. However, remember that in nineteenth-century England, every common man and anthropological authority in the entire Western scientific community had for centuries taught that black people were the lowest form of humanity, so Darwin was only speaking the language that was ubiquitous everywhere in his world until that time. It would be an adequate excuse to say that Darwin was a product of his culture in that era. But as it turns out, Darwin doesn't need that excuse. He was much more progressive than this one paragraph indicates because he challenged both the societal and anthropological status quo. Remember that before Darwin proposed the idea of common ancestry, all the other scientists believed that the races of men counted as different species. This idea was still common even among Darwin's contemporaries like the anthropologist Louis Agassiz. Agassiz was a staunch creationist who said there were eight different races of men.

Remember, *everyone* in Darwin's world thought like this. Even today, Ken Ham reports knowing of a Bible college in Australia that "taught that you don't take the gospel to the Australian aborigines because they're not the same race." Similarly, in Darwin's day, from the clergy throughout academia, virtually every European believed that blacks were "lower" than whites and that different races were different species, so of course we can expect that Darwin believed the same thing. Except that he didn't.

If you read his last book, *Descent of Man*, you see an evolution in his thinking, because early in that book, he's saying the sorts of things you'd expect—but as the work progresses, he begins to question the idea of multiple races. He says the word "race" was inadequately defined and not of any actual value regarding people, and that human demes weren't sufficiently distinct to be considered separate species.

Racist people, then and now, defend their notions of purity by insisting there is some clear division. But Darwin said that our biased judgments against other people were superficial and erroneous—that no matter how distinct other people may appear to European eyes, there was no consistent distinction because some Africans still shared traits with some Caucasians, and the same was true in every other group too. So that every race blends into every other race at some point, such that it is impossible to determine any real division.

To illustrate this, he noticed that even expert anthropologists may disagree and consequently categorize two individuals of the same ethnicity as though they belonged to different races. Most importantly, Darwin tallied the claims of other scientists in his day who couldn't agree on how many races there were.

> Man has been studied more carefully than any other animal, and yet there is the greatest possible diversity amongst capable judges whether he should be classed as a single species or race, or as two (Virey), as three (Jacquinot), as four (Kant), five (Blumenbach), six (Buffon), seven (Hunter), eight (Agassiz), eleven (Pickering), fifteen (Bory de St-Vincent), sixteen (Desmoulins), twenty-two (Morton), sixty (Crawfurd), or as sixty-three, according to Burke. This diversity of judgment does not prove that the races ought not to be ranked as species, but it shows that they graduate into each other, and that it is hardly possible to discover clear distinctive characters between them.

This wasn't the only time that Darwin questioned the judgment of other scientists in his field. He remarked that the people of his day were divided into two schools, monogenists and polygenists, with the polygenists apparently being divided into two groups as well. Darwin described the first group of polygenists as creationists—those who do not accept evolution and must look at different species as separate creations. The other group believed that humans had descended from two or more species as different as orangutans and gorillas. But Darwin argued that if that were the case, there would be clear indications of that in the comparative skeletons of humans. Thus, Darwin predicted that "naturalists . . . who admit the principle of evolution . . . will feel no doubt that all the races of man are descended from a single primitive stock."

He also predicted that once evolution was generally accepted, the dispute between the monogenists and polygenists will "die a silent and unobserved death." That prediction seems to have come true in my lifetime. So far as I know, the last polygenists are racist creationist groups like the Christian Identity, a terrorist group with many subsets, including neo-Nazis and Klansmen. According to the FBI, group members and affiliates were responsible for a number of weapons violations, assaults, murders, and bombings. According to the Anti-Defamation League, these affiliates include the Montana Freemen and Timothy McVeigh, the Christian terrorist who blew up the Murrah Federal Building in Oklahoma City.

The Christian Identity believed that Anglos and Aryans were the true Israelites, and that Jews fall into a class of false people with no soul. One expression of their anti-Semitism is the seedline theory, wherein white people descend from Adam and Eve through their good son, Abel. (No explanation is given for who Abel took as a mate.) They say the seed of Jewish people began when Eve was impregnated by the serpent in the garden. The idea that Eve had sex with the snake has been repeated by individuals in various denominations. Since Christians believe the serpent was the devil in disguise, then Jews, in the "seedline" view, are literally the bastard spawn of Satan.

But the fact is that we never see from Darwin any of the bigotry that was so common from practically everyone else in history until that time, including so many of his critics still living today. Quite the opposite in fact. *On the Origin of Species* offered no support for either racial purity or superiority, despite the fact that it mentions favored races. In that book, he wasn't talking about people.

Darwin was also often accused of having inspired the near annihilation of Australian aborigines. But in *Descent of Man*, Darwin recalls that as a tragic episode, which was already under way years before he got there. Given that it happened decades before he published his theory, he couldn't have had anything to do with it, nor would he have.

He apparently didn't really believe in white supremacy either, because in his first book, *Voyage of the Beagle*, he defined "savages" more by their practice than their lineage, and he said that Europeans should prefer the dark skin of Tahitians to their own comparative pallor. He said the finest people he had ever seen were Tahitians, and that the nicest man he ever knew was a free black military commander stationed in South America. Darwin abhorred every aspect of slavery and wrote extensively about that. He also wrote against the favoritism of Caucasian invaders and opposed the genocide of indigenous tribes, and he often criticized his own race as contrasted against darker tribes, whom he frequently praised.

And yet, if you listen to Ken Ham of Answers in Genesis, somehow it is the creationists who believe that all men are one race, and that Darwin fueled racism. No. I submit this as one more bit of evidence that creationism is driven entirely by lies. Despite Darwin's upbringing in the ethnocentric aristocracy of Imperial England, Darwin was no racist; he was one of the most progressive men of his day, and his critics have done him wrong.

While there are some dishonest scientists out there, there are no hoaxes promoting evolution, and there never have been—not even one. There have, however, been many attempted hoaxes trying to discredit evolution and other associated fields of science. Fortunately as we have seen, the scientific process of peer review seeks out and exposes fraud by design. Antievolutionist arguments on the other hand are withheld from peer review because they are driven entirely by frauds—including misstatements, out-of-context quote mining, and contrived or distorted falsehoods, and terms erroneously redefined into instigative reactionary nonsense unintelligible as anything other than propaganda.

In short, if creationists knew how to expose a fraud, they wouldn't be creationists anymore.

THE 14TH FOUNDATIONAL FALSEHOOD OF CREATIONISM

"Creation Is Evident"

I often hear superstitious people denigrating science and all other forms of progress as well, as if understanding or improving things is somehow bad. At the same time, they really want to claim justification *from* science, as if there was any way their faith in ancient myths could be scientific. Even they know that if they're going to promote the teaching of creationism in public school science classrooms, they have to come up with more than just the fables themselves. They have to be able to cite conditions under which they hope their position could somehow seem plausible even without medication. Since an evidence-based defense of any of the fables themselves is technically impossible (although they try to do that too), they most often try to present some argument for God that they hope will sound scientific. Remember that their goal is to convince people to make believe.

While science cannot technically comment on anything that is neither evident nor testable or falsifiable, the vast majority of scientists do not believe in any sort of god, not even an undefined "higher power." As Christopher Hitchens famously wrote, "Forgotten were the elementary rules of logic, that extraordinary claims require extraordinary evidence and that what is asserted without evidence can also be dismissed without evidence."

According to a poll from 2009, some 94% of the American populace believes in a god, but only about a third of scientists do. If we restrict that only to those associated with the National Academy of Sciences, then only about 7% are believers. It's almost the opposite of the laity.

It seems that among collective atheists who are recognized for their scholarship in any field, there is a predominant impression of what God is and where that notion came from. That hypothesis is that early humans figured out how to make figurines out of wet clay and wondered what it would be like if these clay figures could be brought to life. Their next imaginative assumption was that such could be done and maybe it had already been done, and that that's where we all came from: some previous person with the powers of magical enchantment created people out of clay figurines and breathed into them the breath of life. Immediately though, they had to avoid the question of where that previous person came from; he or she or it or they must have always been there.

This is how most of our modern ideas of God first emerged—from the imaginations of childlike primitives playing in the mud, and making up fables to explain the things they couldn't understand, because they don't know anything about how the world really is. For those who really want to believe those stories of intended purpose, or who want to assert their authority and state it as fact to convince themselves that they're right, the next step is to kill off any skeptics who don't believe in your priest. Voilà! We have a strong selective pressure in population mechanics leading to the emergence of organized religions, all of them without any basis in fact.

As noted earlier, the reason I chose the surname Ra as my *nom de plume* was to call attention to Amen-Ra. He was a composite of the Egyptian air god and sun god, also known as Amun-Re, whom I see as a template for the god of Western monotheism. The archaeology of pre-Judean polytheism shows that the Jewish god, Yahweh (YHWH), was originally part of a Semitic pantheon descended from the father god, El.

Once upon a time some 2,800 years ago, YHWH was even depicted as having a wife, Asherah, although she may have been part of his being melded with El. El's consort Athirat may have become Asherah, just as El and YHWH were merged together into Yahweh/El, whom the Muslims call "Allah" (the god) and the Jews call "Abba" (the father). Composite gods were once fairly common. For example, the trinitarian concept of Jesus shares an identity with El/Allah/Abba/Yahweh.

At one time, all the gods were either magically endowed mortals, like the sixth divine generation from Enuma Elish—or they were anthropomorphized elementals, like the river Apsu plunging into his lover Tiamat, spirit/goddess of the ocean. Amun was both at different times—just like with YHWH. As the deity grew more powerful in the eyes of devotees, the wife became something of an encumbrance, restraining the elemental aspect in human form. Eventually, the wives of both gods were discarded and the deities followed parallel paths, although YHWH was more typically depicted as a volcano god and Amen was an air god. In his full elemental state, Amun became invisible, which meant he could be anywhere, which meant he may as well be everywhere. This is how gods become omnipresent.

We feel the breeze move against our bodies all the time. Since no one yet understood that air was made of particulate matter but everyone knew you would die if you couldn't breathe, then it was believed that the movement of the air was somehow spiritual. YHWH was granted this aspect as well, so when Genesis 1:2 says that only "the Spirit of God moved upon the face of the waters," it's talking about the wind.

The pharaoh Akhenaten is commonly credited with having created the first truly monotheistic religion with the heretical worship of the sun-disk, Aten. This concept evolved out of a polytheistic pantheon that had already become virtually monotheist. By then, all other Egyptian gods were considered manifestations of a single composite of two gods, much like Yahweh-El. A previous pharaoh, Ahmose I, combined the Thebian air god with the solar deity, Ra. Thus, he made Amen-Ra, something that was always looking down on us and whose spirit touched us everywhere in the world. This god could actually be seen and felt, while every other god had to be imagined on faith. So what other deity could compete with him?

There are so many parallels between different gods and heroes that it's obvious the mythmakers borrowed powers and adventures from elder lore. It seems that the Hebrew people also exaggerated however necessary to make their god bigger and badder than everyone else's: "Oh yeah? Well my god can do anything he wants! If he commands a thing to be, he will speak it into being, and it will manifest out of nothing." Something like that.

Desert deities and demons were often depicted like djinni (genies). Early Islamic literature depicts the djinn as devious air elementals. They weren't usually confined to bottles or lamps, but were more often described as free-roaming nomadic spirits. That's why wandering whirlwinds are called "dust devils." There are also strong similarities between the medieval vision of the djinni and our modern impressions of God. Remember how Elijah was taken up to heaven? In a whirlwind.

Such a transition was easy for YHWH, because we supposedly say his name whenever we breathe through our mouths. That's perfect for an air god. Since the earliest creation myths, the gods would "breathe the breath of life" into their clay golems to animate them, and that too is an apparent precedent to Genesis 2. Throughout the time when the Bible was being composed, its authors commonly believed that the first breath of a child was the moment when its body became "infused with the spirit" as a living being. And of course, the flood in Genesis 6 was meant to drown everything that had "the breath of life." In fact, the single wisest comment I could find in the entirety of the Bible again shows better than any other passage how our notions of spirituality actually stem from a misunderstanding of the natural aspects of air.

> I said to myself concerning the sons of men, God has surely tested them in order for them to see that they are but beasts. For the fate of the sons of men and the fate of beasts is the same. As one dies so dies the other; indeed, they all have the same breath and there is no advantage for man over beast, for all is vanity. All go to the same place. All came from the dust and all return to the dust. Who knows that the breath of man ascends upward and the breath of the beast descends downward to the earth? I have seen that nothing is better than that man should be happy in his activities, for that is his lot. For who will bring him to see what will occur after him?

This is Ecclesiastes 3:18–22 according to the New American Standard Bible. The New Revised Standard Version, the American Standard Version, and the King James Version all replace the word "breath" with "spirit." Likewise, if you compare Luke 23:46 in the New American Standard or the New Revised Standard versions of the Bible with the King James or the American Standard Version, you'll see again that "breathed his last" means the same as "gave up the ghost." When the story says that Jesus commits his "spirit" to God, and gives up "the ghost," or "breathed his last," in each case they're talking about the "breath of life" as if that is his "spirit." This translation eloquently illustrates the gaseous origin of man's belief in his own soul. Another example, Ezekiel 37:5–10, has a necromancer going into a bone yard to revive an army of the dead.

> Thus saith the Lord God unto these bones; Behold, I will cause breath to enter into you, and ye shall live: And I will lay sinews upon you, and will bring up flesh upon you, and cover you with skin, and put breath in you, and ye shall live; and ye shall know that I am the Lord. So I prophesied as I was commanded: and as I prophesied, there was a noise, and behold a shaking, and the bones came together, bone to his bone. And when I beheld, lo, the sinews and the flesh came up upon them, and the skin covered them above: but there was no breath in them. Then said he unto me, Prophesy unto the wind, prophesy, son of man, and say to the wind, Thus saith the Lord God; Come from the four winds, O breath, and breathe upon these slain, that they may live. So I prophesied as he commanded me, and the breath came into them, and they lived, and stood up upon their feet, an exceeding great army.

As for the impetus to change YHWH's image from the terrifying volcano god in Exodus to that of a relatively subtle air god, a likely scenario (I think) was illustrated in an old Arnold Schwarzenegger movie. Conan the Barbarian argues that his god is strong, "strong on his mountain." But his companion, who worships the four winds, says his own god is greater: "He is the everlasting sky. Your god lives underneath him."

That is the impression of the Jewish god shared by many scholarly atheists. Given that, and the fact that belief in gods does not provide an explanation for anything, imagine the type of evidence we would need to see in order to believe that such a thing were really real. If that could be done, imagine the breadth of reliable data that would have to be compounded atop that in order to also accept anything of the biblical fables too. As absurd as theism is, it still requires a lot more evidence to believe in the Bible than to just believe in a god. While some may argue that God can never be proved or disproved, we can't say that about the Bible, which has been disproved on every testable claim that it makes. Even if we could prove the existence of God, the Bible would still be wrong.

To my knowledge, there has only ever been one attempt to defend both the notion of God and the Bible in mainstream media. In May 2007, former child actor Kirk Cameron and fellow advocate of idiocy Ray Comfort were on ABC's *Nightline*. Opposite them were Brian Sapient and Kelly O'Connor of the Rational Response Squad. Martin Bashir moderated as they debated the question "Does God Exist?"

In the course of that debate, Cameron and Comfort promised to scientifically prove the existence of God on national television, "100% absolutely, without the use of faith." Of course, that didn't happen. Religious beliefs (as everyone knows) are assumed on faith in lieu of proof and regardless of evidence. Cameron and Comfort had neither. Instead, they presented a series of fallacious absurdities revealing the depth of their own impressive ineptitude.

First they tried to parody cosmology and abiogenesis—as if criticizing them should somehow challenge evolution. As if disproving evolution could imply creation by default. As if God would only be indicated by a failure of science to explain what religion also does not explain. As if evolution did not already adequately explain every aspect of biodiversity very well, where creation offers no explanation at all. As if science hadn't already disproved all the Genesis fables that Cameron and Comfort were still trying to save.

Failing that, they immediately resorted to their usual staple of citations appealing to authority with irrelevant comments made by folks who often meant the very opposite of what these mined quotes implied. Albert Einstein is typically seen as representative of scientific genius, so he is the one most often misquoted by the religious in their attempt to appear scientific. All my life, I have heard that "science without religion is lame, religion without science is blind" and "God does not play dice with the universe." Tonight's citation referred to when Einstein said, "In view of such harmony in the cosmos which I, with my limited human mind, am able to recognize, there are yet people who say there's no God. But what really makes me angry is that they quote me for the support of such views." Not that it really matters what someone else once believed, but I still have to compare and contrast that quote with some other important clarifications that Einstein made in his letters.

> *"It was, of course, a lie what you read about my religious convictions, a lie which is being systematically repeated. I do not believe in a personal God and I have never denied this but have expressed it clearly. If something is in me which can be called religious, then it is the unbounded admiration for the structure of the world so far as our science can reveal it."*
>
> —in reply to Joseph Dispentiere (1954)

"I believe in Spinoza's God, who reveals Himself in the lawful harmony of the world, not in a God who concerns Himself with the fate and the doings of mankind . . ."

—to Rabbi Herbert Goldstein (1929)

"I have never talked to a Jesuit priest in my life and I am astonished by the audacity to tell such lies about me. From the viewpoint of a Jesuit priest I am, of course, and always have been an atheist."

—to Guy H. Raner, Jr. (1945)

"For me, the Jewish religion like all others is an incarnation of the most childish superstitions. . . . The word 'god' is for me nothing more than the expression and product of human weaknesses, the Bible a collection of honourable but still primitive legends which are nevertheless pretty childish. No interpretation, no matter how subtle, can (for me) change this."

—in a letter to philosopher Eric Gutkind (1954)

It doesn't matter whether Einstein believed in the Jewish god, or that he actually believed in some other natural motion that he just liked to refer to as God. In any case, regardless of whatever respect he may have for the modern practice of religious culture and tradition, he still gave no credence to the fables in the Bible, and that's all that might be relevant to me.

Failing on the Einstein front too, Comfort then revealed a profound incompetence in the presentation of evidence. "We'll be simply producing knowledge by looking at three irrefutable evidences for God's existence. . . . Creation is 100% scientific proof there was a creator. You cannot have a creation without a creator."

Now as we explained before, if you're going to "prove" anything outside of mathematics, you're going to have to use the legal definition, that being an overwhelming preponderance of evidence. Comfort only showed that he doesn't know what "evidence" means. His first exhibit was an assumed conclusion based on a logical fallacy: that simply because something exists, it must have been deliberately created by an intelligent being for an intended purpose. This is an unsupported assumption. Worse, it is also the logical fallacy of begging the question (a circular argument in which the conclusion is included in the premise). He says "creation" requires a creator—but if we call it "reality" instead, then we find no need for a realtor.

Comfort then moved on to his second "irrefutable" evidence of God's existence: "Something God has put within each of us: the conscience. You're a self-admitted lying thieving adulterous murderer at heart, so hopefully, your conscience has been stirred by the commandments to show you that you need God's forgiveness."

Again he offers an assumed conclusion, and this time the assumption is based on ignorance of science, including but not limited to sociology, zoology, and (of course) evolution.

My first public speech was the keynote explanation of the evolution of morality. It was at a college in Florida at the invitation of the James Randi Educational Foundation. At that time, I noted how animals that were raised in dependence by a nurturing mother are also the ones who show empathy for others—that man is a social animal, and that as such we have developed an innate sense of compassion for our family, friends, and fellows. This is true of all social animals. Anyone with a dog should be able to see that. Those who collaborate and support the common good are the most beneficial and the most productive, while deviants tend to be ostracized, imprisoned, banished, or killed. So our concept of morality is actually an emergent property of population mechanics driven by strong selective pressure. As a result, we've even developed mirror neurons to enhance our empathy with others. And this is even further enhanced by an overdevelopment of the frontal lobes: where those lacking that tend to be more selfish, cruel, short-sighted, and reactionary.

All social animals, including humans, have this sense of compassion and community that in many species leads to acts of devotion, defense, and even altruistic self-sacrifice on another's behalf.

Even if Ray could find someone in the audience who really felt guilty about something, that wouldn't imply that any god exists or that anything in the Bible is true, particularly not the Ten Commandments. They're not even relevant to this debate. They can only be believed on faith, and Christians don't follow most of the commandments anyway.

For his third "irrefutable" evidence of God's existence, he offers "the radical nature of conversion.

As I explained in the third chapter of this book, what a Christian convert experiences during their imagined rebirth is no different than what converts of many other religions experience. I remember reading the testimony of a woman in Thailand who exclaimed what an overwhelming euphoria she experienced the moment she accepted Buddha into her life. This happens in Hinduism too, as in many cults as well. So again we have an assumed conclusion, this time based on ignorance of religion. Seeking any god, ghost, alien, or other apparition "with your whole heart" means you've already assumed these things exist and have convinced yourself of them without question even without any evidence.

This conversion produces similar results regardless of which religious entity it is—Jesus, Krishna, Buddha, Bast, Xenu, etc. It doesn't matter. Someone in a single-wide trailer in the desert might experience the same sensation when he thinks he's picked up psychic transmissions of a reptilian race from the Sirius star system. The same goes for those who think they've established communication with their dead relatives or some hallucinated remembrance of their past lives. The point is that none of Comfort's "irrefutable" evidence even needs to be refuted because it doesn't even count as evidence.

When all that inevitably failed to impress anyone either, Comfort and Cameron tried to intimidate the audience with a sermon of emotional pleas that relied entirely on fear and ignorance. But this assembly wasn't stocked only with paranoid and superstitious zealots; this was a more intellectually curious group, many of whom already believed in God. Whether they did or not, though, everyone in attendance was sincerely disappointed with Cameron and Comfort's inability to produce anything they promised. As TV and film critic Troy Patterson wrote at *Slate*,

> First, I grew excited at this promise, then began to wonder why no theologian, philosopher, or sitcom star in recorded history had ever done it before—Thomas Aquinas, Immanuel Kant, Tina Yothers, whoever—and I realized I was in for a letdown.

Comfort insulted fellow believers by assuming that if there is a god, then his religion and absurdly narrow interpretation of it was the only acceptable option—a notion his cohort unwittingly described as idolatry. Cameron insulted the rest of the audience by pretending to have once thought as rationalists do, a lie he himself also accidentally exposed when he then accused rationalism of being a belief based on faith, thereby disproving his own argument.

In my opinion, the most surreal of all theist arguments from scripture is this one Comfort cited from Romans 1:20: "The invisible things of him from the creation of the world are clearly seen, being understood by the things that are made." It is a summary of the logical fallacy of question begging. In other words, you already have to believe it before you can believe it, or that it only makes sense if you believe it already. Perhaps this is why every logical fallacy is an argument for God and vice versa.

Finally, Comfort did implicitly admit that his god could only be indicated if one was already determined to believe in it regardless of evidence, and the only claim of evidence the two of them had depended on religious references that they had earlier promised they would neither need nor use. If they knew this going in, then their whole premise was phony because they also knew they didn't have any evidence, much less proof, and would have to rely entirely on assertions of faith and their reverence of scripture. As YouTube vlogger Toffee1000 has noted, "I find it incredulous—if not astounding—that when it comes to knowing all things and having all things at your fingertips, you turn to the Bible, a book riddled with things we know are wrong!"

Amazingly, subsequent interviews showed the duo apparently oblivious to any of their string of utter failures in that forum. They went into this venue as if they actually believed they had something to present, and may even have come out thinking they showed it. But much of the alleged evidence they declared to be "irrefutable" had already been refuted thousands of times.

For example, in the follow-up comments, Cameron said: "You've got adaptation within a species, but you've never seen any animal produce anything other than that type of animal. You have faith that over time, it'll turn into something else, but you've never seen it happen. No one's ever seen it happen, and that is called macroevolution. No one's ever seen it demonstrated." I would invite Kirk Cameron to read the eleventh chapter of this book, where we show that macroevolution has in fact been directly observed and documented dozens of times, both in the lab and in naturally controlled conditions in the field.

Cameron continued: "Science has never found a genuine transitional form—that is, one kind of animal crossing over into another kind—either living or in the fossil record." Again, I would invite Kirk Cameron to read the ninth chapter of this book, or see the original video on YouTube that cites the discovery of a few hundred definite transitions even according to the strictest definition of that word. He should also read subsequent chapters explaining why there is no such thing as a "kind."

In fact in the entire "debate," Comfort only said one thing that was correct, and this single sentence is only true if we leave off the other sentences in that paragraph: "The problem was, I never really took the time to look into the evidence myself, do the research and actually see if the claims that they were making were true."

He still hasn't.

Ray Comfort's comments were merely unsubstantiated assertions that can neither be evidenced nor confirmed. Yet he presented his baseless speculation as though it were certain knowledge. "God is gonna punish murderers. He's good. He's just. He's gonna make sure murderers get what's coming to them. But realize this: that God is so good, he's also gonna punish rapists, adulterers, pedophiles, fornicators, blasphemers, hypocrites, and even thieves and liars."

God is going to punish murderers? Not according to Isaiah 45:5–7. There God describes himself being as much evil as he is good. He brags that he created evil—not as the absence of good, but as a creation that he was proud of. According to Exodus 2:12, Moses was a murderer himself, at least in the 2nd degree, long before the story of the Commandments. He was also guilty of theft, arson, human trafficking, and genocide (among many other crimes). What about Phinehas, son of Eleazar, in Numbers 25? He murdered another Israelite just for being in a mixed-race marriage with a non-Israeli. Not only did God not punish this barbaric crime, he also actually ordered this violent systemic racism and rewarded the murderer with a priesthood. That's how good and just God isn't.

God is going to punish rapists? Not according to Deuteronomy 22:28. Even in the instances where the rapist doesn't get to keep his victim in bondage for the rest of her life, the Bible still says that crimes are to be punished in this life, not the next one. So people are the ones dealing the punishments, not God.

God is going to punish adulterers? Not according to 1 Kings 11:3, because Solomon had 700 wives *and* princesses *and* 300 extramarital concubines, and he was considered a man of God.

God is going to punish pedophiles? Not according to Numbers 31:17–18. That's where Moses ordered whole families to be slaughtered, including little boys, but not preteen girls if they were believed to be virgins. The little girls were awarded to his desert brigands to "keep for yourselves" strictly because of the purity of their sex.

God is going to punish fornicators? You mean except for concubines and as-yet unmolested children orphaned and enslaved as prisoners of war?

God is going to punish blasphemers? Okay, maybe they're in trouble, because any evil thing we do can be forgiven—even offering your innocent children to violent mobs to be brutally gang-raped to death (Genesis 19 and Judges 19) rather than let strangers suffer such indignity. But never *ever* challenge the accuracy or authority of the priests and the dogma they're selling. Because that's the only thing their god will never forgive you for. Now why do you think *that* is? That is not what one should expect of any real being, but it is what we'd expect in the case of an indefensible lie.

God is going to punish hypocrites? What about Kirk Cameron? He expects us to believe that he once thought exactly like rationalists do, meaning that he "believed in evolution" on "faith," and that he only adopted superstitious beliefs after researching a preponderance of evidence, even though he still doesn't know what any of the evidence is? What a hypocrite he is!

God is going to punish liars? Not according to the extravagant revenue of religious organizations like WayoftheMaster.com and Living Waters Ministries, which are both owned by the liar Ray Comfort.

Kelly's closing comment summarized the debate for the entire audience: "I think everybody here can tell that there was not one piece of evidence presented at all for their god."

Obviously Cameron and Comfort hadn't any idea what they were talking about at any point, neither in fact nor fiction. They certainly didn't know what the words "science," "knowledge," "proof," or "evidence" even mean. Creationists typically don't.

In a debate with Michael Shermer, Kent Hovind said, "Everything is just so incredibly complex, there has to be a creator." I saw Canadian creationist Laurence Tisdall say something similar on the Michael Coren show: "Everything that has information and is complex has to have been created." That of course does not account for God, who would have to be infinitely more complex than anything else. It also doesn't account for the fact that all the evidence indicates a progression of emergent patterns and increasing complexity—without any indication of any god.

On a YouTube show called *Creation Guys*, I saw Kent Hovind's son Eric say, "We're both looking at the exact same facts; we're just coming to different conclusions." No, we're not. First of all, facts are objectively verifiable and thus indisputable data. Being verifiably accurate means it can still be shown to be true even to those who don't want to believe it, but dogmatic religious beliefs depend instead on subjective impressions of personal preference, erroneous assumptions, and assertions of logical fallacies.

Second, we could rationalize a few of the facts differently. But mere facts don't qualify as evidence until or unless they positively indicate, or can only be accounted for by, only one available explanation over any other. If a particular fact could still be true whether any contested issue being investigated is true or not, then that fact isn't evidence; it's just a fact. Logically, no single fact can be considered evidence of two different mutually exclusive conclusions at once. Besides, we're obviously not both looking at endogenous retroviruses, atavisms, transitional forms, physiological anatomical and molecular vestiges, ontogeny and developmental biology, protein functional redundancy, convergent phenotypes, mobile genes, observed speciation, or the myriad methods of dating geologic stratigraphy, nor any twin-nested hierarchy of phylogenetic clades. All of these are peer-reviewed and verifiably accurate evidence positively promoting evolution as well as directly disproving creationism.

But you know what we've never seen? We've never seen anything "created." No one has ever seen a complex life-form (or anything else) magically pop out of thin air. But that's what creationists are arguing for! Talismans, incantations, elemental component spells, enchantments, clairvoyance, and prophecies all consistently fail every test.

To confirm this, James "the Amazing" Randi, a former Las Vegas illusionist well-versed in the angles used in supernatural pseudoscience, for nearly twenty years offered a million-dollar prize to anyone who could show testable evidence of the things we should expect would also be true if there were ethereal entities influencing things with molecular structures. In that time, he exposed a few frauds, such as telekinetic James Hydrick, psychic Maureen Flynn, faith-healer Peter Popoff, and spoon-bender Uri Geller. But no one ever produced any actual evidence for faith-healing, telepathy, psionics, precognitive psychic friends with astral bodies, past-life remembrance, or spectral manifestations of any kind.

So where is there any field of study or accurate fact positively promoting a magical creation? Here in the United States there are a number of religious channels, and there are plenty of them claiming that such evidence is out there. For example, the known fraud "Dr." Carl Baugh has a weekly show on the Trinity Broadcast Network, wherein he said, "We're considering this matter of life origins, and there's an incredible body, a pyramid of evidence in support of divine orchestration, divine engineering, divine creation." Okay, great! Where is it? What is it? Because all previous attempts to show it have failed.

Evolution faced a very high-profile formal challenge at the 2005 *Kitzmiller v. Dover* trial held in U.S. District Court. A battery of lawyers working for the Discovery Institute had a much better opportunity to prove their case than Cameron and Comfort did, and they were each much smarter than Cameron or Comfort are. They had been waiting years for this moment, planning well in advance how they were going to present their case to prove intelligent design as implied by the evidence of irreducible complexity. One of the great advantages that they had over Cameron and Comfort was that God's existence was assumed as a given before they even began. Judge John E. Jones III was a Republican conservative Christian nominated by Rick Santorum and appointed by George W. Bush. Even the star witness, textbook author Kenneth Miller, was a self-identified traditional Catholic. It should have been so easy, right?

The argument for irreducible complexity is that there are certain biological structures which cannot be reduced, and therefore could not have developed to become what they are because every incremental step of evolution must be a functional one. Thus, any such structure that could not be functionally reduced could not have evolved either. The intelligent-design proponents had a handful of examples that were championed by Michael Behe, a biochemist with Lehigh University and a fellow of the Discovery Institute. This was one of the most amazing cases in U.S. legal history, and a lot of other people have written brilliant depictions of it. I will offer only the barest summary of the court's response to the defense's arguments, as present by Behe:

Behe:
Removing any one part of the bacterial flagellum will prevent it from acting as a rotary motor.

The Opinion of the Court:
Professor Behe excludes, by definition, the possibility that a precursor to the bacterial flagellum functioned not as a rotary motor, but in some other way, for example as a secretory system.

Behe:
Science will never find an evolutionary explanation for the immune system.

The Opinion of the Court:
He was presented with fifty-eight peer-reviewed publications, nine books, and several immunology textbook chapters about the evolution of the immune system; however, he simply insisted that this was still not sufficient evidence of evolution and that it was not "good enough."

Behe:
Each and every element of the complex cascade of enzymes and cofactors must be in place for blood clotting to work.

The Opinion of the Court:
The evolution of complex molecular systems can occur in several ways. Natural selection can bring together parts of a system for one function at one time and then, at a later time, recombine those parts with other systems of components to produce a system that has a different function. . . . The complex biochemical cascade resulting in blood clotting has been explained in this fashion. . . . Moreover, cross-examination revealed that Professor Behe's redefinition of the blood-clotting system was likely designed to avoid peer-reviewed scientific evidence that falsifies his argument, as it was not a scientifically warranted redefinition.

The Opinion of the Court:
We therefore find that Professor Behe's claim for irreducible

> complexity has been refuted in peer-reviewed research papers and has been rejected by the scientific community at large. Additionally, even if irreducible complexity had not been rejected, it still does not support ID as it is merely a test for evolution, not design.

So again I have to ask, where is this evidence of divine design? Because each of the arguments presented for irreducible complexity (the best arguments creationism ever had) were disproved scientifically and exposed in a court of law—even though God's existence was not even questioned.

There was an earlier court case of *Edwards v. Aguillard* (1987), which made the teaching of creationism illegal in public schools, although some states have since found and exploited loopholes around that. *Kitzmiller v. Dover* was intended to establish whether intelligent design was a defensible scientific concept or whether it was just creationism dressed up in a lab coat. The ID proponents were pushing for schools to accept their textbook, *Of Pandas and People*. Barbara Forrest of the National Center for Science Education was one of the witnesses in the case. She found first edition copies of that book that were written prior to *Edwards v. Aguillard*, when teaching creationism was still legal. She realized that the versions prior to and after the 1987 ruling differed significantly only by one macro command. Wherever the word "creationism" appeared, it was changed to "intelligent design," including the definition. So the definition of creationism *is* the definition of intelligent design. They're exactly the same thing. Despite all their arguments to the contrary, intelligent design = creationism.

The most glaring example of that was where a misspelling confounded the macro command. Wherever the word "creationists" originally appeared, it was replaced by "design proponents"; except in one instance, where it appeared as "cdesign proponentsists." This proved not only that intelligent design is creationism in disguise, but also that the entire case and the preceding events leading to it were a combination of frauds. States Judge John E. Jones III, "It is ironic that several of these individuals, who so staunchly and proudly touted their religious convictions in public, would time and again lie to cover their tracks and disguise the real purpose behind the ID Policy."

Apart from a series of frauds and falsehoods like these, the only arguments antiscience evangelists have ever had seem limited to nothing more than ignorant criticisms of dwindling and already irrelevant gaps in the ever-enveloping advancement of science. But vague criticisms against science still wouldn't count as evidence for creationism even if those arguments weren't all completely wrong; even if there was evidence of gods, it might not be *their* god. Even if it was their god, that wouldn't be evidence of creation because that still wouldn't dismiss any of the evidence for evolution and against mythology. Nor could it change the fact that humans are still apes. Creationism relies on a false dichotomy, rejecting all other options and insisting that there can only be two alternatives so they can imagine that criticizing one will vindicate the other by default.

As an example of that, in a debate between Michael Shermer and Kent Hovind, the audience posed a question for Hovind: "What is your strongest piece of evidence for creationism?" To this, he replied, "I think the evidence for creation would be the absolute impossibility of the contrary."

So convicted fraud and pseudoscience charlatan *Mister* Kent Hovind argues that what has already been directly observed and shown to be certainly true is impossible, and the only option he thinks *is* possible is that an imperceptible mystical imaginary being poofed everything out of nothing by magic. The irony is that what he proposes *is* physically impossible because it defies all natural laws. It's logically implausible too, since it has neither precedent nor parallel anywhere in reality to imply that it could still be true anyway. Where is there evidence anywhere that such a thing actually exists, or that anything even could have any of these abilities?

The evidence of evolution, and even the event of evolution itself, are both directly observed and testable and demonstrably factual. But religious beliefs are none of the above and never have been; they're assumed on faith. Whether or not any of these beliefs might turn out to be correct, they are asserted as true without justification in the form of evidence.

Defenders of pseudoscience scornfully reject scientific methodology and gleefully ignore evidence on purpose, and their leaders—past and present—even admit this openly and publicly. Take, for example, these quotations from Martin Luther, the father of Protestant Christianity,

> There is on earth among all dangers no more dangerous thing than a richly endowed and adroit reason, especially if she enters into spiritual matters which concern the soul and God. For it is more possible to teach an ass

> to read than to blind such a reason and lead it right; for reason must be deluded, blinded, and destroyed.
>
> Whosoever wants to be a Christian should tear the eyes out of his reason.
>
> Faith must trample under foot all reason, sense, and understanding, and whatever it sees it must put out of sight, and wish to know nothing but the word of God.
>
> Reason is the greatest enemy that faith has; it never comes to the aid of spiritual things, but—more frequently than not—struggles against the divine word, treating with contempt all that emanates from God.

Similarly, consider these gems from Henry Morris, the father of Creation Science,

> When science and the Bible differ, science has obviously misinterpreted its data.
>
> The only Bible-honoring conclusion is, of course, that Genesis 1-11 is actual historical truth, regardless of any scientific or chronological problems thereby entailed.

They actually preach that we should not only make positive proclamations of complete conviction even without the slightest indication, but also that we should automatically reject without consideration everything we ever find that doesn't fit into their preconceived bias. As Ken Ham of Answers in Genesis argues,

> We need to be looking at the world through our biblical glasses. We have a revelation from one who says, "I know everything; I've always been there. Here's what happened in the past." So when we take that revelation, put on our set of glasses, and we look at the evidence, we can say, "Aah now I understand! Fossils couldn't have formed before sin. There was no death before sin. There was a global flood. That connects to geology. God made distinct kinds of animals and plants. That connects to biology." And so on.

Notice how Ken Ham admits to bias? He encourages his audience to assume their conclusions, so that the words written by human authors who weren't there are treated as a revelation from God. Then he tells them to adopt confirmation bias (biblical glasses) and simply ignore everything that proves them wrong. That is all creationism is: lying to oneself as well as others. This is how to prove that creationism really is willfully ignorant and deliberately dishonest, because here is another admission that they don't care what the truth really is. They just wanna believe what they wanna believe, and if that doesn't turn out to be true, then they don't want to know what is true, and sometimes they'll even admit that too.

On the YouTube show *Creation Guys*, Kent Hovind's son Eric said, "I'll be the first to admit that creationists view evidence from a biased perspective. Evolutionists, would you be willing to admit you also view evidence from a biased perspective?"

The answer is no. Religion *is* a bias by definition—that's why it relies on propaganda. But science dispels propaganda because it eliminates bias by design. It has to because it's an investigation, not a predetermined conclusion like religion is. Every proposition must be requisitely evidential and potentially falsifiable, and must be subjected to a perpetual battery of independent and unrestricted tests wherein anyone and everyone who thinks they can is welcome to try and find some flaw in it to expose and correct it. Creationists won't subject their beliefs to any of that because they're not interested in finding out what is really true. They want to defend their preferred beliefs, regardless whether they're true or not.

Later in that same episode, Eric explained that it doesn't matter to him what evidence is ever presented because he doesn't care about evidence. He said, "Look, the bottom line is, the creationist answer is God just did it that way." Then one of his buddies chimed in with, "That's good enough for me, man. He just . . . [snaps fingers]."

How could that possibly be good enough for anyone?

Yet Eric agrees, "That's good, yeah. We don't need science to back us up."

That was an eye-opening, jaw-dropping admission! That's the difference between us. I can't choose to believe whatever I want to believe. I am compelled to accept whatever the evidence indicates even if I'd rather not. I have no choice in the matter. Even if I did, I would still reject faith, because I want to know what is really true. So I need evidence. I need science to back me up, because if I don't have that, then how do I know if it's even real? I need that too. I don't want to believe what might not be true; I want to understand what really is true and be able to know the difference.

"One can't believe in impossible things."

"Why, sometimes I've believed as many as
six impossible things before breakfast."

—Alice, speaking to the Red Queen in *Through the Looking Glass*

I saw a Nightline "Faith Matters" spot that really annoyed me. It showed young-earth creationists Billy Jack and Rusty Carter leading hundreds of children on "BC" (biblically correct) tours through natural history museums. It's disturbing to see not just because these guys are constantly lying to impressionable kids, misrepresenting everything; it's also the way they condition very young children to think of these fossils as "boring," because their willfully ignorant guide makes them recite that back to him. How dare he force them to recite his defective opinion and impair their curiosity! Talk about conditioning! How is that not abusive?! If I had been within earshot during one of their tours, I would have made a scene calling him out and correcting him right in front of those kids.

"I've chosen to believe the god of the Bible," Jack said. "Now the evolutionist has chosen not to believe in the god of the Bible. We've chosen to believe. They're both matters of faith." He says this ignoring all the "evolutionists" who are Christian. Professional scientists like:

- Dr. Richard G. Colling, chair of biology at a fundamentalist Christian college
- Dr. Dennis O. Lamoureux, assistant professor of science and religion
- Dr. Keith B. Miller, assistant professor of geology
- Dr. David N. Livingstone, author of *Darwin's Forgotten Defenders*
- Dr. Howard J. Van Till, professor of astronomy,
- Dr. David L. Wilcox, geneticist, author of *God and Evolution*
- Prof. Larry Arnhart, author of *Darwinian Natural Right*

- Prof. Kenneth Miller PhD, author of *Finding Darwin's God*
- Dr. Francis Collins, director of the Human Genome Project
- Prof. Rev. John Polkinghorne PhD, author of *Quarks, Chaos and Christianity*
- Dr. Graeme Finlay, cell biologist and author of *Human Evolution: Genes, Genealogies and Phylogenies*
- Prof. Donald Nield PhD, author of *God Created the Heavens and the Earth*
- Dr. Denis Edwards, author of *The God of Evolution*
- Dr. John Haught, author of *Deeper than Darwin*
- Rev. Stanley Jaki PhD, physicist and author of *Cosmos and Creator*
- Rev. Robert T. Bakker PhD, Pentecostal paleontologist

The next thing that asshole ignored is that science is the opposite of faith, being necessarily rational and empirical. That means that whatever we believe isn't a matter of choice; it's an obligate condition imposed upon us by our knowledge of the evidence, and that position is compelled to change in accordance with our understanding.

What would you do if the truth mattered more than whatever you would rather believe?

- You would question your assumptions.
- You would doubt your convictions.
- You would challenge your biases.
- You would seek out and expose any flaws in your position.
- You would reject or correct those flaws.
- You would be suspicious of authority, testimony, and subjectivity.
- You would believe nothing on faith.
- You would believe only what is positively indicated,
- And you would believe tentatively even then.

A shorter answer is that you would do the very opposite of what religion does. A nice summary of the believer's position comes from an old episode of *Father Bob*, where a clergyman complains,

> Well think about it; very little evidence. Blind faith, that's all we have to go on. There's not a shred of proof anywhere—nothing!

If there's no evidence for a particular notion, then there's no reason to believe it either. It may as well have been imagined out of nothing because it has no basis in fact, and thus no truth to it. We can only proclaim a positive belief if we have sufficient evidence to support only that and no evidence at all against it. Even then, we can only accept it tentatively, because if future evidence ever confirms that we were wrong all along, then we'll be forced to change our minds accordingly even if we didn't want to. That is, if we are honest with ourselves.

The National Academy of Sciences defines science as "the use of evidence to construct testable explanations and predictions of natural phenomena, as well as the knowledge generated through this process." Defense of faith doesn't work that way. It is not testable, not natural, doesn't make predictions, and can't demonstrate knowledge either. It instead relies on apologetics, the branch of theology concerned with the mandatory promotion and reinforcement of faith in a particular religious doctrine, and the compulsory defense of that belief by systematically rationalizing excuses for or denying any or all evidence and arguments that may ever be laid against it.

I'll explain how it works. It's really quite easy: "Goddidit" explains everything by explaining nothing. Since magic is exempt from all rules of nature or logic, they think that means that anything that seems impossible somehow proves them right. Conversely, anything and everything that might imply otherwise can be immediately dismissed as a knee-jerk reaction with the phrase, "That doesn't prove anything!" Virtually all antiscience apologetics are composed of variants of these two thoughtless comments, in addition to the usual propaganda of inflammatory emotional pleas, tall tales, petty bigotry, incredulity, and appeals to authority.

You can see this if you look up the statement of faith for any creationist organization; no matter how scientific, how commonly believed, or how apparently workable or logical it may seem, no evidence of any sort can even be considered if it contradicts their sacred stories, which they insist must take priority in the event of any apparent conflict. Their position is wholly dishonest, and it's everything science isn't, because it's an *a priori* position that must never be seriously questioned, corrected, or rejected.

They have to defend their preconceived notions because they're forbidden on pain of a fate worse than death to even consider that they could be wrong, and they must maintain that belief no matter how wrong they obviously are. Even when they know they're wrong, they still have to make believe anyway. Their position is the definition of a closed mind—it is *not* a search for truth!

I don't make these accusations lightly. I've been arguing with theists online for many years, and they admit these faults openly. I asked a couple of the posters on ChristianForums.com how they would react if they had a time machine and could go back and see that Jesus was never resurrected. Their answers were disappointingly intellectually dishonest and unreasonable.

Liza0315 replied,

> If I found that Jesus did not rise again, then, I am not sure what I would do. Faith requires me to believe anyway. The "faithful" thing to do would be say that you had somehow rigged the system to show me a false version of the events. The logical thing to do would be to stop believing. I sure HOPE (and I know you are going to be disappointed with this answer) that my faith would be strong enough to believe even when my own eyes tell me otherwise.

Inna3, meanwhile, had a more strongly worded response,

> I'll give you the answer here. We don't care about scientific evidence. That's not what our faith is based on. Our faith is built on the Word of God!

Remember that evidence is a body of facts, verifiably accurate data, indicating a particular conclusion. The truth is what the facts are. If you don't care what the facts are, then you don't care what the truth is.

The first creationist video I ever saw on YouTube was "Evolution versus Creation," hosted by Janet Porter, formerly known as Janet Folger. She is the founder and president of Faith2Action, a site dedicated to defending "pro-family" values, but mostly known for opposing the Local Law Enforcement

Hate Crimes Prevention Act of 2007. Her video was the most dishonest production I have ever seen, right from the opening theme song:

I can't get past the evidence,
I can't get past the proof.
I can't get past the evidence.
It's impossible to do.
You are the reason that I can't deny the truth.

By "you," I can only assume the song is referring to God, which is the reason why they *do* deny the truth because they *can* get past the evidence! Every sentence in the entire video was a constant volley of unabashed lies. Not long into it, Folger/Porter says, "For the remainder of this program, we're going to put you right here where I'm sitting in the juror's box. But this time you're going to hear the facts about evolution vs. creation, not some media spin, not wishful thinking, not the biased view of a scientific minority. . . . The truth is you've been misled, scammed, lied to."

Not a word she said was true. She didn't give us any actual facts; she only gave us a biased media spin of wishful thinking from a religious minority objecting to science. That's all apologetics is, making up excuses to rationalize or dismiss whatever you can't accept. It's all lies.

Even when creationist scientists have legitimate degrees, they're still adhering to a prior bias and not observing or practicing science, which means they're still apologists. For example, Jonathan Wells, a fellow of the Discovery Institute and author of *Icons of Evolution*, achieved his PhD for the express purpose of undermining evolution. He was a Moony; his spiritual leader, Rev. Sun Myung Moon, famously financed Well's doctoral education. Moon explained that God wanted Wells to earn a biology degree so that he could "destroy Darwinism from within." Again, it's all about the bias, and the misrepresentation of apologists.

There's no rational need for apologetics and science rejects it. We don't hold evolution sacred. We defend it only because it is evidently true. Superstitious politics have made evolution an icon necessary to the defense of the scientific method, which is the real target of religious fundamentalism. But their underhanded attempts to undermine science are also eroding their parent theology more so than atheism ever could. If you have to lie to defend your truth, then it was never really truth to begin with, and creationism obviously *is* not like the truth and *does* not like the truth. That's also why they don't like free speech. Where science invites critical inquiry in peer review, apologetics depends on one-sided assertions based on nothing and defended only by censorship. To give a famous example, consider the Holy Inquisition's words in judgment of Galileo in 1616,

> The first proposition, that the sun is the centre and does not revolve about the earth, is foolish, absurd, false in theology, and heretical, because expressly contrary to Holy Scripture; and the second proposition, that the earth is not the centre but revolves about the sun, is absurd, false in philosophy, and from a theological point of view at least, opposed to the true faith.

If you look up QuestionEvolution.com, either on the web or in the Talk.Origins newsgroup, you'll see that to their credit, their webmaster once agreed to host a link to an offsite collection of rebuttals from the evolutionary perspective. Despite this, several months later, his home page still said, "These questions remain unanswered by the evolutionist." So we pressed him about correcting that error, and he said he would when he had the time. But that was several years ago, and his site still tells the same lie. All he did was to quietly remove the links to all our answers so he could pretend they didn't exist, rather than correct his false accusations. I showed all of that in a video and I sent him a link to that several years ago, but he has still not replied nor admitted that those questions have been answered.

To give another example based on my own personal experience, someone still stupid enough to promote Kirk Cameron actually bought an Internet domain named after me, with all the tags applicable to me, in an obvious attempt to redirect search engines looking for me. Of course, there was no way to post comments to that site or to contact them to correct it, so they used that page to post idiotic falsehoods deliberately misrepresenting my position. It's so crazy that on some level they have to know how wrong that all is. But they don't care about accuracy, and we've already heard them declare they don't care about evidence either. They've assumed their conclusions without it and admit they'll never change their minds because of it, but sometimes they'll contradict themselves and pretend otherwise if they think they have a few facts on their side.

Laurence Tisdall of the Creation Science Association of Quebec tried to list evidence of creation on *The Michael Coren Show*: "We have footprints of humans with dinosaurs in Dakota. We have dinoglyphs from Lake-on-the-Woods in Ontario all the way down to Peru."

At that point he was interrupted by his opponent, Jason Wiles of the Evolution Education Research Centre at McGill University: "Dinoglyphs being native drawings?"

Tisdall responded, "Native drawings of tyrannosaurs."

"Right," said Wiles, "just like there are native drawings of all kinds of fantastic creatures."

"Yes," Tisdall persisted, "but these aren't fantastic; they're consistent." He finished with the challenge, "it's not that I *don't* know something; it's because I *know* something I'm a creationist." What this creationist didn't know was that he had just cited known forgeries as evidence.

If you look up the artifacts he's talking about, they include drawings of two-headed dragons where the second head grows out of its tail! Even when these things have only one head, they still don't look like dinosaurs; lizards, snakes, crocodilians maybe, but not dinosaurs. The petroglyphs he mentioned have all either turned out to be something else or not a deliberate rendering at all.

Many of the famed Ica stones that he mentioned from Peru have been recently created and artificially aged as part of a hoax financed by a Peruvian dentist seeking to defraud gullible tourists. Javier Carbera Darquea believed, as Erich von Däniken did, that ancient Peruvians were capable of futuristic medical techniques; that they had knowledge of extraterrestrial species and the technology for interstellar travel; and that they rode on domesticated dinosaurs. To prove this, he purchased hundreds of hand-carved stones and clay figurines from villagers and added them to a collection with pre-Columbian artifacts. But the sculptures of dinosaurs were tested in German and Swiss labs and found to be decades old at most, with some so recent that they still contained water.

Some of the hucksters involved have even confessed to their part in this crime. Impoverished villagers have confessed to creating thousands of stones over the years to suit Cabrera's requests. These show no natural erosion but do have traces of sandpaper, and they were given the appearance of advanced age by being soaked in chicken shit. Local farmers claiming to have created the first such depictions in 1975 said they referred to comic books, which is why their dinosaurs are so inaccurate. They look like they came from *The Flintstones*. For example, theropods are shown dragging their tails (the way we used to think they did back in the 1970s) and ceratopsians are depicted with dorsal ridges that they did not really have.

Even other creationists reject this, including cryptozoologists—the ones who most want to believe in persistent existence of extinct paleofauna. Even *they* say they can prove that these artifacts aren't really ancient depictions of dinosaurs. Said Steven Myers, vice president of the Discovery Institute,

> I bought a replica of an Ica stone with a craving of a dinosaur on it from Kent Hovind's website. I was amazed to find out that this carved dinosaur had five fingers and five toes, and a turtle shell on its back, and donut rings on its skin. I do not know of any dinosaur like this.
>
> My theory is that the real stones do not have dinosaurs or men riding dinosaurs, or telescopes, or the planets Neptune or Uranus carved on them. These would be the fake ones. I would like to offer a challenge to Dr. Swift to allow one of his stones with a dinosaur on it to be examined and tested by experts. The first-hand observation by Neil Steede that, even though the stones he examined did have this patina, there was no patina in the grooves. This suggest that while the stones were certainly very old, the carvings were clearly of far more recent origin.

Meanwhile, Ken Ham's Answers in Genesis, which, as you'll recall, owns the Creation Museum, has expressed its disappointment that the stones are fraudulent:

> Unfortunately, some initially plausible evidences for man's contemporaneity with dinosaurs have later turned out to be mistaken. The controversial "Ica stones"—allegedly genuine pre-Inca engravings of dinosaurs from Peru—have since been shown to be a fraud. . . . it turns out that an unscrupulous Peruvian surgeon had purchased the stones from a local artist and installed them in his museum, claiming them to be ancient. The Institute of Geological Sciences in London has since examined one of the stones and confirmed its modern origin.

So much for Laurence Tisdall's evidence. I have investigated every claim pertaining to native drawings of dinosaurs. There is not one that is even remotely plausible. Christian radio-show host Bob Dutko tried to make the same type of claim as Tisdall during a live debate with me. So I challenged him to give me one single example of an artifact that can be scientifically verified to be an authentic pre-Columbian human representation of a dinosaur. In a show of complete confidence, he replied:

> I can give you over thirty of them right now. I could give you over thirty of them. Let me give you, oh just one or two. For example, in the Arizona Historical Society, ancient swords were excavated near Tucson, Arizona, they were excavated in 1924. The swords are referred by the way on page 331 of the book, *Lost Cities of North and Central America*, you can look it up for yourself. They have various artwork designs carved into them. One sword has an exact brachiosaurus carved into it. If you look at this, you can show it to anybody, it looks exactly like it came out of a *Jurassic Park* movie. The Arizona Historical Society owns the sword. You can look at that picture of the brachiosaur. There is nothing ambiguous about it at all. You can't say, "Oh, the ears aren't right," nothing along those lines. That's one example of thirty of them I could give you right now. I would encourage you to look that up.

Dutko should never encourage anyone to look into his claims, because I did, and what I saw was just too funny. He's referring to a collection of sword replicas made out of lead. That's right, lead (Pb). Not steel or bronze or iron, but something that would be utterly useless as a sword. On one of these fireplace decorations is an image of a Flintsones-style sauropod with a forked tongue, because it was drawn back when people thought that dinosaurs were just big lizards. This trait occurs only in monitor lizards and snakes, which are their closest cousins. Lizards are on the wrong side of the reptile family tree from dinosaurs, who would not have had forked tongues. But it gets funnier than that because the Hebrew writing around it belies the fraud.

As Kenneth L. Feder writes in *Encylopedia of Dubious Archaeology*,

> It was difficult to explain how artifacts with a depiction of a dinosaur and Latin and Hebrew writing, perhaps dating to 800, came to be cached in an early twentieth-century lime kiln located just outside Tucson. But this was clarified, sort of, once all the writing had been translated. It appeared that the writing on the artifacts told the story of an ostensible Roman-Jewish colony in Arizona, dating from 775 to 900.

When experts in Latin, Roman history and archaeology, and Southwest prehistory examined the Tucson artifacts, they declared them to be complete nonsense—not just fakes, but bad ones at that. When a University of Arizona professor and highly respected Latin scholar, Dr. Frank Fowler, pored over the inscriptions, he discovered, taken together, the writing made no sense. The inscriptions did not represent a comprehensive or comprehensible message, but were instead a jumble of largely nonsensicle and disconnected, discontinuous phrases. But the argument for fakery seems to have been cinched when Fowler discovered that all of the individual Latin phrases inscribed on the artifacts had been lifted verbatim from three Latin textbooks—Harkness's *Latin Grammar*, the *Latin Grammar* of Allen and Greenough, and Rouf's *Standard Dictionary of Facts*. The earliest publication date for any of these works was 1864, which is rather more recent than the dates inscribed on them. Fowler went on to state that, when whoever inscribed the Latin on the Tucson artifacts attempted to in any way change the phrasing from what appeared in those publications—to change the tense, for example—he or she betrayed the lack of even a rudimentary knowledge of Latin. As a result of Fowler's analysis, it is clear that none of the artifacts can date to before 1864, and therefore none of them date to the purported period for the Roman-colony they mention.

Byron Cummings, director of the Arizona State Museum, was an eyewitness to the removal of at least one of the artifacts. He stated that the artifact in question was embedded in an already existing hole—that it had, in other words, been planted in a hole in the sand and gravel which had then been only imperfectly stamped down around. Cummings included his testimony in a report he presented to the University of Arizona, which at the time was considering purchasing the Tucson artifacts for a substantial sum of money. Geologist James Quinlan maintains that it would not have been difficult to have planted the artifacts in the gravel and suggests that the artifacts were placed there after the lime kiln was abandoned.

The Tucson artifacts are another example of attempts made to connect an ancient Old World culture to America long before the voyages of Christopher Columbus. Like so many of the others, the Tucson artifacts are frauds.

Remember, I asked Bob Dutko for one example on which he would stake his reputation, the best example he had that could be confirmed accurate, and he gave me a laughable example of a failed forgery. Do you see why I say that creationists have no credibility whatsoever?

Fundamentalists have never exposed a single evolutionary fraud or scientific conspiracy since their mindless movement began, but they've perpetrated many of each! Some sites supporting science have long lists of creationists' criminal cons and thoroughly disputed fabrications along with citations of peer-reviewed research proving why every single allegation of evangelical evidence ever examined is either unsubstantiated or entirely erroneous, if not deliberately deceitful. So if you ever need to know how science has answered absolutely every challenge creationism ever offered, look up the Talk.Origins archived index of creationist claims.

Religious fundamentalists seldom correct any of the many flaws in their data, and increasingly desperate wanna-believers keep repeating the same old bogus stories and urban legends long after they know they're not true. That's all they ever had, and before the age of information they could still get away with that. But some of these frauds and rumors are now so blatantly bogus that even apologetics propaganda mills are compelled to admit it because refuting the various fallacies, frauds, fibs, and fakery of creationist claims has become a sort of Internet sport. It's like shooting fish in a barrel; every single thing the fundies ever presented in their defense has always turned out to be either misunderstood and grossly distorted or intentionally misrepresented, if it could ever be tested at all.

This is only because creationism is absolutely wrong about absolutely everything. If there was any truth to it, then creationists could be scientific, and they would be rewarded for that. The first person who comes up with actual evidence of creation would overturn everything, winning not only the Templeton Prize but also a Nobel Prize as well. The reason that hasn't happened is because it *can't* happen—because none of it is real.

Whenever I try to explain to creationists the taxonomic significance of genetic orthologues or comparative morphology, I usually hear comments to the effect that "similarity doesn't imply a common ancestor, it's evidence of a common designer." And they very often use the sequential development of automobiles in their claim that any apparent phylogeny is an illusion. But even if we ignore the fact that big cars don't give birth to compact cars and therefore aren't subject to the laws of population genetics, and even if we ignore the fact that animals reproduce naturally without need of assembly by any manufacturer, there are still plenty of indicators to prove that our biology is definitely a product of common ancestry and does not involve any evident designer.

For example, Adolf Hitler was an intelligent designer. He conceived what was arguably the most versatile production automobile ever made. He called it Volkswagen (VW), the "people's car" (folk's wagon). They were intended to be economical, fuel efficient, and lightweight, and they could even drive on water. They really could! Especially after being modified. The classic VWs could be easily converted into baja bugs, dune buggies, sand rails, all manner of custom kit cars, and even ultralight aircraft, because it was a very simple design. It ran leaded gasoline through a carbureted, rear-mounted, air-cooled aluminum engine in a "popper" configuration with a standard stick-shift transaxle for rear-wheel drive. That was the original Beetle, and it changed very little over nearly forty years. Then it was discontinued for a while and brought back completely different. Now it has a front-mounted, water-cooled, unleaded fuel-injected nickel-and-steel alloy engine in the "in-line" configuration and has an automatic transmission with front-wheel drive.

These are both products of the same designer (Volkswagen, not Hitler), yet the only thing they have in common is that dumb-looking bubble-shaped body style. So they only resemble each other superficially. They share brand names, but otherwise aren't even related.

Through the process of trial and error, a creator can improve his designs of course and usually will a little or a lot, especially as new technology becomes available, or he gets a better idea of what he is doing. But he can also arbitrarily decide to scrap all that and reinvent the original model again while changing everything inside. A creator can choose to create things however he wants, especially if he uses magic powers instead of natural processes because then there are no rules applied. But evolution could never produce the kinds of changes we see in cars, because it is just a matter of population mechanics and is therefore restricted by the laws of physics and phylogeny.

You can't have a whole suite of functional mutations occur all at once, as if choreographed to achieve a certain goal. Nature acts without intent. Usually only one thing changes at a time, and that trait may be carried onto subsequent branches, each of which develop their own unique modifications of the parent template—and that trait marks the next morphological distinction and the subsequent clades. And on it goes, incrementally, indefinitely, with a new branch division at every node. That's what I mean when I say that systematic taxonomy reveals a series of superficial changes being slowly compiled atop compounded tiers of substratal similarities. This is what we consistently see in nature and in the fossil record. It's not just evidence of a single universal common ancestor, but a detailed genealogy of myriad ancestors at many different levels.

This is certainly not what we should expect to see from a common designer, because what you usually get from a creator are things that don't fit into any cladogram. If cars evolved, then we could classify them according to a hierarchy of derived traits, in which case the Chevy Corvair would not be more closely related to Corvettes and Camaros than to the Beetle or Karmann Ghia. Unique traits would emerge in one make or model only, and be carried down or modified in successive models with no dramatic fundamental revisions possible at any point.

But a creator can change the whole design at any time, or any one model might be equipped with different engines, transmissions, or other options that could be swapped with other models too. Multiple creators can steal each other's designs without any ancestral linkage, such that motorcycle manufactures can suddenly start building cars and car companies can decide to design motorcycles, and everyone can build their own SUV.

Again, this sort of thing is impossible in an evolutionary scheme, because if all life-forms evolve at all, then every living thing has to fit into a cladogram of cousins and common ancestors. (And they do.) The implied relationships we describe have to be verifiably accurate according to a suite of specific criteria required in each case. But a magical creator can conjure anything he wants, for any reason or none, and these would consequently defy classification.

That's the nature of created things. Just look at all of our movies or mythology where we've ever invented any fictional creature: they never fit into taxonomy, even when they're supposed to. They always violate the laws of phylogeny. It's just like science fiction shows where there are aliens from other worlds who coincidentally look like humans and can even interbreed with them, but who have two hearts and green blood. They can't all be uniquely original while remaining superficially identical, internally incomparable, and genetically compatible all at the same time. They can if they're created, but not if they're evolved.

In our biological reality, there can be convergence where members of once divergent subclasses later play similar roles and adapt accordingly—

dolphins and ichthyosaurs, dogs and thylocenes, etc.—but they'll never become the same and can never hybridize. It doesn't matter how much they look alike on the outside, because the physiology of the organism determines what it is. The more distantly related they are, the deeper you'll have to look for similarities. But if they're related at all, then the deeper you look, the more similar they'll be. This indicates a common ancestor, not a common designer.

Creationism isn't just wrong scientifically and historically, it's also wrong ethically and morally. The simultaneously saddest and most laughable irony of this whole stupid controversy is that these zealots claim they're opposed to evolution as an issue of morality. Yet while we can cite dozens of examples where politically influential creationists clearly know they're lying about science, there is no such instance wherein evolutionary scientists can be shown to be dishonest in their criticisms of creationism. There's no need to be. Despite all the attempted deception, the baseless assertions and political division produced by the creationism movement, the truth is there has never been a single verifiably accurate argument of evidence indicative of miraculous creation over biological evolution or any other avenue of actual science. Not one—period. Neither has there ever been any credible proponent of creation science, because with Dr. Periannan Senapathy as the one crackpot exception, everyone who has ever published antievolutionary rhetoric to any medium did so only according to a prior religious agenda rather than any amount of scientific comprehension. They've all revealed inexcusable ignorance in the very fields where they claim expertise, and their arguments are all dependent on erroneous assumptions, prejudicial bias, logical fallacies, ridiculous parody, misdefined terms, misquoted authorities, distorted data, fraudulent figures, or out-and-out lies.

Thus, there are only two types of arguments for creationism: those that are untestable, implausible, and indistinguishable from the illusions of delusion, which can neither be indicated nor vindicated, verified or disproved; and those that have already been disproved.

THE 15TH FOUNDATIONAL FALSEHOOD OF CREATIONISM

"Evolution Is a Theory, Not a Fact"

"You, your joys and your sorrows, your memories and your ambitions, your sense of personal identity and free will, are in fact no more than the behavior of a vast assembly of nerve cells and their associated molecules."

—Francis Crick, Nobel laureate and codiscoverer of the DNA molecule, in *The Astonishing Hypothesis*

Comparing polls of the general populace to polls of scientists, we've seen that there is roughly the same percentage of scientists who believe in God as there are common folk who don't, both being a minority. There is a huge difference between educated specialists and the general population in how they understand or think about things. Scientists have a very different perspective than the laity does. For one thing, random people typically say that anything is possible, whereas scientists get excited when it seems that something *might* be possible when it was thought not to be. Because not everything *is* possible. A cow cannot jump over the moon, and monkeys cannot fly out of my ass. Because these things are not just improbable, they're impossible.

Likewise, people very often jump to conclusions, readily accepting whatever they already believe or want to believe with little or no indication, and demanding nothing less than absolutely undeniable proof of whatever they don't want to believe. Then even when they know that such proof exists, they hide from it and don't want to hear it. Or maybe they're inclined to be negative, hateful, or paranoid, and they want to be righteously indignant. So whatever you say, they respond with, "Oh, so in *other* words . . . ," and then they present a gross distortion of things you never said or thought or meant. People tend to be irrational sometimes. Scientists are too on occasion, but science itself is not. The only way to become rich and famous in science is to challenge the status quo, and the methodology behind that is harsh and requires a rational understanding. If you're going to stand up to experts, you'd better be an expert yourself. You'll have to know what they do and more than they do to know something they don't.

Science takes a very different approach than some other human interests. Lies are almost always sold under the guise of absolute truth, so rather than proclaiming "Truth" the way evangelists do, science tends to be tentative, even hesitant, and preferably dispassionate to minimize or eliminate bias and to be as reasonable as possible and open to correction. It's far more important to make sure your understanding and interpretation of the data be defensibly accurate than that you should believe what is probably nonsense and shout it from the mountain to the rooftops the way religion does. That's also why there are so few people who know about transitional species and beneficial mutations and all of that: science is famously intellectual rather than emotional. It's an academic discipline, willing to teach serious students, but it's not about selling sensational headlines. Sensationalism is frowned upon in science, so it's harder for people to get charged up about it the way they might get at a revival meeting.

We're trying to distinguish what we merely believe from what we actually know. We're talking about epistemology—that is, the study of how we know things, also known as the theory of knowledge. Has the theory of knowledge ever been proven? Can it be? Is the idea that people know things "only" a theory?

This is where we have to compare and contrast lay vernacular with the scientific application of terms like hypothesis, fact, law, and (especially) theory. The word "theory" is most commonly confused with a hypothesis, so we'll talk about that aspect first.

The public comprehension of these terms was literally illustrated in an animated *Dilbert* cartoon, as Dilbert and Dogbert were watching the news on TV. "In this episode of *Scientific Truth Journal*, we'll explore the theory of evolution and we'll implicitly mock the people who hold opposing viewpoints."

To which Dogbert gruffed, "Hmph! Evolution, what a crock! Can we change it, please?"

Dilbert replied, "Evolution is a scientific fact—unless you're ignorant."

Then Dogbert revealed the problem most people have. "If it's a fact, why is it called a theory?"

Dilbert answered as one who didn't understand the problem either. "There are scientific reasons; it's all very complicated!"

Let's ignore the fact that no science program would call itself "truth" even though they honestly could and still be accurate, and that religious shows do call themselves "truth" even though there is never any truth to them. Let's ignore also that science shows typically don't mock uneducated people with unrealistic beliefs either. My point is that Dilbert answered that way because the cartoonist who created him, Scott Adams, rejects evolution. That's also why the news talked about scientific truth, as opposed to evidence or a study. How could Dilbert know any better when even his creator doesn't get it? But it's really not that hard to comprehend.

The mission of religion is to make followers believe. Subjective assumptions unsupported by evidence will almost certainly be wrong at the onset regardless of the source, and without any means of regulation and no way to know which changes would be corrections, they will only get wronger over time. (I have seen professional academics use the word "wronger" a few times, but only when they're talking about creationism.) That it's a matter of interpretation only makes it worse; sacred dogma once written is forbidden to be changed, so it can't be rectified either. But science must be amenable to change because its objective is to add to the sum of knowledge and to improve understanding continually, so whatever explanations we ever propose are not to be believed, but to be understood; so they can be tested and corrected, even rejected if necessary, and our explanations must be refined accordingly. As Agent K, played by Tommy Lee Jones, says in the 1997 movie *Men in Black*:

> People are dumb, panicky, dangerous animals, and you know it! Fifteen hundred years ago, everybody knew the earth was the center of the universe; five hundred years ago, everybody knew the earth was flat.

> And fifteen minutes ago, you knew that people were alone on this planet. Imagine what you'll know tomorrow.

Indeed, once upon a time, the religions of Asia and the Near East not only believed the earth was a flat disk (as indicated in Isaiah 40:22, Daniel 4:10–11, and Matthew 4:8), but also that it was divided into four quadrants (mistranslated as "corners" in Isaiah 11:12 and Job 37:3) and that it was enveloped by a giant crystal dome, which was their sky (Job 37:18, Ezekiel 1:22-26 and 10:1). This notion didn't just come from the Bible either. Elements of this perspective were depicted in different sources from Persia through the Indian subcontinent into the Orient. Pythagoras of Samos (569–495 BCE), one of the earliest actual scientists, taught that the earth was a sphere. Eratosthenes of Alexandria (276–196 BCE) deduced its size with a surprising degree of accuracy. Yet hundreds of years later, the authors of scripture still wrote of a disk-shaped world, as did early theologians. According to Saint Augustine of Hippo (354–430 CE),

> As to the fable that there are Antipodes, that is to say, men on the opposite side of the earth where the sun rises when it sets on us, men who walk with their feet opposite ours, there is no reason for believing it. Those who affirm it do not claim to possess any actual information; they merely conjecture that, since the earth is suspended within the concavity of the heavens and there is as much room on the one side of it as on the other, therefore the part which is beneath cannot be void of human inhabitants.
>
> . . . should it be believed or demonstrated that the world is round or spherical in form, it does not follow that the part of the earth opposite to us is not completely covered with water, or that any conjectured dry land there should be inhabited by men.

Later, Saint Procopius of Caesarea (500–565 CE) had this to say,

> If there be men on the other side of the earth, Christ must have gone there and suffered a second time to save them; and therefore there must have been, as necessary preliminaries to his coming, a duplicate Adam, Eden, serpent, and Deluge!

The Dead Sea Scrolls (which contain the book of Isaiah and are the earliest archaeological fragments of the actual Bible) are at their oldest contemporaneous with Eratosthenes, and centuries younger than Pythagoras. It is also centuries earlier than the works attributed to Matthew and Daniel. We know many scholars already knew that the earth was a sphere, but not everyone knew that. As the centuries wore on, some dogmatic believers refused the wisdom of those who knew better and even suppressed or destroyed their knowledge. They held to belief in a flat earth even until Columbus provided the final disproof in 1492.

It was around this time that Hieronymus Bosch painted his masterpiece *The Garden of Earthly Delights*. The exterior panels of this triptych painting show the earth on the third day of creation. The earth is rendered as a flat disk suspended, just as Saint Augustine of Hippo described, in the "concavity" of a spherical crystalline firmament. According to Genesis 1:14–17, this firmament also included the heavens, the sun, moon, stars, and so on. So even after Columbus discovered America, some people still believed the earth was flat. There are plenty of people on the web who do too. Amazingly, flat-earthers and geocentrists abound even now, and they'll cite either the Bible or the Qur'an to justify their beliefs.

Modern creationists oppose evolution and sometimes cosmology the same way flat-earthers reject the theory of geosphericity (the idea that the earth is a sphere) or the same way geocentrists deny the theory of heliocentricity (the idea that the sun is the center of the solar system instead of the earth)! Bill Maher once called attention to this very point in his typical mocking tone,

> President Bush recently suggested that public schools should teach intelligent design alongside the theory of evolution because after all evolution is, quote, 'just a theory.' Then the president renewed his vow to drive the terrorists straight over the edge of the earth.

Rather than swearing in advance to uphold and defend our preconceived notions against all reason, we would do better to suspect and inspect every belief to see if flaws can be found and corrected, and our knowledge consequently increased. As Carl Sagan said in episode four of *Cosmos*, "Heaven & Hell,"

> There are many hypotheses in science which are wrong. That's perfectly alright; it's the aperture to finding out what's right. Science is a self-

> correcting process. To be accepted, new ideas must survive the most rigorous standards of evidence and scrutiny.

Aristotle once proposed that everything was made of earth, air, water, and fire, often represented by the Pythagorean "perfect" solids. The fifth element (represented by the dodecahedron) was considered to be the substance of life. Based on these long-held yet obviously delusive beliefs, George Stahl and other seventeenth-century scientists composed two theories: the theory of vitalism, which held that life was animated by an infusion with an elemental "spirit" (being an evaporous liquid) and the theory of phlogiston.

For decades, European scientists imagined that a nigh-undetectable sort of fiery air called phlogiston was present in everything flammable. Once something burned, the ash left behind was considered to be dephlogisticated material. This explained why combustible materials were lighter after they burned, having liberated their phlogiston into the air. Thus, unburnt wood was considered to be wood-ash with phlogiston still in it, while corroding metals were considered phlogiston plus "calx," which we now describe as oxide. When hydrogen was discovered, it was first thought to be pure phlogiston.

There were some inconsistencies with this. It made sense that burnt objects weighed less, but why did corroded metals weigh more? A series of experiments ensued and some of these men began to rationalize how phlogiston might somehow still account for all the inconsistent data. Finally, more accurate measurements and more critical thinking eventually challenged the status quo. Antoine-Laurent Lavoisier determined that combustion and corrosion weren't adding phlogiston to the air, but that something already in the air was definitely involved. He detected two components in ordinary air, one an as-yet unidentified asphyxiant, and another given to respiration. He had discovered oxygen.

In 1783, Lavoisier was able to demonstrate his proclamation that "phlogiston is imaginary." This prompted the chemical revolution, where the vast majority of scientists abandoned Stahl's theory of phlogiston in favor of Lavoisier's "antiphlogistic" theory, his oxygen theory. Explaining his refutation of phlogiston theory via his discovery of oxygen, Lavoisier said,

> My object is not to substitute a rigorously demonstrated theory but solely a hypothesis which appears to me more probable, more comfortable

to the laws of nature, and which appears to me to contain fewer forced explanations and fewer contradictions.

Ultimately, this brought chemical theory out of the realms of alchemy, with Lavoisier eventually becoming known as the father of modern chemistry.

Vitalism held that organic compounds could be carried from a "vital force" found only in living things. Friedrich Wöhler (1800–1882) synthesized urea artificially, without need of kidneys, thus demonstrating that organic chemistry is a biological process governed by the same laws as those that govern nonliving matter.

So neither of Stahl's theories are valid theories anymore. Phlogiston theory was disproved in 1777, and his theory of vitalism was disproved fifty years later. But after a hundred and fifty years and despite lots of uneducated claims to the contrary, Darwin's theory is still going stronger than ever.

It's hard to explain that to the average guy on the street, or on YouTube. One YouTuber named GEERUP posted a video that argued, "Evolution is still a theory. It's still a theory. We just had two hundred years of evolutionary theory, and it's *still* a theory." By this he meant to imply that it has never been proven. I don't fault people for not understanding this, but even in the 1700s there was a difference between an actual academic theory and a mere hypothesis. Another religious vlogger said, "I heard someone ask one day, 'Well, you know, you haven't disproved the theory of evolution,' speaking about Christians. And my reply was, 'Um, there's nothing to disprove; the theory of evolution has never been proved.'"

You don't have to prove something before it can be disproved, nor should we both prove and disprove the same thing. Science doesn't permit anything to be proven positively, because there's no value in that. Like comedian Darren O'Brien said, "Science knows it doesn't know everything; otherwise it'd stop." Instead, every hypothesis must be potentially falsifiable so they can be disproved to reveal hidden flaws. That means there has to be a way to identify errors: to find out what's wrong with it and to fix it.

It's still possible to falsify evolution too, though it's now so well-supported it will take more than an unsubstantiated anomaly to do it. Sadly, unsubstantiated anomalies are all that creationists usually offer. In Louisiana (where creationism is being taught legally), a public school science textbook says that the existence of the Loch Ness monster disproves evolution! Whoever wrote that doesn't know that the Loch Ness monster doesn't exist. They also don't know that plesiosaurs aren't the same thing as dinosaurs, and they don't know that evolution doesn't require that all dinosaurs be extinct either. After all, birds are still here.

To illustrate part of the problem, some uncritical thinkers believe there's a monster in the loch and they "have a theory" that it is a plesiosaur. Likewise on the side of the Angkor Wat temple in Cambodia, there is a carving of what appears to be a pig. It's one of many stylized animal relief sculptures in the wall of the temple, and all of them have leaflike decorations about them. But in this case, the leaves appear behind the pig as if connected to it. So some young-earth creationists "have a theory" that the leaves are actually growing out of the animal's back and that it's really a Stegosaurus. If it was a stegosaur, it would have a thicker, longer tail with spikes on the end, and it would have a smaller head with a beak instead of a snout; it would *not* by contrast have long pointed ears or hooves.

So it's definitely not a stegosaur, and those who thought otherwise don't have a "theory" about that either. They don't know what a theory means, and they don't know what this theory is, so they don't know how to disprove it. That's why they demand impossible oddities that evolution could never produce. For instance, someone asked me, "If evolution is true, then why don't we grow wings to escape from predators?"

That would actually disprove evolution. If we discover some previously unknown life-form, we can add its branch to the tree of life, but only when we can trace its evolution such that the tree itself already shows where that organism belongs. Systematic taxonomy cannot accommodate created chimeras, like a vertebrate hexopod or a mammal with feathers. It is possible for chickens to have scales and for some mammals to still have scales, because both are derived from a form previously referred to as reptiles. But it is impossible for a mammal to have feathers, and they cannot simply "turn on" any gene that could produce feathers either, because they never had them. Mutations can produce scales that may look like feathers at a glance, but they are not actual feathers. Feathers aren't like hairs or scales; they're a highly complex construct that evidently developed in a series of sequential stages represented by several successive species. It would be a statistical impossibility for an exact duplicate of that level of complexity to coincidentally emerge again in mammals. For example, the typical depiction of angels and the flying monkeys from *The Wizard of Oz* would both destroy the phylogeny currently attributed to evolution because evolution cannot account for primates with two arms, two legs, and two other limbs—especially if they're fully avian wings.

That's why I say that God could make anything he wants any way he wants, but if we ever found a real live Pegasus, it would immediately reduce evolution to horse feathers because we're not just talking about feathers, but the structure beneath. There is no evolutionary process capable of developing a third pair of limbs onto a skeleton incapable of accommodating another set, and of a type that couldn't be formed by any evolutionary process. Mixing animals is common in mythology, but evolution cannot account for the arms of a dinosaur growing out of the body of a horse, nor any other mammal either. It simply cannot happen naturally, so if we ever found such a thing, it would mean that evolution is fundamentally flawed and that life-forms really *are* being created by some other means.

In the 2006 case of *Selman v. Cobb County*, in Marietta, Georgia, one witness testified, "Evolution has not met the test, and deserves only to be treated as theory." Another said, "Darwin considered it a theory, and it is still a theory; it has never been proven, and never will be." As any philologist will readily explain, words may have different meanings in different contexts. Creationists exploit the academic meaning of theory in favor of a colloquial one, as if a theory was only blind speculation like their own position is. But a scientific theory isn't a guess or conjecture. Look it up.

> **Colloquial definition:** a supposition or a system of ideas intended to explain something, especially one based on general principles independent of the thing to be explained.
>
> **Academic definition:** In science, an explanation or model that covers a substantial group of occurrences in nature and has been confirmed by a substantial number of experiments and observations.

In most instances, a theory is a field of academic study. For instance, the high school I went to taught a class called music theory. Being an insufferable jerk then as now, I remember bursting in one day and yelling at the class, "You shouldn't teach music in school. It's just a theory; it's never been proven!" Sadly, I don't think anyone there got the joke. But it does speak to the point I'm making.

If music theory is a field of study, and as such can never be proved, then neither can the theories of evolution or even economics, and for the same reason: the notion is silly. Even if a theory passes every test forever, we still wouldn't say it was proven, because positive proof exists only in matters of mathematics or law wherein evolution actually has been proven. Otherwise, in science, no theory has ever been proved, nor can one ever be. Theories can only be disproved. And when that happens, a theory that doesn't work must be replaced by one that does. We can't discard any theory just because we haven't perfected every part of it yet. You can't trade something that works for nothing that doesn't. If the original theory works at all, you'll still have to use it, and perhaps fix it—but we can't dismiss it until we can replace it with something better. And Darwin's theory is actually better supported than Newton's theory of gravity.

Fact: a point of verifiably accurate data

Law: a statement or equation that has always been shown true.

Hypothesis: a testable, potentially falsifiable explanation of facts/laws.

Theory: the unifying framework of study encompassing all of the above.

Let's look at the facts. Remember that a fact is merely data, a demonstrably accurate observation that is indisputable because it can be objectively verified by either side arguing about it. So if we demonstrate the fact of gravity, we see that things tend to fall down. What's that mean? Well, nothing yet; a fact on its own is meaningless. We need to understand it more specifically. When seen on an astronomical scale, we can determine a universal rule: that matter attracts matter. This is one of the laws of gravity (a law being a general statement of nature that is always true under a specific set of circumstances). Now *why* does matter attract matter? That's the theory!

Now let's look at the fact of evolution. Since the dawn of livestock cultivation and agriculture, we've seen that species diverge, with new races branching out of older family trees, each with their own distinct traits and features that are in turn inherited by new variants diverging thereafter. This is a readily evident fact in any lineage one cares to examine and is implied at every level of taxonomy.

Creationists try to say that this doesn't count because it's only microevolution, but that isn't true because we've actually witnessed the emergence of new species, too. But even if that *were* true, this would still count because it is still evolution according to the laws of population genetics and descent with inherent modification—which is all evolution really is. Here's how the University of California–Berkeley's "Evolution 101" online primer introduces the subject of evolution:

> **The definition**: Biological evolution, simply put, is descent with modification. This definition encompasses small-scale evolution (changes in gene frequency in a population from one generation to the next) and large-scale evolution (the descent of different species from a common ancestor over many generations). Evolution helps us to understand the history of life.
>
> **The explanation**: Biological evolution is not simply a matter of change over time. Lots of things change over time: trees lose their leaves, mountain ranges rise and erode, but they aren't examples of biological evolution because they don't involve descent through genetic inheritance.

Opponents say evolution isn't allelic variance in reproductive populations over successive generations because that doesn't mean spontaneous generation of something coming from nothing, or dirt coming to life, or fish turning into giraffes, or some other ridiculous thing which no scientist could ever endorse. But none of those things are evolution; they're all straw man fallacies created to obfuscate what evolution really is, by those who refuse to admit that it really happens. So they call it something else instead. Here's a classic example of this from John Morris of the Institute of Creation Research,

> Darwin observed many things in in nature. He was a good naturalist, a good observer of information. What he saw was various plants and animals altering somewhat through adaptation, through variation, he saw them change. We never see one basic type of something changing into something else. That has never been observed in science or in in genetics. It just has never been observed. What we see is variety. Variety happens, adaptation happens. Evolution doesn't happen.

Evolution *does* happen, and this twit just admitted it. But he also lied about what it is. Evolution never permits one thing to turn into another, fundamentally different type of anything; that would violate the laws of evolution. Yes, there are natural laws of evolution! But that does not make evolution a law. Creationists have a fundamental misunderstanding about such distinctions. Consider the grossly misinformed take of Terri Leo, antiscience advocate and former chair of the Texas Board of Education,

> They want evolution to be a law (like gravity), not a theory—thus their name change to equate the theory of evolution to laws in science. . . . They do that by downgrading laws of science to now be just "theories." That's why you heard them refer to the "theory" of gravity and "theory" of relativity. They call them all "theories" so as to put evolution on equal grounds with a law. It is still the law of gravity and the law of relativity, and still the "theory of evolution."

There are several laws of gravity, not just one, and the theory of relativity was always called a theory, never the "law" of relativity (although relativity does have its own laws, just like the theory of gravity does). This is one of the reasons that I invited Terri Leo to participate in our debate back in 2005. Isn't it sad when someone like me has to educate the Texas Board of Education? She doesn't know what a theory is, but then she doesn't know what a law is either. So we'll talk about laws now. Contra Texas education, theories do not become laws once they're proven. Laws are less than theories and are enveloped *by* theories. A theory is as good as it gets. In fact, that label has become so exclusive that an idea or explanation kinda has to be proven before it can qualify as a theory.

There are certain rules that nature has to follow. Everything in reality has to conform to the laws of physics. Each of what we call the physical laws of nature are general statements based on repeated experimental observations that are consistently "proven" (shown to be) true under specified circumstances. For example:

> **Kelvin's 2nd law of thermodynamics**: "In all energy exchanges, if no energy enters or leaves the system, the potential energy of the state will always be less than that of the initial state."
>
> **Newton's 3rd law of motion**: "For every action, there is an equal and opposite reaction."

Both of these laws are simply summary sentences that are demonstrably accurate by all observations. Otherwise, laws may be expressed as mathematic equations, such as with Newton's 2nd law of motion, Plank's law, Boyle's ideal gas law, or Einstein's famous mass-energy law, $E=MC^2$. These are laws of physics, but there are other natural laws as well, including some specific to biology—like Mendel's laws of genetics or Karl Ernst von Baer's laws of embryology.

As an aspect of biology, evolution also has its own laws that it has to adhere to just like any other field of science. There are even some laws that are specific to evolution, and we'll talk about those now too. Perhaps the first of these to be formally recognized as such was Dollo's law of irreversibility: "An organism is unable to return, even partially, to a previous stage already realized in the ranks of its ancestors."

In the late 1800s, it was already understood that evolution was a matter of population mechanics, even before the discovery and integration of genetics. Then-paleontologist Lewis Dollo determined that evolution could not move in reverse, meaning that once a population had all inherited particular traits, that population could not then devolve back into a copy of the ancestral stock. To do so would be rather like traveling back to an earlier point in your own life and trying to live every moment in exactly the same way until you get back to the time that you left, without deviation, such that your new present is no different than the one you remember. Obviously, any one of many tiny changes in the interim will complicate and frustrate that attempt. The same is true of evolution, because there would be many peripheral mutations accumulated in modern species that could not all be undone by any attempt to devolve them.

People have tried to reverse engineer the extinct aurochs from existing cattle, and they've tried to breed back the Bullenbeisser from living mastiffs. The best one can get is the appearance of that animal. Paleontologist Jack Horner is attempting to reverse engineer dromaeosaurid dinosaurs using the template of an emu. Emus are Australian ratite birds standing two meters tall, and Dromiceiomimus is a Mesozoic Maniraptor, a dinosaur named for the fact that it looks exactly like an emu except that it still has a long tail and functional arms. Horner wants to reactivate dormant or disabled genes for making teeth in birds and perhaps the genes associated with long tails, although generating their three-fingered hands by this method is highly unlikely. Even if he succeeds in all three of these things, he'll produce an animal we should recognize as the traditional iconic image of a dinosaur—after all, it literally is a dinosaur already. But no matter how perfect his experiments and eventual result might be, it will never be an actual Dromiceiomimus. If we had Dromiceiomimid DNA, it should be very different than that of Horner's emu-saurus.

Sometimes humans are born with small tails, as if we were reverting back to our Old World monkey ancestors; likewise, a dolphin was once found to have hind flippers, making it essentially a four-legged dolphin, just as fossils indicate they all used to be. These are atavisms, traits that were disabled at some point in that organism's evolutionary history but that are recent enough that they might be switched back on in rare occasions. This is also why some people can move their ears where most people can't, and why some horses have canine teeth where most don't. These reactivated genes are still recessive; they're rarely barely expressed, and do not qualify as evolution in reverse.

That's not to say that mutations themselves can't be reversed. Even if a given mutation spreads throughout a particular family group, it may not spread to the point of influencing the entire population. The larger the gene pool, the more it inhibits both the extent and influence of mutations. That's a law too, but I don't know if anyone has bothered to name that one. Let's call it the law of genetic inhibition.

For example, several members of the Vadoma tribe of Zimbabwe have a condition known as ectrodactyly, in which the middle three toes are absent—not just at the visible ends but inside the foot as well. So the foot is divided between two huge toes, like an ostrich. Thus, they are known as the "ostrich-footed" tribe. The remaining toes curve slightly inward, offering some advantage both in running and in climbing trees. But this advantage isn't significant enough for natural or sexual selection in modern human society. Consequently, this novel trait only spread to a few descendants. The more that family blends their genes with the genes of other families, the less likely this characteristic will be expressed in the next generation. The influence of the dominant gene pool is too great. Novel genes have a better chance of spreading throughout smaller isolated populations. This is an important factor of evolution that we'll come back to later.

Another point of confusion worthy of mention is that so many different species show convergent traits, as if they're following a fashion trend rather than inherent genes. The most glaring example of that is "saber" teeth. There is an extinct lineage of several species of machairodont scimitar cats that are known for this feature, as well as nimravids—another extinct lineage of closely related catlike cousins which all share that trait too. But similar saberlike teeth were also present on a variety of other mammals that are not closely related, and therefore they could not have inherited the trait from each other. These oversized canine teeth occurred in an Australian marsupial *Thylacosmilus atrox* and even in gorgonopsids, lion-sized mammallike reptiles from the Permian era. Each of these animals fulfilled the same predatory niche, as if they were all trying to be lions of a different type. However, saber teeth are just a coincidental convergence, an occasionally activated gene that could occur in any cynodont subset, because these oversized canines were also present in Dinocerata (an order of grotesque and vaguely rhinolike ungulates from the Paleocene).

When I was a little boy, I remember cartoonists depicting practically any prehistoric mammal as a saber-toothed version of modern mammals. However this feature wasn't that common, and it wasn't just a prehistoric style; this same trait is also seen in modern species like the Indonesian babirusa as well as walruses, and even in some Asian species of deer. Yes, we have saber-toothed deer! Sometimes a genetic switch gets flipped in mammals and their canines get ridiculous. If you're a predator, that will be more useful to you. If not, figure out what else you can do with them.

The point is that despite the fact that atavisms do occasionally emerge, and certain patterns may seem recurrent, whether they appear in one genus or a convergence of different ones, Dollo's law still holds true. This law has even been demonstrated under laboratory conditions. It is the reason why it is so difficult to reverse-breed ancestral forms, or to reengineer earlier traits now dormant in the modern genome.

Individual mutations may be reversed, colors and proportions may fluctuate, and it is even possible to develop limbs and organs and then lose them again later on. But their initial derivation and subsequent reduction follow different mechanisms, neither of which can be truly reversed.

Famed Harvard professor Stephen Jay Gould expanded on Dollo's principle with a further observation that also qualifies as an evolutionary law. He noted that generalized forms are more versatile, having more assets to adapt, while the more specialized a species is, the more it is limited by the narrow parameters of its own specialty.

For example, whales adapted fins into legs and then turned those back into fins, albeit of different design, but they are now unable to adapt them back the way they were to get back onto land. Likewise, birds made wings out of their arms but cannot "re-evolve" their ancestral hands again, because their fingers have been reduced and reabsorbed and there's nothing to build on. A highly derived lineage may effectively paint itself into a corner, becoming so specialized that it can't adapt to new situations being forced upon it. This leads to an evolutionary dead end. Meanwhile, more primitive and generalized forms should be better able to adapt to any situation or niche new environmental conditions may create.

Raccoons are a good example of this generalized versatility. They're omnivorous basal karyotypes (general form) of the undifferentiated ancestors of all canoidean carnivores (weasels, seals, dogs, and bears). Canoidea is the dog side of the Carnivora family tree. Raccoons are procyonids, which means they're structurally proto-dogs that still have hands. Even though they're so familiar, they could be considered living fossils because of their resemblance to the crown of all canoid Carnivora. (The ancestors of all canoideans were similar in structure to raccoons.) They're still capable of grasping and carrying things with their hands while walking upright, and they can still climb trees and manipulate objects where bears and dogs and such cannot.

The evolution of other canoid carnivores as ground-bound digging and burrowing predators cost them the subtle functions of their former fingers. Dogs took this disadvantage in one respect and turned it into an advantage in a different application. Unlike the rest of the canoids that are plantigrade (plodding about on their whole foot), dogs are digitigrade, always on their toes. This provides a spring to their gate that makes them superpredators capable of running long distances at high speeds. But they can't pick up or carry anything except with their mouth.

Fortunately, humans are still rather generalized in that we retain all five digits on all four extremities, just as our ancestors did when they finally crawled out of the sea for good. But our jaws and our claws are diminished and pathetic. We don't have tails anymore either, and can't get them back, because there's no selective pressure to act on an atavism and build back lost vertebrae. So that's another adaptive attribute we've lost that other species can still employ.

According to Darwin's principle of gradualism, all these changes were slow and slight, because evolution at every level is just a matter of variable proportions, whether chemical or morphological. Although significant mutations appear to occur with almost regular frequency, environmental dynamics rarely apply dramatic selective pressures. When they do, transitions are forced to adapt relatively quickly according to Gould's principle of punctuated equilibrium. This also holds that intermediate states would typically not continue to be favored and would consequently be relatively few and harder to find in the fossil record. Of course, we've still found several hundred of them since then, and scientists have charted macroevolution occurring in real time much faster than previously proposed. However, even in those instances, dramatic responses usually require hundreds of generations, if not tens of thousands, meaning even punctuated equilibrium is still gradual.

Where the laws of evolution really determine phylogeny begins with what I call the principle of monophyly, but which taxonomist Ernst Mayr described as Darwin's second law, the principle of common ancestry. Every new genus, species, or phyla that ever evolved began with speciation, and the newly emerged species is always a modified version of whatever its ancestors were. But we also see them branching into new species that can no longer breed with their ancestral or sister species, even though they're still part of the same clade.

There is no ascension up an evolutionary ladder, no evident direction, and neither is there necessarily any progression toward an implied goal. Things aren't always getting bigger or better either; some get smaller. Others maintain a fashion that seems timeless, but even then there are many different versions. Each adds to an ever-expanding array of different designs with myriad affectations and alterations of a base template established by their ancestry. If anything, natural selection seems to be trying out every course at once and keeps retesting previous options with each new set of descendants. So if there was an intelligent designer for every form of life, it must be feeling its way around in the dark, because we get a wide variety of designs, many of which fail altogether, and only rarely do we see one that is significantly better.

This pattern relates directly to what Mayr described as Darwin's third law, the multiplication of species (more widely known now as biodiversity). It is the understanding that one species may diverge into two or more distinct groups, and their descendants may diverge again, and so on through a branching network of common ancestors.

All of this is vindicated with independent peer-reviewed empirical research. There are several ways of tracing these lineages by genotypes and phenotypes. Phenotypical taxonomy reveals a third method in the form of another trend that could be described as a law, but has instead become a whole field of study unto itself, evo devo.

Earlier I said that as you track any evident lineage backward through fossil sequences, the closer you get to the "crown" or origin of any one clade, the more basal forms will resemble progenitors of their sister clades. Because the further back you go, the more closely related they are. Similarly, the young of two closely related species will look more alike than the adults do. If no one else has claimed that observation yet, I'd like to name it the law of sibling species similarity.

Everything evolutionary science actually demands are things we really do see throughout biology. Consequently, evolution is a unifying theory explaining virtually everything we can prove to be true of biodiversity.

Such is not the case with gravity. Not only has the theory of gravity never been proven, it's demonstrably wrong! One reason is that it incorrectly predicts how Mercury orbits the sun. Albert Einstein's theory of relativity replaced Newtonian mechanics and has even been proven to be more accurate, but it also assumes things about particles that we now know are false. Otherwise, quantum mechanics wouldn't work. Why don't creationists demand that they teach both the strengths and weaknesses of gravity theory? Especially since there *are* holes in that theory!

During his first presidential run, Arkansas governor and Baptist minister Mike Huckabee was asked about evolution. He replied, "Frankly, Darwinism is not an established scientific fact; it is a theory of evolution, that's why it's called the *theory* of evolution."

This is proof that if you're ignorant enough, you could run for president, even multiple times. That's how it was with William Jennings Bryan, and that's how it was with George W. Bush too.

I wish I could explain science to our increasingly embarrassing array of presidential candidates. For one, I would ask whether they would also say that gravity is not an established scientific fact just because it is called the "theory" of gravity. But then, we've already seen that some of them consider such nomenclature to be part of a scientific conspiracy "theory."

I'd also point out that "Darwinism" doesn't exist. At best, it is limited only to Darwin's own postulation, natural selection of mutations, which were not yet understood, and is thus a relic of the nineteenth century. Darwin hypothesized that units of information were passed down from both parents, but he couldn't imagine what those units were, and the mechanisms he devised proved to be wrong. He hypothesized that inherited traits came from "gemmules," essentially the seeds of cells, which were supposed to be shed by cells to reassemble and multiply in a new organism.

Abbot Gregor Johann Mendel (also known as the father of genetics) was a nineteenth-century Austrian Augustinian priest who determined that invisible factors, which came to be called genes, and not gemmules, were responsible for the inheritance of traits in plants. Genetics operate differently than Darwin's gemmules, and, shortly after the publication of *On the Origin of Species*, Mendel identified particular laws of genetics that were later named after him. The significance of his work was not recognized until the turn of the twentieth century. But his discoveries prompted the foundation of genetic study and provided the necessary vessel for Darwin's proposed mechanism of evolution.

With that discovery, Darwinism was replaced by the modern synthesis of Darwinian selection with Mendelian genetics and the subsequent discovery of an additional mechanism, genetic drift. Evolution is not "Darwinism" anymore and never really was. Some now associate it with cell theory.

Theodor Schwann (1810–1882) was a German physiologist who founded modern histology by defining the cell as the basic unit of animal structure that makes elementary parts (such as teeth, bone, muscle, cartilage, nerve tissue, etc.) by cell differentiation. This laid the foundations of the cell theory. Would you also say that cells aren't an established scientific fact because they're called "theory" too?

Huckabee doesn't know the difference between a theory and a fact. We've already covered what facts are, and we'll hit that again at the end, so now we're going to talk about what "theory" really means in science. But first, to make clear what it's not, I'll leave it to Isaac Asimov:

> [Creationists] make it sound as though a "theory" is something you dreamt up after being drunk all night.

This problem persists throughout mundanity; so many times I've seen the Dunning-Kruger effect demonstrated as dim-witted know-nothings think they've somehow outsmarted all the best-educated expert specialists anywhere ever. Case in point, I happened across a video of some unevolved YouTube yokel saying, "So your evolution is umm . . . has got holes all in it; it's just a theory. There's no—you have no way of provin' evolution ever existed. And if we came from apes, why are there still apes? Heh."

A better question is, why are you still an ape?

We can't prove a theory only because that's against the rules imposed by the game of science. But we *can* prove that evolution exists and that it works, just like we can prove that gravity works, even though it too is "just" a theory and has never been proved.

Atomic theory has never been proven either—not even in Hiroshima. But just as evolution is the foundation of modern biology, modern chemistry is completely dependent on atomic theory. And there are huge holes in that theory! Just look at our classic model of atomic structure: it's wrong, and we know it's wrong, but we still teach it in school anyway. Because despite their virtual invisibility and being understood only in theory, atoms are still a matter of undeniable fact. We devised a series of imperfect models (Rutherford-Bohr model, planetary orbit model, electron cloud model, orbital shell model, quantum model) that all have specific applications, but are all inaccurate in other applications. We're still trying to figure out one model that works in all instances.

Physicists are also trying to devise a single theory to blend quantum theory with the theory of relativity, and act as a unifying theory of everything. The closest we've come is string theory, which really isn't an actual theory yet, because it hasn't been vindicated by substantial empirical evidence, and it hasn't borne itself through the battery of critical examinations that every hypothesis must endure before it can graduate to the highest level of confidence science can attain. As I said, nowadays, it is as if something doesn't qualify as a theory until we know so much about it that it's been effectively proved.

Of course those who worship old books still object. I saw an Islamic televangelist, Zakir Naik, denigrating evolution as "a theory, not a fact." He said, "I've not come across any book which says 'fact of evolution.' All the books say, 'theory of evolution.' There's no book I've come across saying 'fact of evolution.'" Well, Naik is welcome to borrow any of my college science textbooks—for example, take *Chemistry* (1st ed.), by Julia Burdge,

> A theory is a unifying principle that explains a body of experimental observations and the laws that are based on them. Theories can also be used to predict related phenomena, so theories are constantly being tested. If a theory is disproved by experiment, then it must be discarded or modified so that it becomes consistent with experimental observations.

As explained earlier, a theory is made of facts. It's an analysis of how reality works, but every theory has holes in it and no theory is complete. That's why science must remain objective.

For example, what some call expanding planet "theory" addresses some compelling points that plate tectonics doesn't adequately account for. Someone cut out all the continents on his world atlas and glued them onto a blue balloon. They seemed to fit all the way around with no oceans. Then he blew up the balloon, and it looked like a normal globe with the oceans looking as they should. That was impressive, but it is still an illusion. For one thing, there is no mechanism to explain how the planet should grow. Where would the additional mass come from? This is not the only problem with this "theory," as described by consulting geologist Timothy Casey,

> The expanding earth theory relies heavily on the assumption that subduction does not occur. Verification of subduction by numerous cosmogenic isotope studies and common direct GPS measurements of subducting plate motion therefore refutes the expanding earth model.

Expanding planet "theory" calls for assumptions that can't be justified, it's refuted by definite observations that are only supportive of plate tectonics, and it can't explain everything that tectonics does. But remember also that Alfred Lothar Wegener, the man who first proposed the theory of plate tectonics, was ridiculed for it until the day he died, though his theory has since gained universal acceptance. He also understood what all good scientists understand, saying,"We have to be prepared always for the possibility that each new discovery, no matter what science furnishes it, may modify the conclusions we draw."

Other theories never had any competition at all. For example, Louis Pasteur (who disproved the supernatural hypothesis of spontaneous generation) also rejected the notion that ailments of the body were of supernatural origin, a belief long promoted by religious leaders. In the early fifth century, Saint

Augustine of Hippo wrote in his *De Divinatione Daemonum*, "All diseases of Christians are to be ascribed to these demons; chiefly do they torment freshly-baptized Christians, yea, even the guiltless new-born infants."

More than a millennium later, Martin Luther, in the sixteenth century, wrote in "On Sicknesses, and of the Causes Thereof,"

> The physicians in sickness consider only of what natural causes the malady preceeds, and this they cure, or not, with their physic. But they see not that often the devil casts a sickness upon one without any natural causes. A higher physic must be required to resist the devil's diseases; namely, faith and prayer, which physic may be fetched out of God's Word. . . . So in spiritual matters, a preacher has much unction and produces more effect upon the conscience than can a layman.

Despite the work of Pasteur and countless others before and after, such beliefs continue to be embraced today. One need look no farther than the Usenet group alt.talk.creationism, where a creationist named Steve Muscat wrote in this very century, "Many diseases are directly caused by demons while the rest can be traced back to sin in the Garden of Eden."

Science can only examine natural explanations, and Pasteur provided that with his proposition that diseases weren't caused by demons, but by germs. Like evolution, germ theory can never be proven, even though we know and can show that it is definitely correct beyond any doubt—so much so that there's no competing scientific theory.

In the last chapter, we mentioned *Kitzmiller v. Dover*. The televised news media of that event is relevant here. Peter Jennings of ABC News reported: "Today the teachers in a rural Pennsylvania town became the first in the country required to tell students that evolution is not the only theory." The *NOVA* special "Judgment Day: Intelligent Design on Trial" shared Dover school board member Bill Buckingham's complaint regarding the newly suggested textbooks.

> In looking at the biology book the teachers wanted, I noticed that it was laced with Darwinism. I think I listed somewhere between twelve and fifteen instances where it talked about Darwin's theory of evolution. It wasn't on every page of the book, but like every couple chapters, there was Darwin in your face again; and it was to the exclusion of any other theory.

There *is* no other theory. Creationists only ever had a few hypotheses, and all of them have been utterly refuted, although they'll never accept that. They're still trying to revive arguments that have been already proven wrong at least a century ago. They include:

> **Vitalism**, promoted (in different versions) by Aristotle, Rene Descartes, and considered theory under George Stahl until his interpretation of it was disproved.
>
> **Platonic geometry inset within planetary orbits**, a hypothesis implying divine design that was both vigorously pursued and eventually disproved by Johannes Kepler.
>
> **Divine Archetypes**, hypothesized by Sir Richard Owen and eventually abandoned even by him—although a similar notion of "created kinds" was adopted by modern creationists using the term "baramins."
>
> **Deluvial Hydroplates**, promoted as "theory" by professional creationist Walt Brown in 1990, was an unsupported and indefensible notion. While it might otherwise have been correctly declared a hypothesis, it failed to account for data previously gathered through magnetostratigraphy of the ocean crust, overlying sediments, and terrestrial lava flows, correlated with radio-isotope dating, ice cores, fossil varves, and myriad other things. Thus it was already disproved before it was presented.

Evolution by natural selection is the only explanation of biodiversity with either evidentiary support or scientific validity. There has only ever been one alternative theory against it, and it was an earlier version of evolution. Centuries before much of the Old Testament was even written, the Greek scientist Anaximander (as well as Aristotle) had already proposed that modern forms of life, including humans, had evolved from simpler forms—albeit on an evolutionary "ladder," which has since been discarded.

This observation seems to have had a far-reaching effect on the religious community of that time. The Bhagavad Gita has been contextually dated to about the same time (sixth century BCE) and Lord Krishna, the central hero of the Gita, appears to be reacting to the "evolutionists" of his day the way modern creationists still do. "The persons of the demonic nature say, the entire cosmic manifestation is unreal, without a creator, without a supreme controller, without cause, originating from natural cohabitation due only to lust, no more than this" (16:8).

Similar ideas were echoed and argued by other figures of the age. The mutability of breeding populations has long been a well-known fact but not at all understood. The first attempted explanation was proposed by Jean-Baptiste Lamarck. He suggested that giraffes stretching their necks to reach higher food would somehow bestow longer necks onto their offspring. His theory of evolution by acquired characteristics proved to be a failed hypothesis almost immediately, but it offered a sort of personal control over racial advancement that appealed to those ambitious in politics. Joseph Stalin and Mao Tse Tung, for example, both embraced Lamarckism and publicly denounced Darwinism, as Hitler did also; yet creationists label all three of them Darwinists and can't distinguish Lamarck's failure from Darwin's success.

Comparing Lamarckian Lysenkoism to Mendelian Darwinism, Lamarckian evolution suggests that if a naturally weak man works out to increase his strength, his children will inherit not only his drive to do the same but they will also be born with increased muscle mass. Darwinian evolution holds that each of our children will vary with different types of strengths as well as liabilities, and that after many generations the environment will "select" certain traits regardless of our ambitions or desires. To secure an illusion of control over human development, Soviet pseudoscientist Trofim Lysenko denounced Mendelian genetics as a capitalist myth. Joseph Stalin wrote and promoted Lamarckian propaganda believing that *Homo sovieticus* (the ideal Communist) could be created through totalitarian psychology rather than breeding. Consequently, Russian dictionaries were rewritten to describe genes as mythical concepts, and those identified as dissenting "Darwinians" were dismissed, gulag'd, or otherwise eliminated.

Flipping through the channels, we find ourselves on a Christian TV station, where someone says, "And when it comes to the subject of origins, there are basically two views: the evolution theory and the creation theory." Except that no branch of creationism has ever met even one of the criteria required of a theory. They can't because science demands both accuracy and accountability. There has to be a way to detect and correct any errors in a given explanation, and determine for certain whether it's wrong in whole or in part, or whether any of it is true to any degree at all. As Stephen Hawking put it in *A Brief History of Time*,

> A theory is a good theory if it satisfies two requirements: It must accurately describe a large class of observations on the basis of a model that contains only a few arbitrary elements, and it must make definite predictions about the results of future observations.

We never see a positive representation of healthy skepticism in entertainment media. A 1981 movie called *Dragonslayer* is one of the best examples of that I can think of. But in that, the skeptic was a murderous villain, and we were expected to side with the believers in magic. Having been accosted and challenged, the wizard's apprentice says, "Ah, so it's a test you're looking for. We don't do tests!"

The villain's response was admirable: "I'm sure you don't. They never do tests. Not many real deeds either. Oh, conversation with your grandmother's shade in a darkened room, the odd love potion or two, but comes a doubter, why, then it's the wrong day, the planets are not in line, the entrails are not favorable, 'we don't do tests!'"

That was a beautiful indictment of mysticism. Why did he have to be the bad guy?

A theory has to be tested indefinitely, which is another reason why it can never be proved. Religion demands complete conviction, but science advises against that. It demands understanding instead of belief, so it must be based on verifiable evidence; it must explain related observations with a measurable degree of accuracy; it must withstand continuous critical analysis in peer review; and it must be falsifiable too. If it doesn't fulfill all these conditions at once, then it isn't science. If it meets none of them, it could be religion.

Getting back to the common misconception, we look back at *Selman v. Cobb County* and the public testimony to have intelligent design taught in the county's schools. One of the people involved in that described his interpretation of intelligent design on an episode of Penn & Teller's Showtime series *Bullshit!*. He said, "Intelligent design is a theory that there was some, there is some master plan, some creator of some type that put together the world as it is."

This guy is trying to portray intelligent design as if it were something different from creationism, even though earlier testimony already asked for creationism and identified ID as such. He then admitted that intelligent design is a "theory" (by which he means "blind guess") that a creator created creation. In other words, intelligent design is creationism.

As we've seen in the previous chapter and throughout this book, intelligent design isn't a theory at all; it's a scam, a scheme conceived solely to undermine legitimate science. It doesn't even count as a hypothesis, because it isn't based on evidence, offers no mechanism, and isn't falsifiable either. It is backed by nothing and produces nothing because it is nothing but untestable conjecture. None of it has been shown to be right and lots of it has been proven wrong, so it's useless in any field because only accurate information can have practical application. That's why AIDS researchers use evolutionary principles in their treatment regimens. They have to, because the virus evolves.

For another example, the venom of the male Sydney funnel-web spider is only lethal to one line of vertebrates: it is deadly to all primates, and *only* to primates. Other large vertebrates are not at risk, but any human bitten would be foolish to disregard our taxonomic classification. If you know this thing kills only monkeys, then you should know that it can kill you too.

That's why we have billion-dollar industries in medicine, toxicology, agriculture, and biotechnology, where we have Nobel Prize–winning research that is all dependent on the functionality of evolution and that would only work if evolution were factually correct. Those who argue that evolution is "just" a theory don't realize all the other theories they diminish along with that, including Big Bang theory. To quote Stephen Hawking again, this time from his Ted Talk titled "Questioning the Universe":

> Up until the 1920s, everyone thought the universe was essentially static and unchanging in time. Then we discovered that the universe is expanding. Distant galaxies were moving away from us. This meant they must have been closer together in the past. If we extrapolate back, we find that they must all have been on top of each other about fifteen billion years ago. This was the Big Bang, the beginning of the universe.

Now the very earliest stages of the universe are as yet unknown because our current theories are insufficient to explain it, so we can't yet be certain what "the singularity" is or where or what it came from. Creationists like to distort that by saying that first there was nothing and then nothing exploded, but it's not like that at all. The current understanding is that the singularity was not a point in space, that it was the single point where all matter, energy, and space-time existed at once, and that there was nothing, not even a void of empty space outside of that until the universe inflated from that point.

I should mention a couple relevant hypotheses. Because time slows down both as one approaches the speed of light and also under the influence of extreme gravity, and the singularity would be as extreme as it gets, then imagine a Cartesian coordinate system where time is an asymptote such that one second equals infinity when T = zero. A curious feature of that is that the universal singularity could still be eternal even though the universe has a beginning.

Another simultaneous possibility is this: we know that gravity bends space-time. A good illustration is the way a bowling ball distorts a trampoline. In a classroom demonstration, a tennis ball can appear to orbit around the bowling ball. But really extreme gravity doesn't just bend the fabric of space; it punches a hole in it, and that is apparently why we have black holes. Well, if the singularity is the single unification of all matter in the universe, then that would be some freakishly extreme gravity, and maybe it could punch a hole into another source, say a multiverse or a fourth spacial dimension. Whatever the catalyst is wouldn't require or imply intent. Fourth-dimensional content flooding into three-dimensional space would erupt through a singular pinpoint in the time-space continuum and inflate the universe from there with no need of a god, an incantation spell, or a talking snake. But that's just how I see it anyway.

Big Bang theory has a lot of compelling support behind it according to all the smartest people alive. So how do creationists contend with that? How do they explain the evident expansion and its reverse implication? How do they explain galactic acceleration? How do they account for the cosmic microwave background radiation that was predicted by the theory and discovered later on?

They don't. Creationists delight in saying that scientists have "proven" the universe had a beginning, and they use that argument against the Big Bang. But the discovery of the Big Bang is our proof that the universe had a beginning! That's one example of how bewilderingly inane creationism is.

To quote another of the Cobb County yahoos, "We must require that evolution agree with all the facts if it is to be promoted from theory to truth. Evolution as an explanation of the origin of man cannot pass this test."

Without mentioning the fact that there is no truth to creationism, but there are a whole lotta lies behind it, I have to explain that nothing in science would ever be promoted to "truth," because truth implies that there's nothing more to learn. (Unless the truth is what the facts are, in which case theories would have to be *de*moted to truth, not *pro*moted.) That's why science, being objective, demands that everything be considered theory no matter how proven it seems to be.

Evolution has survived every test the greatest minds of the modern age have ever been able to pit against it. It's been demonstrated myriad ways with lab and field experiments, and is further enhanced by compounded revelations in paleontology and systematics, as well as by developments in embryology and advances in genomic research and bioengineering. Evolution is now one of the strongest theories in science. There is no fact it doesn't agree with, and it's never failed any test. But sadly, those controlling education in many parts of the world, including in parts of the United States, don't want students to know that. So what can you expect?

One of the Cobb County creationists said, "I don't believe personally that the evolution itself is anything more than a theory." This is an important point: there is nothing more than a theory. Theory is the highest level of confidence science can get. A theory is a field of study and a body of knowledge enveloping a collection of facts, hypotheses, and natural laws. A theory is not "just" a hypothesis; it is made of facts. Theories never graduate to become laws either; laws are subordinate components of theories. According to the National Academy of Sciences (and indeed the entire global scientific community), evolution is both a fact and a theory.

I mentioned that the first video I ever saw on the evolution-creationism debate, "Evolution versus Creation," was also the most dishonest production I had ever seen. I need to close by addressing something the host of that video said, "Well, you see, evolution is a theory, not a scientific fact, as it's generally considered to be."

Wrong.

It is a fact that evolution happens, that biodiversity and complexity does increase, and that both occur naturally only by evolutionary means.

It is a fact that alleles vary with increasing distinction in reproductive populations, and that these are accelerated in genetically isolated groups.

It is a fact that natural selection, sexual selection, and genetic drift have all been proven to have predictable effect in guiding this variance.

It is a fact that significant beneficial mutations do occur and are inherited by descendant groups, and that multiple independent sets of biological markers exist to trace these lineages backward over many generations.

It is a fact that birds are a subset of dinosaurs the same way humans are a subset of apes, primates, eutherian mammals, and vertebrate deuterostome animals.

It is a fact that the collective genome of all animals has been traced to its most basal form, and that those forms are also indicated by comparative morphology, physiology, and embryological development.

It is a fact that everything on earth has definite relatives either living nearby or evident in the fossil record.

It is a fact that the fossil record holds hundreds of transitional species even according to its strictest definition, and that both microevolution and macroevolution have been directly observed.

Evolution is a *fact!*

ABOUT THE AUTHOR

Aron Ra is a science educator, secular activist, debater, and public speaker. He cohosts the Ra-Men podcast and writes at Reason Advocates on the Patheos network. He is the former Texas state director of American Atheists and current president of Atheist Alliance of America. He is also heading the Phylogeny Explorer Project, an effort to render the entire evolutionary tree of life as a navigable online encyclopedia. He is best known for his popular video series exposing the foundational falsehoods of creationism.